AF361653

Horticultural Crop Response to Different Environmental and Nutritional Stress

Horticultural Crop Response to Different Environmental and Nutritional Stress

Editor

Stefano Marino

MDPI • Basel • Beijing • Wuhan • Barcelona • Belgrade • Manchester • Tokyo • Cluj • Tianjin

Editor
Stefano Marino
University of Molise
Italy

Editorial Office
MDPI
St. Alban-Anlage 66
4052 Basel, Switzerland

This is a reprint of articles from the Special Issue published online in the open access journal *Horticulturae* (ISSN 2311-7524) (available at: https://www.mdpi.com/journal/horticulturae/special_issues/Horticultural_Crop_Response).

For citation purposes, cite each article independently as indicated on the article page online and as indicated below:

LastName, A.A.; LastName, B.B.; LastName, C.C. Article Title. *Journal Name* **Year**, *Volume Number*, Page Range.

ISBN 978-3-0365-1948-7 (Hbk)
ISBN 978-3-0365-1949-4 (PDF)

Cover image courtesy of Stefano Marino

© 2021 by the authors. Articles in this book are Open Access and distributed under the Creative Commons Attribution (CC BY) license, which allows users to download, copy and build upon published articles, as long as the author and publisher are properly credited, which ensures maximum dissemination and a wider impact of our publications.
The book as a whole is distributed by MDPI under the terms and conditions of the Creative Commons license CC BY-NC-ND.

About the Editor

Stefano Marino is Professor in Precision farming at the Department of Agricultural, Environmental and Food Sciences since 2021. He obtained his master's degree at University of Molise, Italy, in 2002, and Ph.D. degrees in Protection and quality of agro-food production at University of Molise, Itraly, in 2006. His research areas cover the fields of precision farming, irrigation, plant nutrition, horticulture, Proximal and remote sensing, Crop management; Abiotic stresses

 *horticulturae*

Editorial

Horticultural Crop Response to Different Environmental and Nutritional Stress

Stefano Marino

Department of Agricultural, Environmental and Food Sciences (DAEFS), University of Molise, Via De Sanctis, 86100 Campobasso, Italy; stefanomarino@unimol.it

Abstract: Environmental conditions and nutritional stress may greatly affect crop performance. Abiotic stresses such as temperature (cold, heat), water (drought, flooding), irradiance, salinity, nutrients, and heavy metals can strongly affect plant growth dynamics and the yield and quality of horticultural products. Such effects have become of greater importance during the course of global climate change. Different strategies and techniques can be used to detect, investigate, and mitigate the effects of environmental and nutritional stress. Horticultural crop management is moving towards digitized, precision management through wireless remote-control solutions, but data analysis, although a traditional approach, remains the basis of stress detection and crop management. This Special Issue summarizes the recent progress in agronomic management strategies to detect and reduce environmental and nutritional stress effects on the yield and quality of horticultural crops.

Citation: Marino, S. Horticultural Crop Response to Different Environmental and Nutritional Stress. *Horticulturae* **2021**, *7*, 240. https://doi.org/10.3390/horticulturae7080240

Received: 3 August 2021
Accepted: 10 August 2021
Published: 11 August 2021

Publisher's Note: MDPI stays neutral with regard to jurisdictional claims in published maps and institutional affiliations.

Copyright: © 2021 by the author. Licensee MDPI, Basel, Switzerland. This article is an open access article distributed under the terms and conditions of the Creative Commons Attribution (CC BY) license (https://creativecommons.org/licenses/by/4.0/).

1. Introduction

Food and agriculture systems may follow alternative pathways, depending on the evolution of a variety of factors, such as population growth, dietary choices, technological progress, income distribution, the state and use of natural resources, climatic changes and efforts to prevent and resolve conflicts. These pathways can and will be impacted by strategic choices and policy decisions. Swift and purposeful actions are needed to ensure the sustainability of food and agriculture systems in the long term [1].

Climate change is considered as one of the future challenges that either directly or indirectly affect all sectors negatively [2]. Environmental interactions also affect sectors that have a direct reliance on natural resources for production, highlighting their significance for national socio-economic development. The agriculture sector, in turn, has about 2.5 billion livelihoods that are dependent on it. The quality and yield of horticultural crops need to be improved for their production, cultivation management, and biotic/abiotic resistances. Biotic and abiotic factors are the main factors limiting production in agricultural systems [3,4]. Abiotic stresses such as temperature (cold, heat), water (drought, flooding), irradiance, salinity, nutrients, and heavy metals can strongly affect plant growth dynamics, increase crop yield losses and the yield and quality of horticultural products.

Such effects become more and more important in the course of global climate change. Water scarcity, climate change, and drought are the main hurdles in our efforts to make our agri-food systems resilient and sustainable [4]. Different strategies and techniques can be used to detect, investigate, and mitigate the effects of environmental and nutritional stress. Horticultural crop management is moving towards digitized, precision management through wireless remote-control solutions, but data analysis, although a traditional approach, remains the basis of stress detection and crop management.

2. Special Issue Overview

This Special Issue collects current research findings that deal with a wide range of topics to detect environmental and nutritional stress effects on the yield and quality of horticultural crops.

1

The effect of environment conditions (temperature, rainfall, altitude, soil types, hail) [5–9], nutritional strategies [5,6,10–14], water (stress, salinity) [5,8,12,15], container substrate of cultivation [16] on yield, yield traits and quality were analyzed in different crops in an open field [8,9,13,14,17] and greenhouses [5–7,10–12,15,16].

Different innovative strategies were presented that included protected structures, container type, cultivation techniques [5,10,16], symbiotic relationships [6,7], fertigation strategies [10] new agricultural management technologies (remote sensing, smartphone) [11] novel hybrids for breeding [12] spraying of Gibberellic acid [17], water supplies [8,15] and substrate [16] cover crops management [13] and stand reduction [9].

2.1. Nutrient Concentration of African Horned Cucumber (Cucumis metalliferous L.) Fruit under Different Soil Types, Environments, and Varying Irrigation Water Levels

This study was conducted during the 2017/18 and 2018/19 growing seasons, under the greenhouse, shade net, and open-space environment at the Florida science campus of the University of South Africa [5]. The aim was to determine the effect of different water stress levels, soil types, and growing environments (greenhouse, shade net, and open field) on the nutrient concentration of the African horned cucumber fruit. Total soluble sugars, crude proteins, β-carotene, vitamin C, vitamin E, total flavonoids, total phenols and micro-nutrients were analyzed. The results showed that African horned cucumber fruits are nutrient-dense when grown under moderate water stress treatment on a loamy or sandy loam substrate in shade-net and open-field environments. Quality parameters (total flavonoids, total phenols, micro-nutrients and vitamins metabolites) seem to be treatment-imposed. The data show that this crop can grow well under protected structures, which eliminates the potential damage caused by higher rainfall, hail, and extreme heat in summer.

2.2. Growth and Competitive Infection Behaviors of Bradyrhizobium japonicum and Bradyrhizobium elkanii at Different Temperatures

Growth and competitive infection behaviors of two sets of *Bradyrhizobium* spp. strains were examined at different temperatures to explain strain-specific soybean nodulation under local climate conditions [6]. Each set consist of three strains—*B. japonicum* Hh 16-9 (Bj11-1), *B. japonicum* Hh 16-25 (Bj11-2), and *B. elkanii* Hk 16-7 (BeL7); and *B. japonicum* Kh 16-43 (Bj10J-2), *B. japonicum* Kh 16-64 (Bj10J-4), and *B. elkanii* Kh 16-7 (BeL7)—which were isolated from the soybean nodules cultivated in Fukagawa and Miyazaki soils, respectively. The authors compared growth and infection behaviors at different temperatures in Japan (Fukagawa and Miyazaki soils) and elucidated the reason why the species-specific nodule compositions are present in the Fukagawa and Miyazaki soils and locations. The experiments performed in liquid cultures revealed better growth of *B. japonicum* at lower temperatures and *B. elkanii* at higher temperatures, and therefore it can be assumed that the temperature of soil affects rhizobia growth in the rhizosphere and could be a reason for the different competitive properties of *B. japonicum* and *B. elkanii* strains at different temperatures. In addition, competitive infection was suggested between the *B. japonicum* strains.

2.3. Effects of Fertigation Management on the Quality of Organic Legumes Grown in Protected Cultivation

The experimental trial was carried out in a protected cultivation certified organic farm in the province of Almería, South-East Spain on four legume cultivars of *Phaseolus* and *Pisum* [10]. The objective of the study was to determine the effects of two fertigation treatments, normal (T100) and 50% sustained deficit (T50), on the physico-chemical quality of legumes. The fertigation treatments had significant effects on the morphometric traits (width for mangetout and French bean; fresh weight for French bean; seed height for Pea cv. Lincoln). Furthermore, only French bean plants had significantly lowered productivity under 50% fertigation conditions. Furthermore, mangetout became the highest source of total soluble solids, reaching higher content at 50% fertigation treatment. Fertigation treatments did not significantly affect the antioxidant compounds (total polyphenols and ascorbic acid), minerals and protein fraction contents of the legumes studied.

2.4. Latitudinal Characteristic Nodule Composition of Soybean-Nodulating Bradyrhizobia: Temperature-Dependent Proliferation in Soil or Infection?

To examine the possible reasons for the temperature-dependent distribution of soybean-nodulating rhizobia, competitive inoculation experiments at different temperatures were conducted [7]. The study chose three locations Fukagawa, Matsue and Miyazaki which are considered to have different climatic conditions in Japan. The aim was to elucidate the possible reasons for the latitudinal characteristic distribution of soybean-nodulating rhizobia in local climate conditions. Rainfall might affect the soil conditions during the winter season, though the difference in the soil storage did not seem to be serious on the composition of rhizobia and the difference in rainfall would not significantly affect the nodule composition. The most interesting result was that the soil temperature mainly affected the dominant nodule composition in the different environmental conditions.

2.5. A Novel Method for Estimating Nitrogen Stress in Plants Using Smartphones

The objectives of the research were to (i) test the hypothesis that the ratio of blue light reflectance to that of combined reflectance in the visible band can be used as an index for N stress, (ii) study the association between the N stress index and the physiological pathways in plants, and (iii) develop a smartphone application to measure the N stress index in species with differences in plant architecture [11]. The experiment was conducted in a greenhouse at Purdue University, West Lafayette, IN, USA. Nitrogen stress was provided by supplying a fertilizer solution with an EC of 0.75 dS·m^{-1} and maintaining a θ level of 0.48 m^3·m^{-3}. The study developed an index, calculated as the ratio of reflectance of blue relative to the reflectance of combined wavelengths in the visible wavelengths band. The index value decreased when plants were exposed to nitrogen stress relative to optimal conditions. Furthermore, the index value decreased gradually with increasing N stress in plants. Therefore, the continuous measurement of the index can aid in the timely detection of N stress in plants.

2.6. Cucurbita Rootstocks Improve Salt Tolerance of Melon Scions by Inducing Physiological, Biochemical and Nutritional Responses

An experiment was conducted to evaluate whether grafting with hybrid *Cucurbita maxima* × *Cucurbita moschata* rootstocks could improve the salt tolerance of melon and to determine the physiological, biochemical, and nutritional responses induced by Cucurbita rootstocks under hydroponic salt stress [12]. Results indicated that the shoot and root growths of grafted and nongrafted melon plants were detrimentally affected by salt stress. Significant reductions were recorded in some agronomic and physiological plant traits. Susceptible plants responded to salt stress by increasing leaf proline and malondialdehyde (MDA), ion leakage, and leaf Na$^+$ and Cl$^-$ contents. The highest plant growth performance was exhibited by Citirex/Nun9075 and Citirex/Kardosa graft combinations. These Cucurbita cultivars had a high rootstock potential for melon, and their significant contributions to salt tolerance were closely associated with inducing the physiological and the biochemical responses of the scions. These traits could be useful for the selection and breeding of salt-tolerant rootstocks for sustainable agriculture in the future.

2.7. The Effects of Gibberellic Acid and Emasculation Treatments on Seed and Fruit Production in the Prickly Pear (Opuntia ficus-indica (L.) Mill.) cv. "Gialla"

The author tested the application of two methods (injection and spraying) of gibberellic acid (GA$_3$) on the prickly pear cactus both at pre- and post-blooming in order to obtain well-formed seedless fruits in emasculated flowers [17]. The experiments were conducted in the Apulia region, Italy. Different application methods (injection and spraying) and concentrations of GA$_3$ (0, 100, 200, 250, and 500 ppm) combined with floral-bud emasculation were applied to a commercial plantation to evaluate their effects on the weight, length, and diameter of the fruits, total seed number, hard-coated viable seed number, and seed weight per fruit. The application of 500 ppm GA_3 sprayed on emasculated floral buds was the most effective method for reducing the seed numbers of prickly pear fruits (-46.0%). The injection method resulted in a very low number of seeds (-50.7%) but produced unmarketable fruit. The spraying of the GA_3 (both at low and high levels) enhanced the growth performance of all analyzed variables of the treated fruits, while the application of these treatments in an industrial-scale requires support to evaluate the processes.

2.8. Water Use and Yield Responses of Chile Pepper Cultivars Irrigated with Brackish Groundwater and Reverse Osmosis Concentrate

Freshwater availability is declining in most of the semi-arid and the arid regions across the world [15]. The study evaluated the effects of natural brackish groundwater and RO concentrate irrigation on the water use, leaching fraction, and yield responses of Chile pepper cultivars (*Capsicum annuum* L.). The study was conducted in a greenhouse located at the New Mexico State University (NMSU). Saline irrigation caused a reduction in the water uptake of the Chile peppers and increased LFs. The four saline water treatments used for irrigation were tap water with an electrical conductivity (EC) of 0.6 dS m^{-1} (control), groundwater with EC 3 and 5 dS m^{-1}, and an RO concentrate with EC 8 dS m^{-1}. The WUE was not substantially different but decreased significantly in the other two higher salinity treatments. Therefore, irrigating Chile peppers with up to 3 dS m^{-1} brackish water could be possible by maintaining appropriate leaching fractions to sustain Chile pepper production in freshwater-scare areas where brackish groundwater is the only available source of irrigation. The yield response curves showed that the yield reductions in the Chile peppers irrigated with natural brackish water were less, compared to those of NaCl-dominant solution studies. Low yield reductions could be related to significant Ca^{2+} concentrations in the brackish groundwater and RO concentrate.

2.9. Alterations in the Chemical Composition of Spinach (Spinacia oleracea L.) as Provoked by Season and Moderately Limited Water Supply in Open Field Cultivation

The study shows the relationship of the irrigation water supply with that of the chemical composition of the Spinach (*Spinacia oleracea* L.) [8]. Trials of the study recorded a slight effect on the chemical composition of the plant from providing a moderate water supply which ultimately influenced the product quality of field-grown spinach plants. In the reduced water supply treatment, the total amount of supplied water, including both irrigation and natural precipitation, amounted to 90%, 94% and 96% in 2015, 2016 and 2017, respectively, of the full optimal water supply treatment. The study was carried out on Spinach cv. 'Silverwhale' grown under open field conditions at Geiseheim University, Germany. The chemical composition of both the dry and the fresh biomass of spinach was shown to be strongly influenced by the climatic conditions and/or the water supply. Fresh biomass-related levels of ascorbic acid, potassium, nitrogen, phosphorous as well as total flavonoids and carotenoids increased upon limiting the water supply. Considering the composition of the dry biomass itself, authors demonstrated that even mild water supply reductions led to significant increases of inositol, zinc and manganese levels, while malic acid, phosphate and chloride levels decreased. The nutritional composition of spinach was sensitive to even moderately reduced water supply, but the overall quality of fresh spinach did not suffer regarding the levels of health-promoting constituents such as minerals, trace elements, flavonoids and carotenoids.

2.10. Container Type and Substrate Affects Root Zone Temperature and Growth of 'Green Giant' Arborvitae

The objective of this research was to evaluate the combined effects of the container type and the substrate on RZT and growth of *Thuja standishii* × *plicata* 'Green Giant' [16]. Two separate studies were conducted concurrently at the Tennessee State University and the Auburn University Ornamental Horticulture Research Center, USA. Trade gallon arborvitae were transplanted into black, white, or air pruning containers filled with pine bark (PB) or 4 PB: 1 peatmoss (*v:v*) (PB:PM). Plants grown in PB:PM were larger and had greater shoot and root biomass than plants grown in PB, likely due to the increased volumetric water content. Plant growth response to container type varied by location, but white containers with PB:PM produced larger plants and greater biomass compared with the other container types. Root zone temperature was greatest in black containers and remained above 38 °C and 46 °C for 15% and 17% longer than white and air pruning containers, respectively. Utilizing light color containers in combination with substrates containing peatmoss can reduce RZT and increase substrate moisture content thus improving crop growth and quality.

2.11. Effects of Non-Leguminous Cover Crops on Yield and Quality of Baby Corn (Zea mays L.) Grown under Subtropical Conditions

The objective of this study was to evaluate the effects of non-leguminous cover crops and increments in chopping time versus Days After Planting (DAP) on the yield and quality of no-till baby corn (*Zea mays* L.) [13]. The experiment was carried out during kharif seasons under the subtropical climatic conditions. The experiment was conducted at the Punjab Agricultural University, Ludhiana, India. Three cover crops (pearl millet (*Pennisetum glaucum* L.), fodder maize (*Zea mays* L.), and sorghum (*Sorghum bicolor* L.)) and the control (no cover crop) were in the main plots and chopping time treatments (25, 35, 45 days after planting (DAP)) in the subplots. The yield (cob and green fodder yield) and dry matter accumulation of baby corn following cover crop treatments were significantly higher than the control (no cover crop) and improved with increment in chopping time. Increment in chopping time (from 25 DAP to 45 DAP) had a significant effect on the protein and sugar content of the baby corn cob. Chopping of cover crops at 45 DAP showed the highest yield and dry matter. Non-leguminous cover crops and their times of chopping evaluated in this study could be used for a sustainable maize crop production system to improve baby corn growth and yield, baby corn quality, and topsoil quality.

2.12. Effect of Stand Reduction at Different Growth Stages on Yield of Paprika-Type Chile Pepper

The goal of this study was to understand how a simulation of population losses by four levels of stand reduction at three different growth stages affected the yield and yield components of the paprika-type red Chile [9]. Two trials, one per year, were conducted in southern New Mexico. 'LB-25', a standard commercial cultivar, was direct seeded on 29 March 2016 and 4 April 2017. Field experiments were conducted at the New Mexico State University, USA. Plants were thinned at three different growth stages; early seedling, first bloom, and peak bloom at four different levels at each phenological stage: 0% stand reduction (control; ~200,000 plants ha^{-1}), 60% stand reduction (~82,000 plants ha^{-1}), 70% stand reduction (~60,000 plants ha^{-1}), and 80% stand reduction (~41,000 plant ha^{-1}). The timing of stand reductions (growth stage) for paprika-type Chile did not impact the marketable red yields. Paprika-type Chile has some capacity to recover and compensate for stand reduction losses. Data show that a farmer could lose up to 70% of their paprika-type Chile stand due to hail damage and experience minimal to no impact on their yields. Furthermore, the paprika-type Chile crop losses can be estimated based on percentage of stand losses instead of growth stage.

2.13. Fertilization and Soil Nutrients Impact Differentially Cranberry Yield and Quality in Eastern Canada

The objective of the research activities was to support site-specific nutrient man-agement decisions in cranberry agroecosystems [14]. A 3-year trial was conducted on permanent plots at four production sites in Quebec, Canada. This paper quantified the trade-off between berry yield and quality as driven primarily by N fertilization. Berry yield was closely related to the number of fruiting uprights (r = 0.92), berry counts per fruiting upright (r = 0.91), number of reproductive uprights (r = 0.83), and fruit set (r = 0.77). Nitrogen increased berry yield nonlinearly but decreased berry firmness, total anthocyanin content (TAcy), and total soluble solids content (°Brix) linearly, indicating a trade-off between berry yield and quality. Fertilizer dosage at a high-yield level ranged between 30 and 45 kg N ha^{-1} in both conventional and organic farming systems. Berry yield could be predicted most accurately from berry counts per fruiting upright. Nitrogen fertilization increased berry yield nonlinearly and decreased fruit quality-based indices in a linear trend. As shown by redundancy analysis (RDA), cranberry performance was related to soil pH and soil test nutrients. The K and Ca were negatively correlated between them, indicating an upper limit for K additions. The RDA indicated close relationships between cranberry performance indices and soil properties, and thus supported the need for further soil test calibration.

Funding: This research received no external funding.

Institutional Review Board Statement: Not applicable.

Informed Consent Statement: Not applicable.

Acknowledgments: We gratefully acknowledge all the authors that participated in this Special Issues.

Conflicts of Interest: The authors declare no conflict of interest.

References

1. FAO. *The Future of Food and Agriculture—Alternative Pathways to 2050*; FAO: Rome, Italy, 2018; p. 60.
2. Francini, A.; Sebastiani, L. Abiotic Stress Effects on Performance of Horticultural Crops. *Horticulturae* **2019**, *5*, 67. [CrossRef]
3. FAO. *The Impact of Disasters and Crises on Agriculture and Food Security: 2021*; FAO: Rome, Italy, 2021. [CrossRef]
4. Dongyu, Q. *High Level Political Forum Special Event: Global Acceleration Framework for Sustainable Development Goal (SDG)*; FAO: New York, NY, USA, 2020; Available online: http://www.fao.org/director-general/speeches/detail/es/c/1297450/ (accessed on 11 August 2021).
5. Maluleke, M.K.; Moja, S.J.; Nyathi, M.; Modise, D.M. Nutrient Concentration of African Horned Cucumber (Cucumis metuliferus L) Fruit under Different Soil Types, Environments, and Varying Irrigation Water Levels. *Horticulturae* **2021**, *7*, 76. [CrossRef]
6. Hafiz, M.H.R.; Salehin, A.; Itoh, K. Growth and Competitive Infection Behaviors of Bradyrhizobium japonicum and Bradyrhizobium elkanii at Different Temperatures. *Horticulturae* **2021**, *7*, 41. [CrossRef]
7. Hafiz, M.H.R.; Salehin, A.; Adachi, F.; Omichi, M.; Saeki, Y.; Yamamoto, A.; Hayashi, S.; Itoh, K. Latitudinal Characteristic Nodule Composition of Soybean-Nodulating Bradyrhizobia: Temperature-Dependent Proliferation in Soil or Infection? *Horticulturae* **2021**, *7*, 22. [CrossRef]
8. Schlering, C.; Zinkernagel, J.; Dietrich, H.; Frisch, M.; Schweiggert, R. Alterations in the Chemical Composition of Spinach (*Spinacia oleracea* L.) as Provoked by Season and Moderately Limited Water Supply in Open Field Cultivation. *Horticulturae* **2020**, *6*, 25. [CrossRef]
9. Joukhadar, I.; Walker, S. Effect of Stand Reduction at Different Growth Stages on Yield of Paprika-Type Chile Pepper. *Horticulturae* **2020**, *6*, 16. [CrossRef]
10. García-García, M.D.C.; Font, R.; Gómez, P.; Valenzuela, J.L.; Fernández, J.A.; Del Río-Celestino, M. Effects of Fertigation Management on the Quality of Organic Legumes Grown in Protected Cultivation. *Horticulturae* **2021**, *7*, 28. [CrossRef]
11. Adhikari, R.; Nemali, K. A Novel Method for Estimating Nitrogen Stress in Plants Using Smartphones. *Horticulturae* **2020**, *6*, 74. [CrossRef]
12. Ulas, A.; Aydin, A.; Ulas, F.; Yetisir, H.; Miano, T.F. Cucurbita Rootstocks Improve Salt Tolerance of Melon Scions by Inducing Physiological, Biochemical and Nutritional Responses. *Horticulturae* **2020**, *6*, 66. [CrossRef]
13. Singh, A.; Deb, S.K.; Singh, S.; Sharma, P.; Kang, J.S. Effects of Non-Leguminous Cover Crops on Yield and Quality of Baby Corn (*Zea mays* L.) Grown under Subtropical Conditions. *Horticulturae* **2020**, *6*, 21. [CrossRef]
14. Jamaly, R.; Parent, S.-É.; Parent, L.E. Fertilization and Soil Nutrients Impact Differentially Cranberry Yield and Quality in Eastern Canada. *Horticulturae* **2021**, *7*, 191. [CrossRef]
15. Baath, G.S.; Shukla, M.; Bosland, P.W.; Walker, S.J.; Saini, R.K.; Shaw, R. Water Use and Yield Responses of Chile Pepper Cultivars Irrigated with Brackish Groundwater and Reverse Osmosis Concentrate. *Horticulturae* **2020**, *6*, 27. [CrossRef]
16. Witcher, A.L.; Pickens, J.M.; Blythe, E.K. Container Type and Substrate Affect Root Zone Temperature and Growth of 'Green Giant' Arborvitae. *Horticulturae* **2020**, *6*, 22. [CrossRef]
17. Marini, L.; Grassi, C.; Fino, P.; Calamai, A.; Masoni, A.; Brilli, L.; Palchetti, E. The Effects of Gibberellic Acid and Emasculation Treatments on Seed and Fruit Production in the Prickly Pear (*Opuntia ficus-indica* (L.) Mill.) cv. "Gialla". *Horticulturae* **2020**, *6*, 46. [CrossRef]

horticulturae

MDPI

Article

Fertilization and Soil Nutrients Impact Differentially Cranberry Yield and Quality in Eastern Canada

Reza Jamaly [1], Serge-Étienne Parent [1,2] and Léon E. Parent [1,3,*

[1] Department of Soil and Agri-Food Engineering, Université Laval, Québec, QC G1V 0A6, Canada; reza.jamaly.1@ulaval.ca (R.J.); Serge-Etienne.Parent@USherbrooke.ca (S.-É.P.)

[2] Département de Médecine de Famille, Faculté de Médecine et des Sciences de la Santé, Université de Sherbrooke, Sherbrooke, Québec, QC J1K 2R1, Canada

[3] Departamento dos Solos, Universidade Federal de Santa Maria, Av. Roraima, 1000 Camobi, Santa Maria 97105-900, Brazil

* Correspondence: Leon-Etienne.Parent@fsaa.ulaval.ca

Abstract: High berry yield and quality of conventionally and organically grown cranberry stands require proper nutrient sources and dosage. Our objective was to model the response of cultivar "Stevens" to N, P, K, Mg, Cu, and B fertilization under conventional and organic farming systems. A 3-year trial was conducted on permanent plots at four production sites in Quebec, Canada. We analyzed yield predictors, marketable yield, and fruit quality in response to fertilization and soil properties. Cranberry responded primarily to nitrogen fertilization and, to a lesser extent, to potassium. Berry yield was closely related to the number of fruiting uprights ($r = 0.92$), berry counts per fruiting upright ($r = 0.91$), number of reproductive uprights ($r = 0.83$), and fruit set ($r = 0.77$). Nitrogen increased berry yield nonlinearly but decreased berry firmness, total anthocyanin content (TAcy), and total soluble solids content (°Brix) linearly, indicating a trade-off between berry yield and quality. Fertilizer dosage at a high-yield level ranged between 30 and 45 kg N ha^{-1} in both conventional and organic farming systems. Slow-release fertilizers delayed crop maturity and should thus be managed differently than ammonium sulfate. Berry weight increased with added K. Redundancy analysis showed a close correlation between marketable yield, berry quality indices, and soil tests, especially K and Ca, indicating the need for soil test calibration.

Keywords: Brix; TAcy; nitrogen; potassium; compositional data; cranberry yield parameters; firmness; local diagnosis; redundancy analysis

Citation: Jamaly, R.; Parent, S.-É.; Parent, L.E. Fertilization and Soil Nutrients Impact Differentially Cranberry Yield and Quality in Eastern Canada. *Horticulturae* **2021**, *7*, 191. https://doi.org/10.3390/horticulturae7070191

Academic Editor: Stefano Marino

Received: 8 June 2021
Accepted: 9 July 2021
Published: 13 July 2021

Publisher's Note: MDPI stays neutral with regard to jurisdictional claims in published maps and institutional affiliations.

Copyright: © 2021 by the authors. Licensee MDPI, Basel, Switzerland. This article is an open access article distributed under the terms and conditions of the Creative Commons Attribution (CC BY) license (https://creativecommons.org/licenses/by/4.0/).

1. Introduction

Cranberry (*Vaccinium macrocarpon* Ait.) is grown in low-fertility acidic (pH 4.0 to 5.0) sandy or peaty soils located in low-lying landscape positions to facilitate water transfer [1]. Wisconsin, Quebec, and Massachusetts are the world's largest cranberry producers [2]. Quebec is the world leader in organically grown cranberries. Berry yield and quality depend on site, fertilization, cultivar, maturity, harvest date, and temperature [3,4]. While the fertilization of conventional systems has been documented to some degree, information on the fertilization of organic systems is scanty.

Fruit set, number of uprights, relative abundance of reproductive uprights (% of total uprights), number of flowers per reproductive upright, and berry weight are useful yield predictors [5,6]. A cranberry plant typically produces one to three fruits per reproductive upright from two to seven flowers. Depending on the cultivar and weather, berries take 60 to 120 d or even more to reach maturity and deep coloring [7]. Berry firmness, size, soluble solids, and ascorbic acid and anthocyanin content are the most important traits for the industry [8–10]. Berry moisture before harvest is also crucial to reduce the economic loss in terms of yield, quality, and drying cost [11,12].

7

Fertilization timing, source, and dosage must be managed to balance berry yield and quality [13]. Nitrogen fertilization is the most effective means to stimulate cranberry growth but it may affect floral induction negatively [14]. Excessive N dosage can produce intraspecific competition among neighboring plants [15]. The N source, timing, and rate should thus be managed carefully [16–18]. If applied in excess, N decreases berry yield and quality [16,19] and increases the risk of overgrowth of vegetative parts [20]. Ammonium sulfate is the most common N fertilizer in cranberry production. Slow-release nitrogen fertilizers are other sources of N fertilizer that may delay berry maturity depending on the N release rate [21]. Decision on proper N dosage and sources involves a trade-off between berry yield and quality that must be addressed locally but has not yet been modeled under the climatic conditions of eastern Canada.

Soil tests have been little documented in cranberry production. Potassium and phosphorus showed variable effects on berry yield and quality [22]. Soil test P and P fertilization were found to be poorly correlated with crop yield [23,24]. The K dosage was found to depend on soil test K, cultivar, and site, hence requiring local calibration [4]. Tissue testing is the most frequent method to diagnose other elements [4,7,25,26].

On the other hand, redundancy analysis provides a means to explore the relationship between cranberry performance and soil tests. To run a multivariate analysis such as redundancy analysis (RDA), soil compositions should be log-ratio transformed to remove false correlations between components that resonate on each other within the constrained sample space of nutrient compositions [27,28].

We hypothesized that (1) berry yield parameters and berry quality are impacted by N, P, K, Mg, B, and Cu dosage and the N source in conventional and organic cranberry agroecosystems, and (2) cranberry performance indices are correlated with soil test. The first hypothesis was tested using a mixed model. The second hypothesis was addressed by RDA. Our objective was to support site-specific nutrient management decisions in cranberry agroecosystems.

2. Materials and Methods

2.1. Experimental Site and Design

One hundred and forty-four permanent plots of cultivar 'Stevens' were delineated in four commercial fields in southern Quebec, Canada (Figure 1). 'Stevens' is characterized by high yield and moderate color development [29]. The stands established in 1995, 1999, 2004, and 2007 at sites 9, 45, A9, and 10, respectively, were monitored from the spring of 2014 to the fall of 2016. Site A9 was under organic farming. Others were under conventional farming. Meteorological data during the 2014 to 2016 period (Appendix A, Figure A1) were obtained from the closest weather stations of Environment Canada [30]. Cranberry stands were irrigated to maintain soil matric potential between -3 and -7 kPa [31].

2.2. Soil Analysis

Soil samples (0 to 15 cm) were collected in each plot before fertilization in June 2014 and every year thereafter. Samples were air-dried and sieved to less than 2 mm. Grain-size distribution was determined by sedimentation [32] and sand composition by hand-sieving. Grain-size distribution is presented in Table 1. Minerals were extracted using the Mehlich III method [33] and quantified by ICP-OES (inductively coupled plasma optical emission spectrometry). The C and N contents were quantified by combustion (Leco CNS-2000 analyzer, St. Joseph, MI, USA). Soil pH was reported as pH_{cacl2}. Results of soil analyses at the onset of the experiment are presented in Table 1.

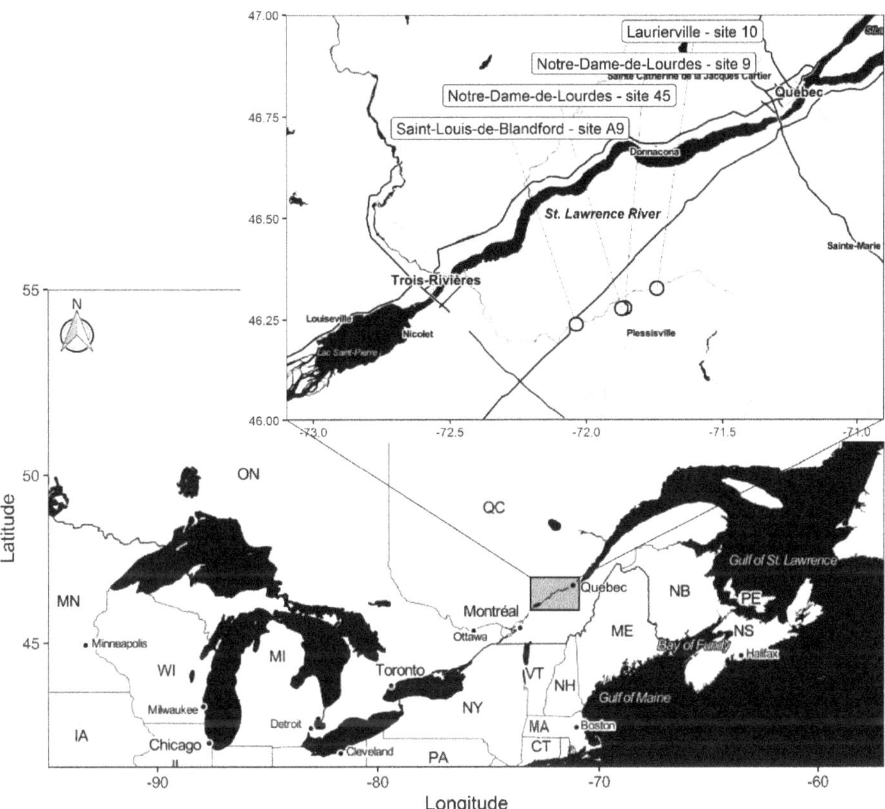

Figure 1. Map of the four study sites located on the south shore of the St. Lawrence River in Quebec, Canada.

Table 1. Soil particle size distribution and mean and standard deviation (SD) of soil data at each experimental site across 144 plots in June 2014.

Site	10	45	9	A9
		g kg^{-1}		
Clay	23	26	28	36
Silt	40	32	41	36
Sand total	937	943	932	928
• 1–2 mm	6.1	19.1	34.1	13.2
• 0.5–1.0 mm	28.3	70.4	161.5	72.1
• 0.25–0.5 mm	332.9	283.5	405.3	417.3
• 0.1–0.25 mm	466.5	478.5	261.8	380.7
• 0.25–0.01 mm	103.2	91.5	69.2	44.7
		mean ± standard deviation		
pH (0.01 M CaCl$_2$)	4.2	4.3 ± 0.1	4.3 ± 0.1	4.1 ± 0.1
		g kg^{-1} (mean ± standard deviation)		
Total C	3942 ± 664	12,170 ± 4218	10,615 ± 1875	10,528 ± 1836
Total N	162±88	479 ± 182	436 ± 173	400 ± 183

Table 1. *Cont.*

Site	10	45	9	A9
Mehlich-3 analysis		mg kg^{-1} (mean ± standard deviation)		
P	63 ± 15	100 ± 21	164 ± 29.6	91 ± 19.2
K	8.5 ± 1.5	26.1 ± 6.8	16.6 ± 2.8	10.7 ± 3.2
Ca	31.6 ± 5.5	70.3 ± 24	106.1 ± 24.8	14.7 ± 4.1
Mg	10.7 ± 2	6.6 ± 2.6	9.8 ± 2.1	3.9 ± 1.3
Cu	1.8 ± 0.6	1.1 ± 0.6	1.7 ± 0.5	0.9 ± 0.3
Zn	0.6 ± 0.1	1.0 ± 0.2	2.0 ± 0.5	0.4 ± 0.1
Mn	1.5 ± 0.3	0.7 ± 0.3	1.2 ± 0.5	0.4 ± 0.3
Fe	177 ± 25	3043 ± 4	282 ± 30	239 ± 46
Al	594 ± 175	1240 ± 22	1510 ± 175	1488 ± 114

2.3. Fertilization

The experiment comprised 18 treatments (Appendix A, Table A1) arranged as randomized block designs and replicated twice at each site, for a total of eight observations per treatment per year. Plot size was 12 m^2 (4 by 3 m). Fertilizer treatments are presented in Table 2. The transition to organic farming at site A9 started in 2015. The N treatments (0, 15, 30, 45, or 60 kg N ha^{-1}) comprised ammonium sulfate (21% N) or sulfur-coated urea (SCU = 24% N, 5% P$_2$O$_5$, 11% K$_2$O) on conventional sites and certified fish emulsions (6% N, 1% P$_2$O$_5$, 1% K$_2$O) or amino acids of plant origin (8% N) on the organic site (Table 2). The K doses (0, 40, 80, or 120 kg K ha^{-1}) were applied as potassium sulfate (50% K) or sulfate of potassium and magnesium (18% K and 9% Mg). Phosphorus was supplied as triple superphosphate (46% P$_2$O$_5$) on the conventional sites or as bone meal (13% P$_2$O$_5$) on the organic site. Two Mg doses (0 or 12 kg Mg ha^{-1}), two Cu doses (0 or 2 kg Cu ha^{-1}), and two B doses (0 or 1 kg B ha^{-1}) were applied as Epsom salt (9% Mg), copper sulfate (25% Cu), and sodium borate (20% B), respectively.

Where one element was varied, other elements were applied at rates of 45 kg N ha^{-1}, 15 kg P ha^{-1}, 80 kg K ha^{-1}, 12 kg Mg ha^{-1}, 2 kg Cu ha^{-1}, and 1 kg B ha^{-1}. Fertilizers were applied uniformly by hand during the growing season. The N fertilizer was applied within a 3- to 4-week window that coincides with fruit set and initial bud formation [34]. From early June to mid-June, B, Cu, and Mg fertilizer were applied during bud break and bud elongation. Thereafter, NPK were applied at four occasions as follows: 15% at early flowering (29 June to 2 July), 35% at 50% flowering (8 to 11 July), 35% at 50% fruit set (16 to 19 July), and 15% 1 to 2 weeks after the third application.

Table 2. Fertilization treatments during the 2014–2016 period at conventional and organic sites.

	2014	2015		2016	
	Conventional	Conventional	Conventional	Organic	Conventional	Organic
Nutrient			kg element ha^{-1}			
N	0, 15, 30. 45, 60	0, 15, 30. 45, 60	0, 15, 30. 45, 60	0, 15, 30, 45, 60	0, 15, 30. 45, 60	0, 15, 30. 45, 60
P	0, 15, 30	0, 15, 30	0, 15, 30	0, 15, 30	15	15
K	0, 40, 80, 120	0, 40, 80, 120	0, 40, 80, 120	0, 40, 80, 120	0, 40, 80, 120	0, 40, 80, 120
Mg	0 or 12	0 or 12	0 or 12	0 or 12	12	12
Cu	0 or 2	0 or 2	0 or 2	0 or 2	2	2
B	0 or 1	0 or 1	0 or 1	0 or 1	1	1

2.4. Plant Measurements

Yield parameters used as yield predictors were flower counts, number of reproductive uprights, number of flowers per reproductive upright, berry counts, number of fruiting uprights, berry counts per fruiting upright, and fruit set (ratio of berry counts to flower counts). The counts of flowers and reproductive uprights were measured in 2014 and 2015 at the end of June on four representative areas totaling 0.37 m^2 per plot while fruits and fruiting uprights were counted in mid-August. Berries were hand-harvested in four areas totaling 0.37 m^2 in each plot, one to two weeks before starting the planned flooding operation at the beginning of October, to avoid too early flooding. Berries were counted and weighed to derive average berry weight (g) and marketable yield (Mg ha^{-1}).

2.5. Berry Quality

Fruit quality followed commercial criteria of the USDA shipping point and market inspection instructions for fresh cranberries (USDA, 2007). One kilogram of randomly selected berries was weighed to determine berry quality after discarding unmarketable fruits. Berry quality was determined as moisture content, total soluble solid concentration (Brix), total anthocyanin concentration, acidity, and firmness [10].

Berries were frozen at $-10\,°C$ and analyzed for moisture content, TAcy [35], and °Brix (refractometry) for soluble solids at the Ocean Spray quality department in Warren, Wisconsin. As berries were harvested before commercial harvesting, average TAcy indices [24] could be lower than market requirements of 350 to 450 mg kg^{-1} for bonus payments reachable at harvest. Berry firmness was quantified using the TA.TX2 Texture Analyzer (Texture Technologies Inc., Scarsdale, NY, USA) [36]. Fifty berries per treatment were refrigerated overnight then maintained at room temperature for 1 to 2 h before performing the test. Pre-test speed was 1 mm s^{-1}, test speed was 2 mm s^{-1}, post-test speed was 10 mm s^{-1}, and trigger force was 0.1 N. Firmness was reported in N mm^{-1} as the mean and standard deviation of 50 samples.

2.6. Statistical Analysis

Statistical analyses were conducted in the R statistical environment version 4.0.5 [37]. We used the R meta-package tidyverse version 1.3.0 [38] for generic data analysis, weathercan [39] for historical weather data, and ggmap [40] for spatial visualization. There were 13 dependent variables including seven yield predictors, as follows: flower counts, number of reproductive uprights, number of flowers per reproductive upright, berry counts, number of fruiting uprights, berry counts per fruiting upright, and fruit set. Other dependent variables were quality indices (TAcy, Brix, firmness, and berry moisture), marketable yield, and berry weight. The experimental setup was analyzed as a mixed model with treatments as fixed factors and years, sites, and replications as random factors [41]. Ammonium sulfate (21-0-0) was set as the reference fertilizer treatment to run the nlme model. Outliers were removed by Z-score test [42] if they exceeded 5 times the standard deviation (5.85% of total observations).

Tests of significance ($p = 0.05$) were used to reject the null hypothesis, but not to accept it as true [43]. Non-significant results did not mean that there was no difference between groups or there were no treatment effects [44]. For each primary outcome, we computed 95% compatibility intervals [44].

There were twelve soil properties and six cranberry performance indicators. The dataset of matrix Y (cranberry yield and quality indices) and explanatory matrix X (pH, N, P, K, Mg, Cu, Ca, Zn, Mn, Fe, Al, and C) was explored by redundancy analysis (RDA) to analyze the impact of "explanatory variables" on "response variables". The R packages to run RDA were the vegan version 2.5-7 [45] for RDA, R meta-package for ordination, and compositions for *clr* transformations to avoid spurious correlations. A permutation procedure was performed (anova.cca function in vegan package) [45] to test the significance of RDA models. Soil test nutrient concentrations were transformed into centered log-ratios [46] before conducting RDA due to Euclidean geometry. The centered log-ratios (clr) were computed as follows [27,28]:

$$clr(x_i) = ln\left(\frac{x_i}{g(x_i)}\right)$$

where x_i is the ith nutrient soil concentration and $g(x_i)$ is the geometric mean. The *clr* transformation allows for computing Euclidean distances between any two compositions [29]. Redundancy analysis (RDA) related matrix Y (cranberry performance) to explanatory matrix X (soil test) based on Euclidean distance between observations [47,48]. Indeed, due to closure to the bounded measurement unit, compositional data should be log-ratio transformed before running linear univariate or multivariate statistical analyses [49].

3. Results

3.1. Effect of Nitrogen Source on Berry Yield and Quality

Berry quality was impacted significantly ($p < 0.05$) by N sources (Figure 2). The effects of SCU and fish emulsions differed compared to ammonium sulfate depending on the variable tested. Berry counts, flower counts, number of fruiting and reproductive uprights, and percentage of fruit set tended to decrease adding amino acids (8-0-0). Organic fertilizers (6-1-1 and 8-0-0) had similar effects on berry yield ($p > 0.05$). Fish emulsions and SCU decreased TAcy by four to six units while fish emulsions increased firmness by 11%, indicating delayed maturity. Compared to ammonium sulfate, SCU increased berry yield by 13% ($p < 0.05$) but reduced TAcy ($p < 0.05$). TAcy responded differently than firmness and °Brix to fish emulsions. Overall, average anthocyanin content decreased (7.23 mg TAcy 100 g^{-1}) significantly with the organic fertilizer (6-1-1). The °Brix and berry moisture showed no significant differences between nitrogen sources.

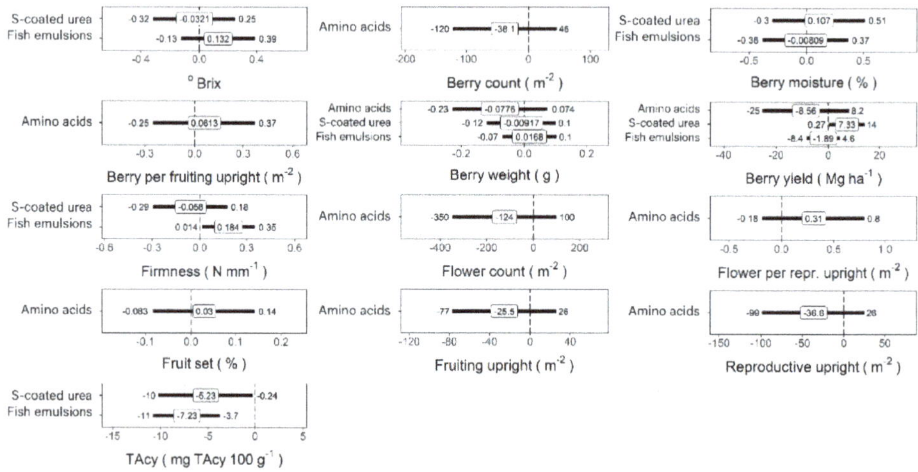

Figure 2. Coefficients of the linear mixed model on the x axes showing the effects on berry yield and quality of amino acids (8-0-0), sulfur-coated urea (24-5-11), and fish emulsions (6-1-1) compared to ammonium sulfate (21-0-0) (y axes). The in-box black line shows 95% confidence intervals. The on-line values are the mean and lower and upper limits for 95% confidence intervals.

3.2. Effect of N, P, and K Regimes on Yield Parameters and Fruit Quality

The N and K fertilization impacted significantly berry yield and quality while the effect of P fertilization was not significant (Figures 3 and 4). The effects of Mg, B, and Cu regimes were also not significant (Appendix A, Figures A2 and A3). Berry yield responded non-linearly to N fertilization ($p < 0.05$) and tended to plateau between 30 and 60 kg N ha^{-1}. The highest yield of 33 Mg ha^{-1} was reached at 45 kg N ha^{-1} under organic farming. The highest yield average of 48 Mg ha^{-1} was reached at 30 to 45 N ha^{-1} under conventional farming.

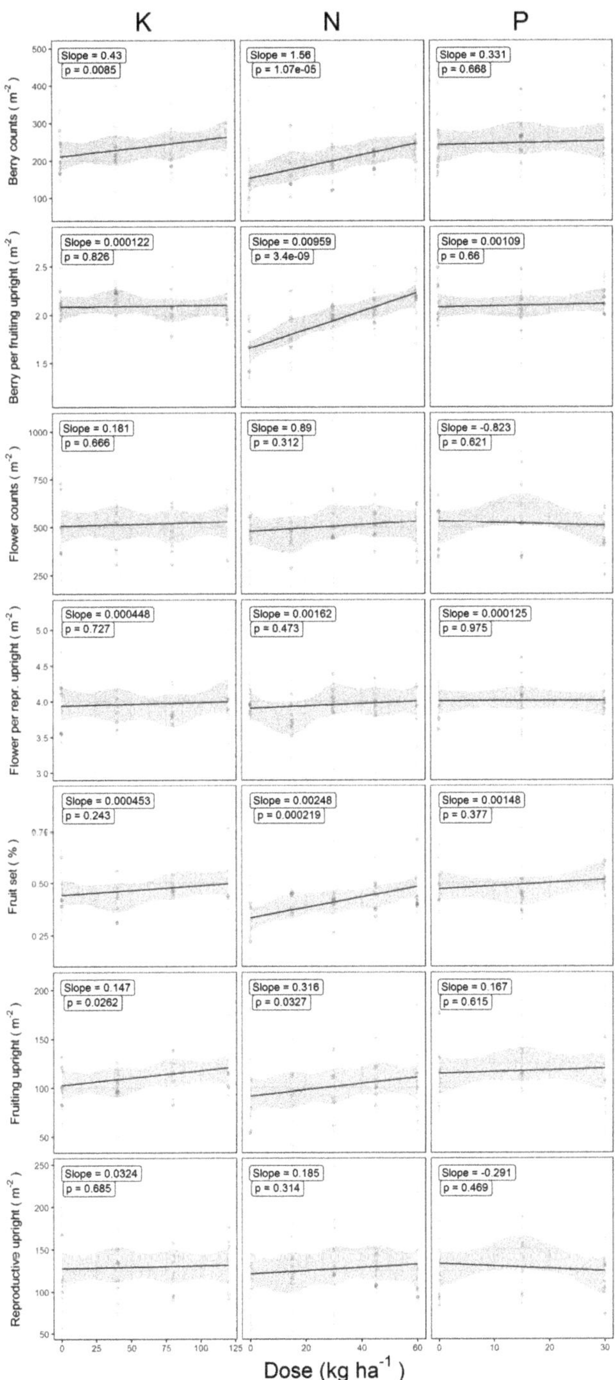

Figure 3. Response of cranberry yield components to added N, P, and K. The solid line represents the model fit with the slope and intercept of the line; the shaded area represents the 95% confidence interval.

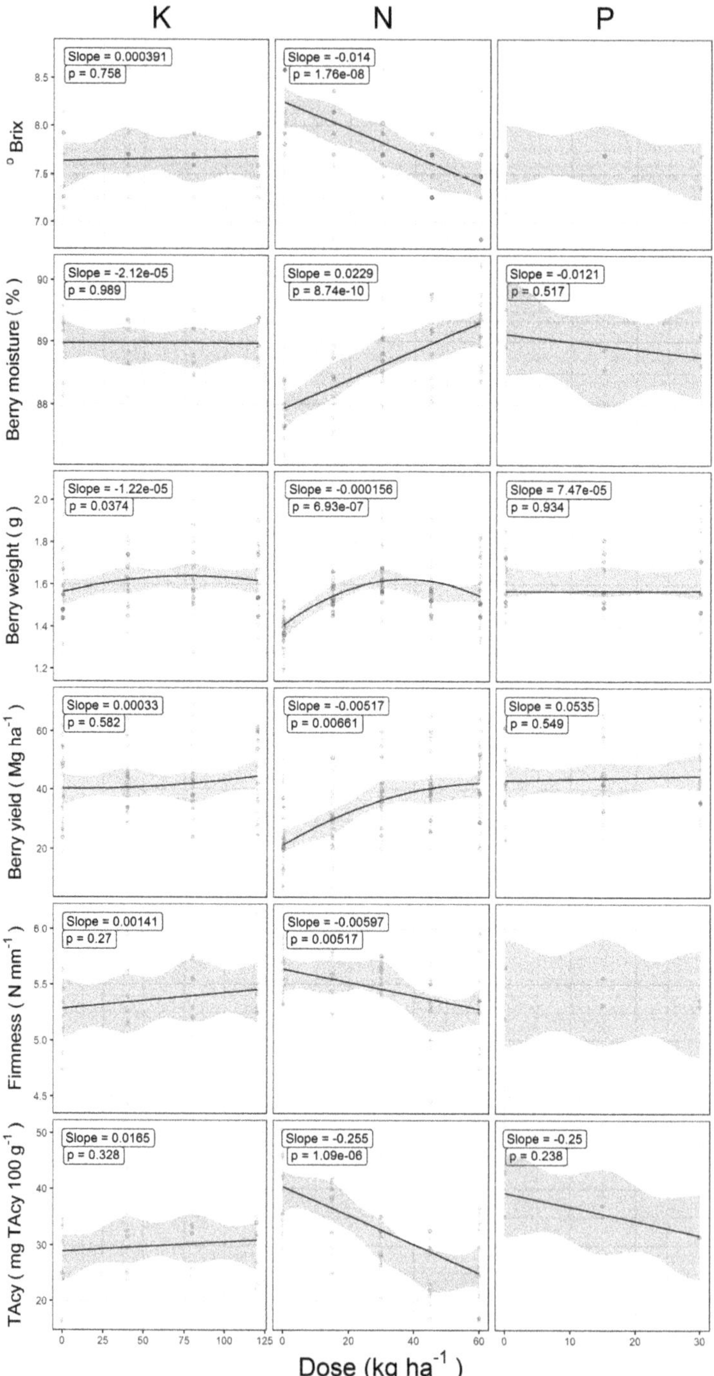

Figure 4. Response of berry yield, weight, and quality indices to added N, P, and K. The solid line represents the model fit with the slope and intercept of the line; the shaded area represents the 95% confidence interval.

Fruit set increased linearly between 0 and 60 kg N ha^{-1} and between 0 and 120 kg K ha^{-1} (Figure 3). Likewise, counts of fruiting uprights increased linearly from 95 to 123 by adding N and K. Berry count per fruiting upright increased from 205 to 257 with K additions between 0 and 120 kg K ha^{-1}. Berry weight was highest with 30 to 45 kg N ha^{-1} and 40 to 120 kg K ha^{-1}. There was no significant yield response ($p > 0.05$) to added K in 2015, where the maximum yield was 36 Mg ha^{-1}, but there was a significant response ($p < 0.05$) in 2014 and 2016 where yields reached 54 and 44 Mg ha^{-1}, respectively. Hence, yield response to K fertilization was apparently related to yield level.

There were 3.9 flowers per reproductive upright, 2.1 berries per fruiting upright, and 48% of fruit set at N application rate of 45 kg N ha^{-1}. Each kg of N per ha increased berry moisture by 0.023% unit in the range of 0 to 60 kg N ha^{-1} (Figure 4). Cranberry responded negatively to added N for °Brix, TAcy, firmness, and positively for the percentage of berry moisture (Figure 4). There was a non-linear response to added N for berry weight and berry yield. The K positively impacted °Brix, TAcy, and firmness, increasing gently between 0 and 120 kg K ha^{-1}. Response to added P was not significant. Response to Mg, B, and Cu regimes was also not significant (Appendix A, Figures A2 and A3).

3.3. Correlations among Berry Yield and Quality Parameters

Relationships among yield parameters are presented in Figure 5. There were close correlations between the number of reproductive uprights, flower counts, number of fruiting uprights, and berry counts (Figure 5). Flower counts and the number of reproductive uprights ($r = 0.95$), as well as berry counts and number of fruiting uprights ($r = 0.93$) were closely related with marketable yield. Fruit set showed moderate correlation ($r = 0.77$) with berry counts and flower counts. Fruit set fluctuated between 33% to 59% in 2014 and 36% to 41% in 2015. The relationship between fruit set and berry yield was thus inconsistent across years.

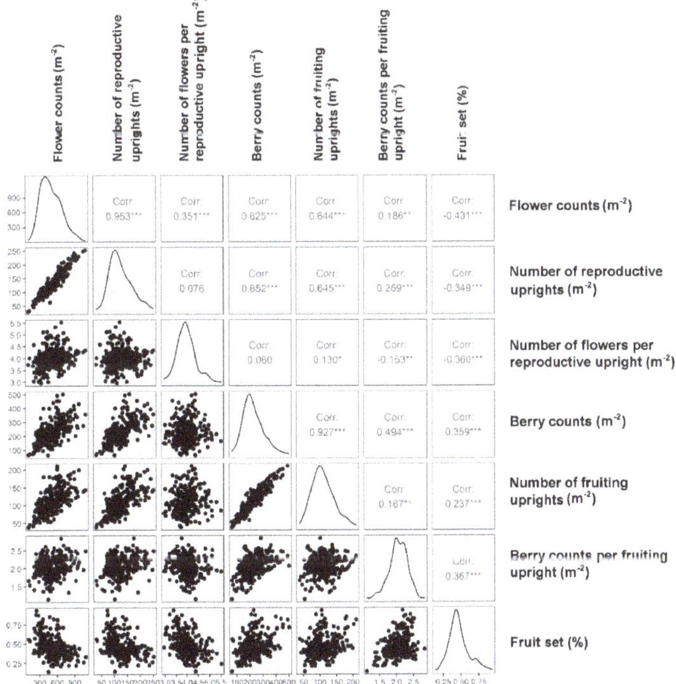

Figure 5. Matrix of correlations among yield components in the cranberry data set. The diagonal shows graphs of the original data after adjusting for all other variables. Upper: Pearson correlation. ** Correlation significant at the 0.05 level; *** Correlation significant at the 0.01 level.

3.4. Redundancy Analysis

Relationships between soil test and cranberry performance are illustrated in Figure 6. The first RDA axis was significant (RDA1: F = 19.44, p = 0.001), and explained 85.9% of the total variation. The anova.cca ranking was Ca (F = 11.98, $p < 0.001$), K (F = 3.92, $p < 0.05$), Fe (F = 2.57, $p < 0.05$), and pH_{CaCl2} (0.01 M $CaCl_2$) (F = 4.53, $p < 0.05$). Soil pH_{CaCl2} was the most important soil factor affecting cranberry performance. The optimum pH_{CaCl2} was 4.14 ± 1.66. Among soil nutrient tests, Ca ranked first, explaining 56.3% of the cumulative variance. The second key factor was K with 18.4% of the cumulative variance, followed by Fe, and Mg. The K and Ca were negatively related.

Berry yield and quality indices were related to soil tests. Arrows in RDA illustrated the complex relationships between them (Figure 6). Berry weight was related negatively to TAcy. The °Brix was negatively related to berry moisture. Berry moisture was positively related to pH. There was a positive relationship between soil test Ca and TAcy. Soil test K, Fe, and Mg were related positively to °Brix. Berry yield was negatively related to soil test K, Fe, and Mg. Berry weight was related negatively to soil test Ca and Zn. Yield and berry moisture were located in the lower right quadrant, and were positively related to pH. Berry firmness and weight were closely related to each other. The Ca and TAcy located in the upper right quadrants were positively related. In contrast with K fertilization applied as potassium sulfate (Appendix A, Table A1), soil test K was positively related to °Brix but negatively related to berry yield.

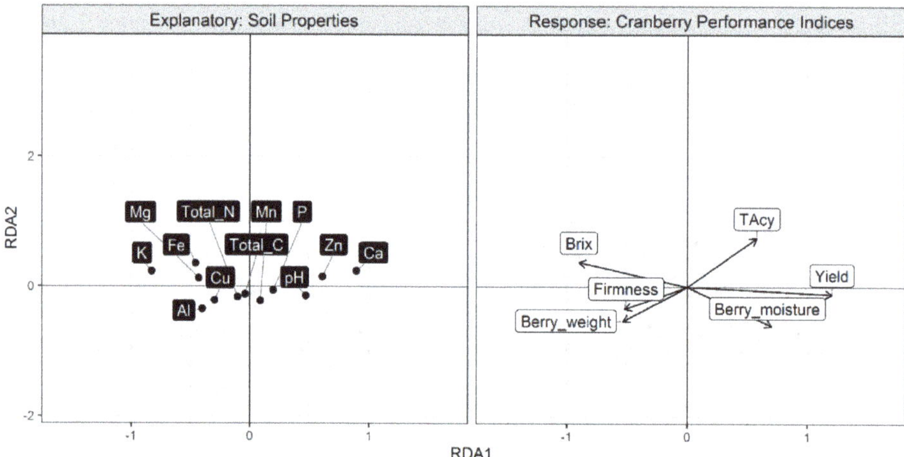

Figure 6. Redundancy analysis relating soil properties on the left to cranberry performance indices on the right. Distances were computed using the Euclidean distance.

4. Discussion

4.1. Impact of Fertilization on Berry Quality

Fruit quality characteristics are of prime importance in cranberry breeding programs [29]. Fruit quality depends on genetics, management, and the environment [50,51]. Berry quality traits comprise anthocyanin content (color), fruit texture characteristics (crispness, hardness, juiciness, and mealiness), fruit anatomy (skin, flesh, or air pocket), and fruit external appearance (size and shape) [52]. Cultivar 'Stevens' showed the smallest average berry size (1.51 cm) among commercial cultivars [8]. Stevens [53] first suggested that factors such as temperature and rainfall could impact fruit keeping quality through fungal infection and disturbed fruit physiology.

The N and K fertilization regimes can also influence cranberry production. Increased N dosage resulted in a linear decrease in °Brix, firmness, and TAcy, and increased moisture content. In general, a high N dosage was found to reduce TAcy [4,54], firmness, [55] and °Brix [29]. In contrast, Davenport [56] found no significant effect on TAcy by applying up to 44 kg N ha^{-1}. As anthocyanins are located primarily in the fruit epidermal layers [57], TAcy decreases as fruit size increases [58]. While TAcy increased [59], fruit firmness was found to decrease as the fruit ripened [10]. Bourne [60] found that similar to cranberry, apple firmness decreased with N additions. In contrast with previous research showing a linear response of hybrid cultivars [61], we found a quadratic relationship between added N and berry weight of 'Stevens'. Cranberry quality indices showed a small but significant response trend adding 80 to 120 kg K ha^{-1}. Crop response to added K was found to be related to yield level in eastern Canada, which was not necessarily the case across cultivars and sites in North America [19,56].

4.2. Impact of Fertilization on Yield Parameters

The numbers of flowers per reproductive upright and number of berries per fruiting upright were within the range of published values [7,62]. As the number of reproductive uprights and flower counts m^{-2} are related to fruit yield, they provided performance indices a few months before harvest [63]. Berry count m^{-2} was found to be the yield parameter most closely related to berry yield. Fruit set is also an important indicator of yield variation [5,64] as related to sunlight [65], pollination [64], and yield [66], but the relationship between carbohydrate concentration, fruit set, and yield can be inconsistent [64]. Carbon allocation between reproductive and vegetative parts [67] depends on temperature [63,68] and is affected by excessive rainfall or drought [69].

Marketable yields were higher by 10% to 25% under the conventional vs. organic systems as reported elsewhere [70,71]. In both conventional and organic farming systems, the effect of N dosage on yield plateaued between 30 and 60 kg N ha^{-1}, within the 20 to 65 kg N ha^{-1} range reported in Davenport [56] and the 39 to 56 kg N ha^{-1} range reported in DeMoranville and Ghantous [21]. While 'Stevens' yield was found to plateau at 20 Mg ha^{-1} after adding 22 to 44 kg N ha^{-1} [56], 'Stevens' yields up to 40 Mg ha^{-1} decreased and stolon weight increased where N dosage exceeded 34 kg N ha^{-1} [20]. The N dosage is site-specific. A too high N dosage (60 kg N ha^{-1}) lead to plant crowding [15,50] and overgrowth of the vegetative parts [72].

Slow-release N fertilizers such as SCU in conventional cranberry production and certified fish emulsions in organic production should be managed differently than ammonium sulfate. Berry firmness increased by applying an organic source likely due to delay in berry maturity. Anthocyanin content tended to decrease where ammonium sulfate was replaced by organic nitrogen or SCU as also reported in [18]. Compared to ammonium sulfate, SCU and fish emulsions should be applied earlier in the season to sustain N release during the whole season and avoid delaying berry maturity and the reddening finish close to harvest time.

4.3. Ranking of Soil Test Variables

Soil test calibration has been little addressed in cranberry production except for P [23,24]. As first shown by Bray [73,74], soil tests for nutrients showing low mobility in the soil as well as nutrient source and placement must show different coefficients of efficiency. Crop response to fertilization dosage and soil test value are generally addressed separately then assembled into a modified Mitscherlich equation. In this paper, we addressed crop response to added nutrients separately using a mixed model and soil test using RDA.

As shown by RDA, soil test K appeared to be the most discriminant nutrient for cranberry performance, followed by Ca, Fe, and Mg. Soil test K is often low in cranberry soils because K is easily leached at soil pH values less than 5.5 [75], and cation-exchange capacity is low in sandy soils [76]. Soil K can also be supplied as non-exchangeable K by primary and secondary soil minerals such as feldspar, mica, and illite, which are common in soils of eastern Canada [77–79]. Mica K is released much faster compared to phlogopite, biotite, and muscovite [80]. The reactivity of soil minerals also depends on the grain-size distribution, pH [81], and rhizosphere exploration of the soil [82]. Owing to the low clay and high sand contents in cranberry soils, soil minerals are assumed to contribute little to cranberry K requirements. It is thus difficult to maintain high soil test K values in acidic cranberry sandy soils. Low soil test K results in low berry yield [19,22], requiring fertilization.

The K fertilization should meet annual K requirements, at a rate that avoids affecting the uptake of other cations. Added K may trigger Ca leaching and reduce Ca uptake [17,22, 26]. Since tissue K increases with K dosage or soil test K level while tissue Mg might decline [4,17], the tissue K–Mg interaction should also be monitored [83]. The RDA supported calibrating soil tests to provide minimum critical soil test values to sustain the cranberry production on sandy soils. Nevertheless, our results supported the present K recommendation of 54 to 92 kg K ha^{-1} at a low soil test K [84]. The minimum soil test K value to be maintained in cranberry soils should be addressed in future research.

The RDA indicated that soil pH played a key role in cranberry nutrient management [18,85]. High levels of Ca in cranberry soils can reduce the absorption of Mn, Fe, and Zn, potentially reducing berry yield and size [83]. To address nutritional balance in cranberry crops and predict cranberry yield and nutrient requirements, tissue testing is complementary to soil testing [86].

5. Conclusions

This paper quantified the trade-off between berry yield and quality as driven primarily by N fertilization. Berry count per fruiting upright, fruit set, and berry weight responded consistently to N treatments. Berry yield could be predicted most accurately from berry counts per fruiting upright. Nitrogen fertilization increased berry yield nonlinearly and decreased berry quality indices linearly. The SCU and fish emulsions delayed berry maturity and should thus be managed differently than ammonium sulfate through earlier applications to account for the slow-release patterns.

Cranberry responded moderately to K where yield potential was high. As shown by RDA, cranberry performance was related to soil pH and soil test nutrients. The K and Ca were negatively correlated between them, indicating an upper limit for K additions. The RDA indicated close relationships between cranberry performance indices and soil properties, and thus supported the need for further soil test calibration.

Author Contributions: Conceptualization, R.J., S.-É.P. and L.E.P.; data curation, R.J., S.-É.P. and L.E.P.; formal analysis, R.J., S.-É.P. and L.E.P.; funding acquisition, S.-É.P. and L.E.P.; methodology, R.J., S.-É.P. and L.E.P.; project administration, S.-É.P.; Software: R.J. and S.-É.P.; supervision, S.-É.P. and L.E.P.; validation, R.J., S.-É.P. and L.E.P.; writing—original draft preparation, R.J., S.-É.P. and L.E.P.; writing—review and editing, R.J., S.-É.P. and L.E.P. All authors have read and agreed with the published version of the manuscript.

Funding: This collaborative research project was supported by Les Atocas de l'Érable Inc., Les Atocas Blandford Inc., La Cannebergière Inc., the Natural Sciences and Engineering Research Council of Canada (RDCPJ-469358-14).

Informed Consent Statement: Not applicable.

Conflicts of Interest: The authors declare no conflict of interest.

Appendix A

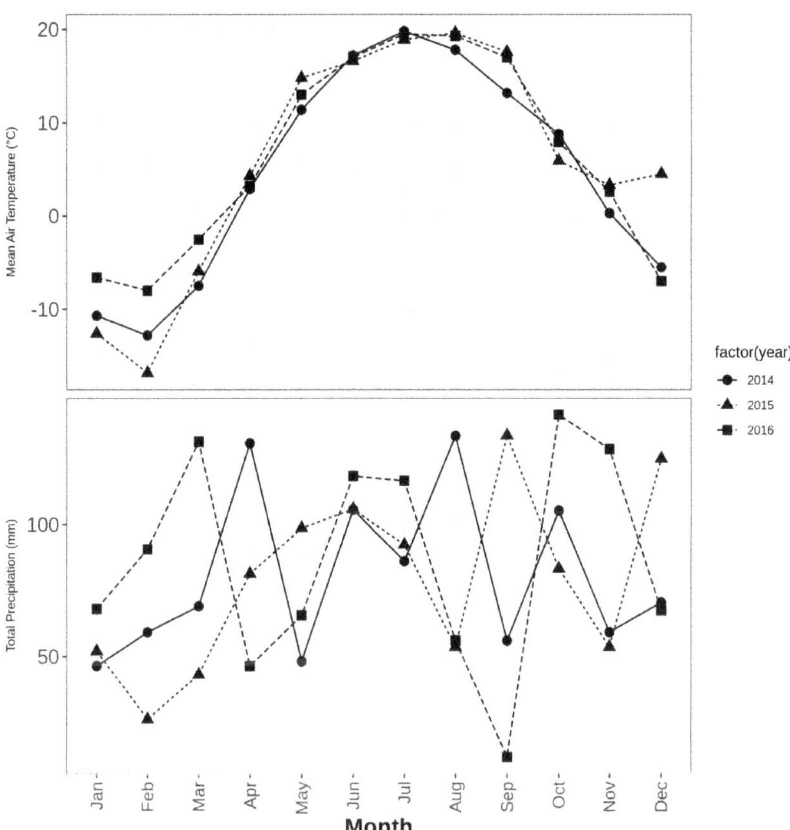

Figure A1. Monthly mean air temperature and total precipitation at the four experimental sites in Quebec, Canada.

Table A1. Fertilizer regimes applied during the 3-year experimental period.

	Cropping Systems	2014	2015	2016
N	Conventional	Ammonium sulfate (21-0-0)	Ammonium sulfate (21-0-0)	Ammonium sulfate (21-0-0) Sulfur coated urea (24-5-11)
	Organic		Amino acids (8-0-0)	Fish emulsions (6-1-1)
P	Conventional	Triple superphosphate (0-46-0)	Triple superphosphate (0-46-0)	Triple superphosphate (0 16 0)
	Organic		Bone meal (0-13-0)	Bone meal (0-13-0)
K	Conventional	Potassium sulfate (0-0-50)	Potassium sulfate (0-0-50)	Potassium sulfate (0-0-50)
	Organic			
Mg	Conventional	Epsom salt (9% Mg)	Epsom salt (9% Mg)	Epsom salt (9% Mg)
	Organic	Sulfate of potassium and magnesium (0-0-22)	Sulfate of potassium and magnesium (0-0-22)	Sulfate of potassium and magnesium (0-0-22)

Table A1. *Cont.*

	Cropping Systems	2014	2015	2016
Cu	Conventional	Copper sulfate	Copper sulfate	Copper sulfate
	Organic	(25% Cu)	(25% Cu)	(25% Cu)
B	Conventional	Solubor	Solubor	Solubor
	Organic	(20% B)	(20% B)	(20% B)

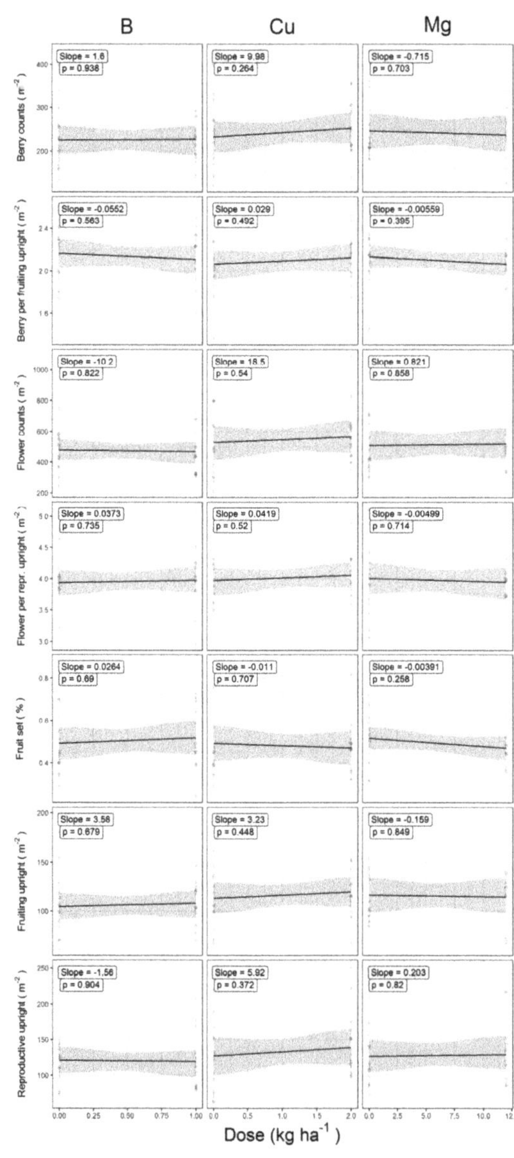

Figure A2. Response of cranberry yield components to added B, Cu, and Mg. The solid line represents the model fit with the slope and intercept of the line; the shaded area represents the 95% confidence interval.

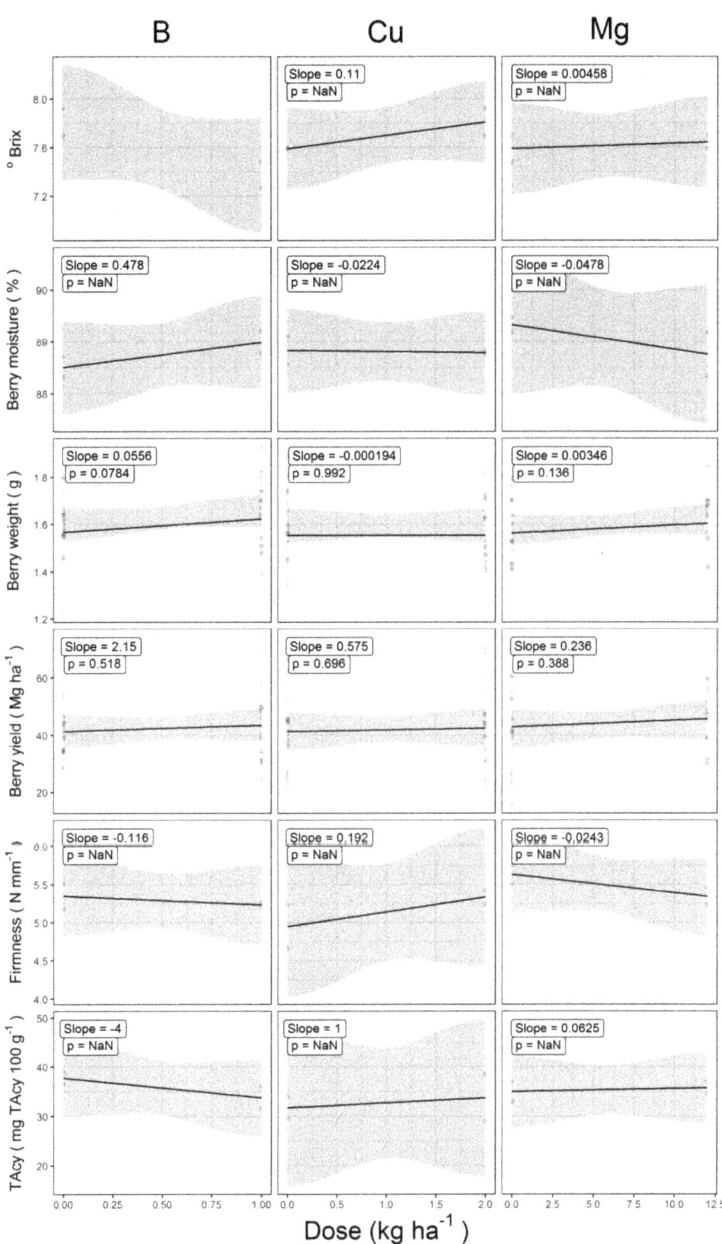

Figure A3. Response of berry yield, weight, and quality indices to added B, Cu, and Mg. The solid line represents the model fit with the slope and intercept of the line; the shaded area represents the 95% confidence interval.

References

1. Sandler, H.A.; DeMoranville, C.J. Cranberry production: A guide for Massachusetts—Summary edition. In *Cranberry Production Guide*; Station, C., Ed.; University of Massachusetts: Wareham, MA, USA, 2008; Volume 5, p. 37.
2. APCQ. Association des Producteurs de Canneberges du Québec (Quebec Cranberry Growers Association). Available online: http://www.notrecanneberge.com/Content/Page/Stats (accessed on 1 May 2021).
3. Vorsa, N.; Polashock, J.; Cunningham, D.; Roderick, R. Genetic inferences and breeding implications from analysis of cranberry germplasm anthocyanin profiles. *J. Am. Soc. Hortic. Sci.* **2003**, *128*, 691–697. [CrossRef]
4. Hart, J.M.; Strik, B.C.; DeMoranville, C.; Davenport, J.R.; Roper, T. Cranberries: A nutrient management guide for south coastal Oregon. In *OSU Extension Catalog*; Service, E., Ed.; Oregon State University: Corvallis, OR, USA, 2015; p. 52.
5. Baumann, T.E.; Eaton, G.W. Competition among berries on the cranberry upright. *J. Am. Soc. Hortic. Sci.* **1986**, *111*, 869–872.
6. Eaton, G.W. Floral induction and biennial bearing in the cranberry. *Fruit Var. J.* **1978**, *32*, 58–60.
7. Dana, M.N. *Cranberry Management*; Prentice-Hall: Englewood Cliffs, NJ, USA, 1990; pp. 334–362.
8. Diaz-Garcia, L.; Rodriguez-Bonilla, L.; Phillips, M.; Lopez-Hernandez, A.; Grygleski, E.; Atucha, A.; Zalapa, J. Comprehensive analysis of the internal structure and firmness in American cranberry (*Vaccinium macrocarpon* Ait.) fruit. *PLoS ONE* **2019**, *14*, e0222451. [CrossRef] [PubMed]
9. Berezina, E.V.; Brilkina, A.A.; Veselov, A.P. Content of phenolic compounds, ascorbic acid, and photosynthetic pigments in *Vaccinium macrocarpon* Ait. dependent on seasonal plant development stages and age (the example of introduction in Russia). *Sci. Hortic.* **2017**, *218*, 139–146. [CrossRef]
10. Forney, C.F.; Kalt, W.; Jordan, M.A.; Vinqvist-Tymchuk, M.R.; Fillmore, S.A.E. Blueberry and cranberry fruit composition during development. *J. Berry Res.* **2012**, *2*, 169–177. [CrossRef]
11. Rennie, T.J.; Mercer, D.G. Effect of blanching on convective drying and osmotic dehydration of cranberries. *Trans. ASABE* **2013**, *56*, 1863–1870. [CrossRef]
12. Sinha, N.; Sidhu, J.; Barta, J.; Wu, J.; Cano, M.P. *Handbook of Fruits and Fruit Processing*; John Wiley & Sons: Hoboken, NJ, USA, 2012.
13. DeMoranville, C.J.; Sandler, H.A.; Jeranyama, P.; Averill, A.L.; Caruso, F.L.; Sylvia, M.; Ghantous, K. *2014 Chart Book: Table of Contents*; University of Massachusetts Amherst: Amherst, MA, USA, 2014.
14. McArthur, D.A.J.; Eaton, G.W. Cranberry growth and yield response to fertilizer and paclobutrazol. *Sci. Hortic.* **1989**, *38*, 131–146. [CrossRef]
15. Engels, C.; Kirkby, E.; White, P. Mineral nutrition, yield and source–sink relationships. In *Marschner's Mineral Nutrition of Higher Plants*; Elsevier Ltd.: Amsterdam, The Netherlands, 2012; pp. 85–133. [CrossRef]
16. Vanden Heuvel, J.E.; Davenport, J.R. Growth and carbon partitioning in cranberry uprights as influenced by nitrogen supply. *Hortscience* **2006**, *41*, 1552–1558. [CrossRef]
17. Eaton, G.W. Effect of N, P, and K fertilizer applications on cranberry leaf nutrient composition, fruit color and yield in a mature bog. *J. Am. Soc. Hortic. Sci.* **1971**, *96*, 430–433.
18. Rosen, C.J.; Allan, D.L.; Luby, J.J. Nitrogen form and solution pH influence growth and nutrition of 2 *Vaccinium* clones. *J. Am. Soc. Hortic. Sci.* **1990**, *115*, 83–89. [CrossRef]
19. Davenport, J.; McMoranville, J.; Hart, J.M.; Patten, K.; Peterson, L.; Planer, T.; Poole, A. Cranberry tissue testing for producing beds in North America. In *OSU Extension Catalog*; Service, E., Ed.; Oregon State University: Corvallis, OR, USA, 1995.
20. Davenport, J.; DeMoranville, C.J.; Hart, J.; Roper, T. *Nitrogen for Bearing Cranberries in North America*; Extension Service, Oregon State University: Corvallis, OR, USA, 2000.
21. DeMoranville, C.J.; Ghantous, K. *2018–2020 Chart Book: Nutrition Management*; University of Massachusetts, Amherst: Amherst, MA, USA, 2018.
22. Roper, T.R. Mineral nutrition of cranberry: What we know and what we thought we knew. In Proceedings of the IX International Vaccinium Symposium, Corvallis, OR, USA, 14–18 July 2008; Volume 810, pp. 613–625.
23. DeMoranville, C. Reducing phosphorus use in cranberry production: Horticultural and environmental implications. In Proceedings of the X International Symposium on Vaccinium and Other Superfruits, Maastricht, The Netherlands, 17–22 June 2012; VanKooten, O., Brouns, F., Eds.; Volume 1017, pp. 447–453.
24. Parent, L.E.; Marchand, S. Response to phosphorus of cranberry on high phosphorus testing acid sandy soils. *Soil Sci. Soc. Am. J.* **2006**, *70*, 1914–1921. [CrossRef]
25. Davenport, J.R.; Provost, J. Cranberry tissue nutrient levels as impacted by three levels of nitrogen fertilizer and their relationship to fruit yield and quality. *J. Plant Nutr.* **1994**, *17*, 1625–1634. [CrossRef]
26. Eaton, G.W.; Meehan, C.N. Effects of N application and K application on leaf composition, yield, and fruit-quality of bearing Mcfarlin cranberries. *Can. J. Plant Sci.* **1976**, *56*, 107–110. [CrossRef]
27. Aitchison, J. *The Statistical Analysis of Compositional Data*; Chapman and Hall Ltd.: New York, NY, USA, 1986; p. 416.
28. Pawlowsky-Glahn, V.; Buccianti, A. *Compositional Data Analysis: Theory and Applications*; Wiley: Hoboken, NJ, USA, 2011.
29. McCown, B.H.; Zeldin, E.L. 'HyRed', an early, high fruit color cranberry hybrid. *Hortscience* **2003**, *38*, 304–305. [CrossRef]

30. National Climate Data and Information Archive of Environment. *The Meteorological Service of Canada*; Government of Canada: Ottawa, ON, Canada, 2021.

31. Caron, J.; Pelletier, V.; Kennedy, C.D.; Gallichand, J.; Gumiere, S.; Bonin, S.; Bland, W.L.; Pepin, S. Guidelines of irrigation and drainage management strategies to enhance cranberry production and optimize water use in North America. *Can. J. Soil Sci.* **2017**, *97*, 82–91. [CrossRef]

32. Kettler, T.A.; Doran, J.W.; Gilbert, T.L. Simplified method for soil particle-size determination to accompany soil-quality analyses. *Soil Sci. Soc. Am. J.* **2001**, *65*, 849–852. [CrossRef]

33. Mehlich, A. Mehlich-3 soil test extractant a modification of Mehlich-2 extractant. *Commun. Soil Sci. Plant Anal.* **1984**, *15*, 1409–1416. [CrossRef]

34. Kennedy, C.D.; Alverson, N.; Jeranyama, P.; DeMoranville, C. Seasonal dynamics of water and nutrient fluxes in an agricultural peatland. *Hydrol. Process.* **2018**, *32*, 698–712. [CrossRef]

35. Fuleki, T.; Francis, F.J. Quantitative methods for Anthocyanins.1. Extraction and determination of total anthocyanin in cranberries. *J. Food Sci.* **1968**, *33*, 72–77. [CrossRef]

36. Lamikanra, O.; Kueneman, D.; Ukuku, D.; Bett-Garber, K.L. Effect of processing under ultraviolet light on the shelf life of fresh-cut cantaloupe melon. *J. Food Sci.* **2005**, *70*, C534–C539. [CrossRef]

37. R Development Core Team. *R: A Language and Environment for Statistical Computing*; R Foundation for Statistical Computing: Vienna, Austria, 2021.

38. Wickham, H.; Averick, M.; Bryan, J.; Chang, W.; McGowan, L.D.A.; François, R.; Grolemund, G.; Hayes, A.; Henry, L.; Hester, J. Welcome to the Tidyverse. *J. Open Source Softw.* **2019**, *4*, 1686. [CrossRef]

39. LaZerte, S.E.; Albers, S. Weathercan: Download and format weather data from Environment and Climate Change Canada. *J. Open Source Softw.* **2018**, *3*, 571. [CrossRef]

40. Kahle, D.; Wickham, H.; Kahle, M.D. ggmap: Spatial Visualization with ggplot2. *R J.* **2013**, *5*, 144–161. [CrossRef]

41. Pinheiro, J.; Bates, D.; DebRoy, S.; Sarkar, D.; R Core Team. nlme: Linear and Nonlinear Mixed Effects Models. *R J.* **2019**, *3*, 139.

42. Clark-Carter, D. z Scores. In *Wiley StatsRef: Statistics Reference Online*; Wiley: Hoboken, NJ, USA, 2014. [CrossRef]

43. Makowski, D.; Ben-Shachar, M.S.; Chen, S.H.A.; Lüdecke, D. Indices of effect existence and significance in the Bayesian framework. *Front. Psychol.* **2019**, *10*, 2767. [CrossRef] [PubMed]

44. Amrhein, V.; Greenland, S.; McShane, B. Scientists rise up against statistical significance. *Nature* **2019**, *567*, 305–307. [CrossRef] [PubMed]

45. Oksanen, J.; Blanchet, F.G.; Kindt, R.; Legendre, P.; Minchin, P.R.; O'hara, R.; Simpson, G.L.; Solymos, P.; Stevens, M.H.H.; Wagner, H. Package 'vegan'. *Community Ecol. Package Version* **2013**, *2*, 1–295.

46. Greenacre, M. Compositional data analysis. *Annu. Rev. Stat. Appl.* **2021**, *8*, 271–299. [CrossRef]

47. Borcard, D.; Gillet, F.O.; Legen, P. *Numerical Ecology with R*, 2nd ed.; Springer: Cham, Switzerland, 2018. [CrossRef]

48. Legendre, P.; Legendre, L. *Numerical Ecology*, 3rd ed.; Elsevier: Amsterdam, The Netherlands, 2012.

49. Filzmoser, P.; Hron, K.; Reimann, C. Univariate statistical analysis of environmental (compositional) data: Problems and possibilities. *Sci. Total Environ.* **2009**, *407*, 6100–6108. [CrossRef]

50. Roper, T. *The Physiology of Cranberry Yield*; Wisconsin Cranberry Crop Management Newsletter: Madison, WI, USA, 2006; p. 21.

51. Parent, L.E.; Rozane, D.E.; de Deus, J.A.L.; Natale, W. Diagnosis of nutrient composition in fruit crops: Major developments. In *Fruit Crops*; Elsevier: Amsterdam, The Netherlands, 2020; pp. 145–156.

52. Gallardo, R.K.; Klingthong, P.; Zhang, Q.; Polashock, J.; Atucha, A.; Zalapa, J.; Rodriguez-Saona, C.; Vorsa, N.; Iorizzo, M. Breeding trait priorities of the cranberry industry in the United States and Canada. *HortScience* **2018**, *53*, 1467–1474. [CrossRef]

53. Stevens, N.E. Thickness of cuticle in cranberry fruits. *Am. J. Bot.* **1932**, *19*, 432–435. [CrossRef]

54. Boucher, V. *Effet des Doses de Fertilisant Azoté sur le Rendement et la Qualité des Fruits de Canneberge Produits au Québec*; Université Laval: Quebec City, QC, Canada, 1999.

55. Jamaly, R.; Marchand, S.; Parent, S.-E.; Gumiere, S.J.; Deland, J.-P.; Parent, L.-E. Impact of Fertilization on the Firmness of Cranberry (*Vaccinium macrocarpon* AIT). In Proceedings of the North American Cranberry Researcher and Extension Workers Conference, Plymouth, MA, USA, 27–30 August 2017.

56. Davenport, J.R. The effect of nitrogen fertilizer rates and timing on cranberry yield and fruit quality. *J. Am. Soc. Hortic. Sci.* **1996**, *121*, 1089–1094. [CrossRef]

57. Sapers, G.M.; Phillips, J.G.; Rudolf, H.M.; Divito, A.M. Cranberry quality selection procedures for breeding programs. *J. Am. Soc. Hortic. Sci.* **1983**, *108*, 241–246.

58. Vorsa, N.; Welker, W.V. Relationship between fruit size and extractable anthocyanin content in cranberry. *Hortscience* **1985**, *20*, 402–403.

59. Vorsa, N.; Johnson-Cicalese, J. American cranberry. In *Fruit Breeding*; Springer: Boston, MA, USA, 2012; Volume 8, pp. 191–223.

60. Bourne, M.C. Texture of temperate fruits. *J. Texture Stud.* **1979**, *10*, 25–44. [CrossRef]

61. Davenport, J.R.; Vorsa, N. Cultivar fruiting and vegetative response to nitrogen fertilizer in cranberry. *J. Am. Soc. Hortic. Sci.* *JASHS* **1999**, *124*, 90. [CrossRef]

62. Brown, A.O.; McNeil, J.N. Fruit production in cranberry (Ericaceae: *Vaccinium macrocarpon*): A bet-hedging strategy to optimize reproductive effort. *Am. J. Bot.* **2006**, *93*, 910–916. [CrossRef]

63. DeVetter, L.; Colquhoun, J.; Zalapa, J.; Harbut, R. Yield estimation in commercial cranberry systems using physiological, environmental, and genetic variables. *Sci. Hortic.* **2015**, *190*, 83–93. [CrossRef]
64. Birrenkott, B.A.; Stang, E.J. Selective flower removal increases cranberry fruit set. *HortScience* **1990**, *25*, 1226–1228. [CrossRef]
65. Roper, T.R.; Klueh, J.; Hagidimitriou, M. Shading, timing and intensity influences fruit-set and yield in cranberry. *Hortscience* **1995**, *30*, 525–527. [CrossRef]
66. DeMoranville, C.J.; Davenport, J.R.; Patten, K.; Roper, T.R.; Strik, B.C.; Vorsa, N.; Poole, A.P. Fruit mass development in three cranberry cultivars and five production regions. *J. Am. Soc. Hortic. Sci.* **1996**, *121*, 680–685. [CrossRef]
67. Burd, M. "Excess" flower production and selective fruit abortion: A model of potential benefits. *Ecology* **1998**, *79*, 2123–2132. [CrossRef]
68. Degaetano, A.T.; Shulman, M.D. A statistical evaluation of the relationship between cranberry yield in New-Jersey and meteorological factors. *Agric. For. Meteorol.* **1987**, *40*, 323–342. [CrossRef]
69. Franklin, H.J.; Stevens, N.E. *Weather and Water as Factors in Cranberry Production*; University of Massachusetts: Amherst, MA, USA, 1946.
70. Reganold, J.P.; Wachter, J.M. Organic agriculture in the twenty-first century. *Nat. Plants* **2016**, *2*, 15221. [CrossRef]
71. Ponti, D.T.; Rijk, H.C.A.; Ittersum, V.M.K. The crop yield gap between organic and conventional agriculture. *Agric. Syst.* **2012**, *108*, 1–9. [CrossRef]
72. De Moranville, C.J. *2017 Chart Book: Nutrition Management*; University of Massachusetts Amherst: Amherst, MA, USA, 2017; p. 15.
73. B Bray, R.H. A nutrient mobility concept of soil-plant relationships. *Soil Sci.* **1954**, *78*, 9–22. [CrossRef]
74. Bray, R.H. Soil-plant relations: I. The quantitative relation of exchangeable potassium to crop yields and to crop response to potash additions. *Soil Sci.* **1944**, *58*, 305–324. [CrossRef]
75. Haynes, R.J.; Swift, R.S. Effect of soil amendments and sawdust mulching on growth, yield and leaf nutrient content of highbush blueberry plants. *Sci. Hortic.* **1986**, *29*, 229–238. [CrossRef]
76. Phillips, I.; Burton, E. Nutrient leaching in undisturbed cores of an acidic sandy podosol following simultaneous potassium chloride and di-ammonium phosphate application. *Nutr. Cycl. Agroecosyst.* **2005**, *73*, 1–14. [CrossRef]
77. Al-Kanani, T.; MacKenzie, A.; Ross, G. Mineralogy of surface soil samples of some Quebec soils with reference to K status. *Can. J. Soil Sci.* **1984**, *64*, 107–113. [CrossRef]
78. Kodama, H. Clay minerals in Canadian soils their origin, distribution and alteration. *Can. J. Soil Sci.* **1979**, *59*, 37–58. [CrossRef]
79. Simard, R.R.; Dekimpe, C.R.; Zizka, J. The kinetics of nonexchangeable potassium and magnesium release from Quebec soils. *Can. J. Soil Sci.* **1989**, *69*, 663–675. [CrossRef]
80. Feigenbaum, S.; Edelstein, R.; Shainberg, I. Release rate of potassium and structural cations from micas to ion-exchangers in dilute-solutions. *Soil Sci. Soc. Am. J.* **1981**, *45*, 501–506. [CrossRef]
81. Mc Lean, E.O.; Watson, M. Soil Measurements of Plant-Available Potassium. In *Potassium in Agriculture*; Rde, M., Ed.; ASA CSSA and SSSA: Madison, WI, USA, 1985; pp. 277–308. [CrossRef]
82. Kuchenbuch, R.; Jungk, A. Influence of potassium supply on the availability of potassium in the rhizosphere of rape (Brassica-Napus). *Z. Fur Pflanzenernahrung. Und Bodenkd.* **1984**, *147*, 435–448. [CrossRef]
83. Marchand, S.; Parent, S.E.; Deland, J.P.; Parent, L.E. Nutrient Signature of Québec (Canada) cranberry (*Vaccinium macrocarpon* AIT.). *Rev. Bras. Frutic.* **2013**, *35*, 292–304. [CrossRef]
84. Parent, L.E.; Gagné, G. *Guide de Référence en Fertilisation*; Les Impressions STAMPA Inc.: Québec, QC, Canada, 2010; p. 473.
85. Davenport, J.R.; DeMoranville, C.J.; Hart, J.; Kumidini, S.; Patten, K.; Poole, A.; Roper, T.R. Spatial and temporal variability of cranberry soil pH. *Acta Hortic.* **2003**, *626*, 315–327. [CrossRef]
86. Parent, L.E.; Jamaly, R.; Atucha, A.; Parent, E.J.; Workmaster, B.A.; Ziadi, N.; Parent, S.-É. Current and next-year cranberry yields predicted from local features and carryover effects. *PLoS ONE* **2021**, *16*, e0250575. [CrossRef]

 *horticulturae*

Article

Nutrient Concentration of African Horned Cucumber (*Cucumis metuliferus* L) Fruit under Different Soil Types, Environments, and Varying Irrigation Water Levels

Mdungazi K Maluleke [1],*, Shadung J Moja [2], Melvin Nyathi [3] and David M Modise [4]

[1] Department of Environmental Sciences, College of Agriculture and Environmental Sciences, University of South Africa, Tshwane 0002, South Africa

[2] Council of Geosciences, Water and Environment Business Unit: Geological Resource Division, Silverton, Tshwane 0002, South Africa; sjmoja@geoscience.org.za

[3] Agricultural Research Council, Tshwane 0002, South Africa; Mnyathi@arc.agric.za

[4] Faculty of Natural and Agricultural Sciences, School of Agricultural Sciences, North-West University, Potchefstroom 2520, South Africa; david.modise@nwu.ac.za

* Correspondence: malulm@unisa.ac.za

Abstract: The nutrient concentration of most crops depends on factors such as amount of water, growing environment, sunlight, and soil types. However, the factors influencing nutrient concentration of African horned cucumber fruit are not yet known. The objective of the study was to determine the effect of different water stress levels, soil types, and growing environments on the nutrient concentration of African horned cucumber fruit. Freeze-dried fruit samples were used in the quantification of *β-carotene* and total soluble sugars. The results demonstrated that plants grown under the shade net, combined with severe water stress level and loamy soil, had increased total soluble sugars (from 8 to 16 °Brix). Under the shade-net environment, the combination of moderate water stress level and loamy soil resulted in increased crude protein content (from 6.22 to 6.34% °Brix). In addition, the severe water stress treatment combined with loamy soil, under greenhouse conditions, resulted in increased *β-carotene* content (from 1.5 to 1.7 mg 100 g^{-1} DW). The results showed that African horned cucumber fruits are nutrient-dense when grown under moderate water stress treatment on the loamy or sandy loam substrate in the shade-net and open-field environments.

Keywords: biochemical constituents; *β-carotene*; vitamins; micro-nutrients; growing environments

Citation: Maluleke, M.K; Moja, S.J; Nyathi, M.; Modise, D.M Nutrient Concentration of African Horned Cucumber (*Cucumis metuliferus* L) Fruit under Different Soil Types, Environments, and Varying Irrigation Water Levels. *Horticulturae* **2021**, *7*, 76. https://doi.org/10.3390/horticulturae7040076

Academic Editor: Stefano Marino

Received: 18 February 2021
Accepted: 18 March 2021
Published: 10 April 2021

Publisher's Note: MDPI stays neutral with regard to jurisdictional claims in published maps and institutional affiliations.

Copyright: © 2021 by the authors. Licensee MDPI, Basel, Switzerland. This article is an open access article distributed under the terms and conditions of the Creative Commons Attribution (CC BY) license (https://creativecommons.org/licenses/by/4.0/).

1. Introduction

In Sub-Saharan Africa, indigenous crops have been a source of food for rural resource-poor households who experience nutritional food insecurity [1]. However, deficiencies in micronutrients, such as zinc, iron, and *β-carotene*, have been described as a major nutritional challenge faced by many rural households [2]. Several researchers claimed that the benefits of indigenous crops are that (i) they grow naturally in the wild [3]; (ii) are resistant to most pests and diseases; (iii) have better environmental stress tolerance; (iv) require low agricultural inputs, such as irrigation and fertilizers; and (v) have a shorter period to mature and are readily available for consumption [2]. However, most indigenous fruits and vegetables have not yet been commercialized, particularly in Southern Africa, because they are not produced under well-defined agronomic practices, and there is a lack of market value chain, since they do not have a high demand [2,4]. The nutritional composition of these crops has not been widely investigated, despite their usefulness to the communities. There appears to be scanty knowledge about their nutritional content, particularly when grown under different growing conditions. This knowledge could aid in influencing policymakers in the commercialization and products innovation in many countries, since the crop is adaptable in various growing environments. Ref. [2] reports that most of these crops have the potential to supplement several nutrients needed by the human body, in both smaller and larger quantities. Ref. [1] iterated that there is a need to promote the consumption of indigenous crops, and that can be achieved by the investigation of their agronomical viability and qualities, such as nutritional content. The African horned cucumber fruit is palatable, with a similar taste of a mixture of banana and pineapple [5]. The internal part of the fruit contains a high moisture content, which can aid body hydration [6]. Ref. [7] the other benefits of consuming this fruit: (i) It is a source of vitamin C, and (ii) it contains biochemical compounds such as phenols, which help the body to eliminate toxins; thus, growing this crop is of the outmost important in terms of promoting biodiversity and stewardship of the natural heritage and ecosystem of the Sub-Sahara region. The objective of the study was to determine the effect of different water stress levels, varying soil types, and growing environments on the nutrient concentration of the African horned cucumber fruit.

2. Material and Methods

This study was conducted during the 2017/18 and 2018/19 growing seasons, under the greenhouse, shade net, and open-space environment at the Florida science campus of the University of South Africa (26°10′ 30″ S, 27°55′ 22.8″ E). Before plant cultivation, gravimetric water content (GWC) was carried out, determining the field water capacity of the soils. Briefly, dry soil was filled in a 30 cm depth planting pot, weighed, and then watered to filled capacity (3000 mL). The pots were then weighed after 72 h, when drainage was completed. The process was repeated until the soil reached permanent wilting point. Water stress levels was then determined by using the formula (e.g., 3000 mL–filled capacity $\times 75 \div 100 = 2250$ mL moderate stress, while 3000 mL $\times 35 \div 100 = 1050$ mL severe water stress). Soil samples (loamy soil and sandy loam) were analyzed for mineral and/or chemical content (Table 1), using the method followed by [8]. The above analysis was conducted at the Agricultural Research Council, Institute for Soil, Climate and Water (ARC-ISWC) in Pretoria (25° 44′ 19.4″ S28° 12′ 26.4″ E). Sterilized growth media (loamy soil and sandy loam) were used. In addition, certified seeds of African horned cucumber were purchased from Seeds for Africa, Cape Town. A factorial experiment with two factors, i.e., soil (loamy soil and sandy loam soil) and irrigation water levels (no water stress, moderate water stress, and severe water stress), was conducted. The pot experiment was a completely randomized design with nine (9) replicates per treatment. The pots were spaced 1 m apart, and an up-rope vertical trellising was used to support the plants. On each site, pots were either filled with loamy soil or sandy loam. Each block comprised 18 plants in pots, resulting in 54 plants per site. A total of 162 plants were used for the experiment. Each site had plants used as guard plants, in order to separate the plants from the external effects outside the experimental plot. Well-established, uniform, and healthy African horned cucumber seedlings, germinated from peat substrate, that were 30 days old, were transplanted into 30 cm depth $\times$ 30 cm width. Briefly Area (depth $\times$ width) 30 cm $\times$ 30 cm = 900 cm^2, A = $\pi \left(\frac{d}{2}\right) \times 2$ d = 286.5 cm^2 planting pots, and the treatments were imposed four (4) weeks later, after establishment. Plants were well irrigated prior to imposition of the treatments. Granules fertilizers (potassium phosphate), 10 g per plant pot, were applied once every 7th day of the week during the experimental period.

The impact of soil, water, and growing environment on the nutrient composition of African horned cucumber fruit was evaluated at 12 weeks after planting during 2017/18 and 2018/2019. Prior to fruit analysis, optimization analysis of crude protein and total soluble sugars of fruit was carried out before the actual fruit analysis, whereby fruit were harvested from each irrigation water level (no-water-stress control, moderate water stress, and severe water stress), soil type, and growing environment. The goal for the fruit optimization analysis was to find the optimum value for one or more target variables among African horned cucumber fruit harvested under different treatments.

Table 1. Soil analysis for the experiment (mineral/chemical analysis).

	Chemical Analysis (Micro-Minerals)				
	Fe	Mn	Cu	Zn	pH
	mg kg^{-1}	mg kg^{-1}	mg kg^{-1}	mg kg^{-1}	
SL	30.3	59.4	1.24	9.36	7.69
L	33.2	59.8	1.27	8.96	7.74
	P	Ca	Mg	K	Na	Total N
	mg kg^{-1}	mg kg^{-1}	mg kg^{-1}	mg kg^{-1}	mg kg^{-1}	%
SL	35.16	1900	141	243	35.5	0.105
L	34.4	1810	133	217	28.7	0.113

2.1. Determination of β-Carotene

The analysis of *β-carotene* was carried out with a Prominence-i High-Performance Liquid Chromatography–PDA model system equipped with a sample cooler LC-2030C (Shimadzu, Japan), with slight modifications (triplicate), as described by [4], since most of the compounds measured were expected to be similar to those of the current study. A mixture of approximately 0.1 g/mL of extracted sample with ice-cold hexane:acetone (1:1, *v/v*) was vortexed for two (2) minutes, before being centrifuged at 2000 rpm for two (2) minutes. The organic phase was decanted into a tube containing saturated sodium chloride solution and placed on ice. The remaining residue was similarly re-extracted until the extract was colorless. Each time, the extract separated organic phase was filtered through 0.45 μm syringe filtered before injection into the HPLC. Chromatographic separation was achieved, using a C$_{18}$ Luna$^{®}$ column (150 × 4.6 mm, 5μ) maintained at 35 °C.

An isocratic mobile phase which consisted of acetonitrile:dichloromethane:methanol (7:2:1) was used, with a flow rate of 1 mL/min, an injection volume of 20 μL, and the detection was at 450 nm. Peak identification and quantification of the compound (*β-carotene*) were both achieved based on authentic *β-carotene* standard, which was used for plotting the calibration curves [9].

2.2. Determination of Total Soluble Sugars

The African horned cucumber fruit harvested from the greenhouse, shade net, and open field, irrigated with different water levels and soil types, were analyzed for total soluble sugars concentration (°Brix) following the method by [10]. The fruit was cut into two portions, then juice was squeezed from a fruit portion by hand to release about 0.03 mL juice onto the aperture of the hand refractometer (HI 96801 Refractometer, USA) and readings were taken immediately. About 18 fruits were measured per treatment. The aperture was washed between different juices samples, with distilled water, and dried with a soft paper towel.

2.3. Determination of Vitamin C and E

The fruit samples were freeze-dried for 72 h, using a freeze drier (HARVEST-RIGHT, Barcelona). The freeze-dried fruit slices were rigorously homogenized, using a sterilized food blender, and mixed with dried powder before nutritional analysis. The method described by [4] was followed with slight modifications (triplicate). Individual samples were weighed (1 g) into tube, followed by the addition of 5% metaphosphoric acid (10 mL). It was sonicated 15 min before centrifuged and then filtrated in the ice-cold water bath. The analysis was carried out on the model system described above, Prominence-i HLCP–PDA. A C18 Luna $^{®}$ column (150/4.6 mm, 5 μL) held at 25 μC was used to achieve chromatographic separation. A water-based isocratic mobile phase: acetonitrile: formic acid (99:0.9:0.1) was used at a flow rate of 1 mL/min. The volume of injection was 20 μL and 245 nm of detection was set. Depending on the calibration curve plotted by using L-ascorbic acid, sample quantification was achieved.

2.4. Determination of Total Flavonoids

The African horned cucumber fruit samples were quantified, using the aluminum chloride colorimetric method described by [4]. Catechin was used as a standard for calibration curve, and total flavonoids content was expressed in mg catechin equivalents (CEs) per dry weight.

2.5. Determination of Total Phenolic Content

Total phenolic content of the fruit samples was carried out, using [4], with a slight modification (triplicate). Garlic used as standard for plotting curve. Total phenolic content was expressed in mg garlic acid equivalents (GAEs) per g dry weight (DW).

2.6. Determination of Micro-Nutrients

Freeze-dried fruit samples were digested in a diffused microwave system (MLS 1200 Mega; Milestone S.r. L, Sorisole, Italy), and the samples were further congelated–dried, following the procedure described by [4] with minor modifications. The modifications were that samples were measured in three (3) replicates per treatment (around 15–25 mg) weighed into polytetrafluoroethylene vessels and 2 mL HNO_3 (67%, analphur) and 1 mL H_2O_2 (30%, analytical grade) added in the vessels [4f]. Every solution was diluted to 15 mL, in a deionized-water test tube, after digestion, and analyzed by Inductively Coupled Plasma–Mass Spectrometry (ICP–MS). An ICP–MS (Agilent 7700; Agilent Technologies, Tokyo, Japan) based on quadrupole mass analyzer and octapole reaction system (ORS 3) was used to conduct the analysis. Nutrient elements, such as zinc (Zn), iron (Fe), molybdenum (Mo), copper (Cu), and manganese (Mn), were analyzed.

The calibration solution was prepared by appropriate dilution of the single element certified reference material with 1.000 g/L for each element (Analytika Ltd., Czech Republic) with deionized water (18.2 $M\Omega{\cdot}cm$, Direct-Q; Millipore, France). Measurement of accuracy was verified by using certified reference material of water TM-15.2 (National Water Research Institution, Ontario, Canada).

2.7. Statistical Analysis

Analysis of variance (ANOVA) was performed with a three-way ANOVA), to determine the main and interaction effects of all studied variables (crude protein, total soluble sugars, *Beta carotene*, vitamin C, vitamin E, total phenols, total flavonoids, and macro- and micro-nutrients). Homogeneity and uniformity tests were carried to determine the difference and similarities between variance. Mean separation was done by using the Fischer's unprotected least significance difference test at 5% significance level. Treatment means for each measured parameter were compared, and differences were noted. All statistical analyses were done, using GenStat (version 14, VSN, Rothamstead, UK).

3. Results

3.1. Total Soluble Sugars

Figure 1 presents the treatment interaction effect on total soluble sugars content of African horned cucumber fruit grown at different environments (greenhouse, shade net, and open field), soil types (loamy soil and sandy loam), and water stress levels (no water stress, moderate water stress, and severe water stress). The results indicated that there was no significant ($p > 0.05$) interaction between location, different water stress levels, and soil types on total soluble sugars content of African horned cucumber fruit during both growing seasons. However, fruit total soluble sugars ranged from 8.0 to 16 °Brix. In addition, the study revealed that there was a significant ($p \leq 0.05$) difference in total soluble sugars under varying water levels. Total soluble sugars among different water levels ranged from 11.4 to 14.4 °Brix. Furthermore, the results illustrated that the severe-water-stress level obtained the highest total soluble sugar content (14.4 °Brix), while the lowest content was observed from the no-water-stress (control) water level, with 11.4 °Brix.

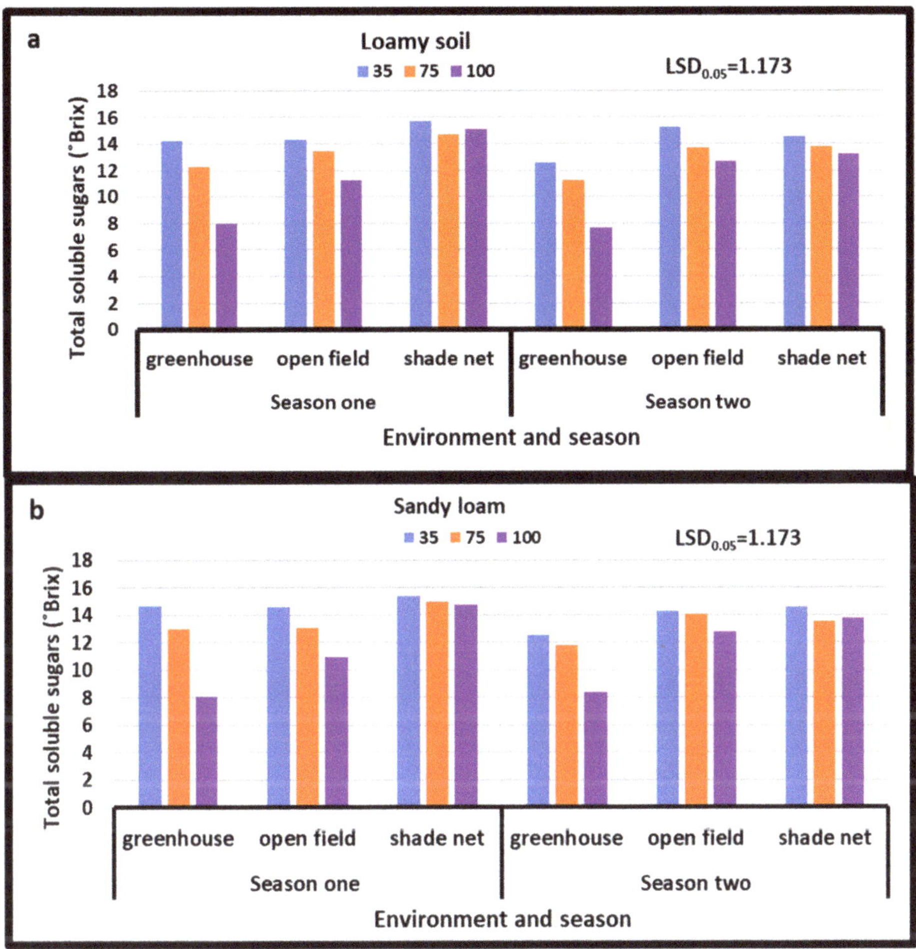

Figure 1. Treatment effect on the total soluble sugars content of African horned cucumber fruit grown in different environments; (**a**) effect of different water stress levels and loamy soil, in different environments, during different seasons (2017/18, season one; and 2018/19, season two); (**b**) effect of different water stress levels and sandy loam, in different environments, during different seasons (2017/18, season one; and 2018/2019, season two); 35 means severe water stress, 75 means moderate water stress, and 100 means no water stress (control). LSD$_{0.05}$ is the least significant difference of means.

3.2. Crude Proteins

For crude protein content, the results of the study showed that there was no significant ($p > 0.05$) difference in crude protein content between interaction of growing environment, water stress levels, and soil types (Figure 2). However, the results delineated that fruit crude protein ranged from 6.22 to 6.29%. Moreover, the results of the study demonstrated two extremes: The treatment of no water stress and severe water stress combined with both soil types (loamy soil and sandy loam) at growing conditions (greenhouse and shade net) during both seasons decreased crude protein content from 6.29 to 6.22% (Figure 2a,b), whereas the treatment of severe water stress combined with loamy soil at shade-net conditions increased crude protein content from 6.22 to 6.29% (Figure 2a). In addition, results showed evinced that there was a significant ($p \leq 0.05$) difference for crude protein content under different growing environment. Crude protein under varying growing environments ranged from 6.24 to 6.28%. Moreover, results showed that shade-net growing environment obtained the highest crude protein, at 6.28%, while the greenhouse environment expressed the lowest content, at 6.24%.

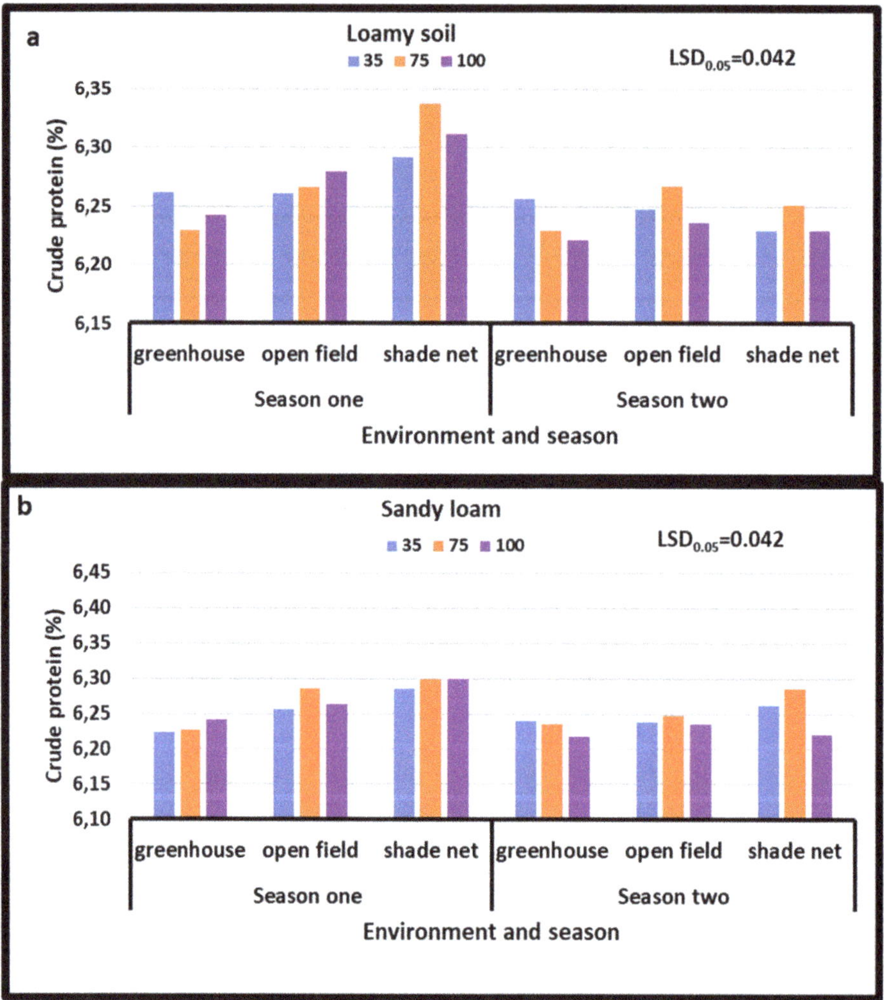

Figure 2. Treatment interaction effect on crude protein content of African horned cucumber fruit; (**a**) interaction effect of different water stress levels and loamy soil, in different environments, during season one (2017/2018); (**b**) interaction effect of different water stress levels and sandy loam, in different environments, during season two (2018/19); 35 means severe water stress, 75 means moderate water stress, and 100 means no water stress (control). $\text{LSD}_{0.05}$ is the least significant difference.

3.3. β-Carotene

Table 2 presents the treatment effect on the *β-carotene*, vitamin C, vitamin E, total flavonoids, and total phenols of African horned cucumber fruit under different growing environments. For the greenhouse, shade, and open-field environment, the results illustrated that there was a significant ($p \leq 0.05$) difference between the interaction of water stress levels and soil types. *β-carotene* ranged from 1.5 to 17 mg 100 g^{-1} DW. In addition, the results demonstrated that the severe water stress combined with sandy loam slightly decreased *β-carotene* from 1.7 to 1.5 mg 100 g^{-1} DW, whereas the treatment of severe water stress combined with loamy soil increased it from 1.5 to 1.7 mg 100 g^{-1} DW. For the shade-net environment, *β-carotene* ranged from 1.5 to 1.6 mg 100 g^{-1} DW. In addition, the results showed that water stress levels (moderate and severe water stress) combined with both soils slightly decreased *β-carotene* from 1.6 to 1.5 mg 100 g^{-1} DW, whereas moderate water stress treatment combined with loamy soil increased it from 1.5 to 1.6 mg 100 g^{-1} DW. Under the open-field environment, *β-carotene* increased from 1.5 to 1.6 mg 100 g^{-1} DW. The treatment of water levels (moderate and severe water stress) indicated a decrease from 1.6 to 1.5 mg 100 g^{-1} DW, whereas no-water-stress treatment (control) combined with both soils expressed an increase from 1.5 to 1.6 mg 100 g^{-1} DW.

Table 2. Treatment effect on biochemical constituents of African horned cucumber fruit harvested from different growing environments.

Treatment	*β-carotene* (mg 100 g^{-1} DW)	Vitamin C (mg 100 g^{-1} DW)	Vitamin E (mg 100 g^{-1} DW)	Total Flavonoids (CE g^{-1} DW)	Total Phenols (GAE g^{-1} DW)
Greenhouse					
W1S1	1.6(0.0)	26.6(0.2)	11.7(1.1)	0.66(0.03)	3.1(0.1)
W2S1	1.6(0.0)	24.3(0.2)	29.8(0.1)	0.56(0.03)	5.2(0.2)
W3S1	1.7(0.01)	23.5(0.5)	24.4(13.4)	0.25(0.1)	4.5(0.1)
W1S2	1.5(0.01)	23.8(1.9)	9.3(5.5)	0.26(0.1)	4.4(0.1)
W2S2	1.6(0.01)	30.3(0.9)	31.7(0.5)	0.55(0.02)	5.8(0.2)
W3S2	1.5(0.01)	23.2(0.9)	35.1(0.5)	0.21(0.0)	4.2(0.0)
Grand mean	1.6	25.3(0.1)	23.7(0.9)	0.4	4.5
LSD0.05	0.020	1.528	12.95	0.060	0.196
p-value	0.001	0.001	0.204	0.001	0.001
Shade net					
W1S1	1.5(0.0)	33.1(0.5)	18.1(16.9)	0.75(0.01)	4.2(0.1)
W2S1	1.6(0.1)	30.2(0.4)	16.9(3.7)	0.84(0.0)	5.3(0.1)
W3S1	1.5(0.0)	28.2(0.0)	11.3(5.8)	0.54(0.1)	3.6(0.0)
W1S2	1.5(0.0)	31.7(16.8)	10(3.4)	0.63(0.02)	4.3(0.3)
W2S2	1.5(0.1)	22.6(0.1)	12.5(2.9)	0.77(0.03)	4.4(0.1)
W3S2	1.5(0.0)	27.2(0.1)	14.3(1.1)	0.49(0.0)	3.5(0.1)
Grand mean	1.5	28.8	13.9	0.670	4.2
LSD0.05	0.009	12.58	19.35	0.038	0.205
p-value	0.001	0.658	0.29	0.009	0.001
Open field					
W1S1	1.6(0.0)	17.0(0.7)	10.7(0.5)	0.73(0.03)	6.4(0.01)
W2S1	1.5(0.0)	19.0(0.4)	11.8(2.7)	0.47(0.02)	4.1(0.1)
W3S1	1.5(0.01)	16.6(0.6)	8.3(3.0)	0.85(0.02)	4.8(0.1)
W1S2	1.6(0.01)	18.7(0.6)	13.4(3.2)	0.42(0.02)	5.4(0.1)
W2S2	1.5(0.0)	27.5(0.9)	13.5(0.4)	0.41(0.02)	5.1(0.0)
W3S1	1.5(0.01)	15.5(0.7)	9.7(0.3)	0.65(0.04)	3.1(0.2)
Grand mean	1.5	19.03	11.2	0.59	4.8
LSD0.05	0.009	1.231	6.079	0.057	0.207
p-value	0.001	0.001	0.809	0.001	0.001

W1 means no water stress (control); W2 means moderate water stress; W3 means severe water stress. S1 is loamy soil, and S2 is sandy loam soil. Numbers in brackets represent the standard deviations of the mean. LSD$_{0.05}$ is the least significant difference of means. The *p*-values in bold are lower than 0.05. Note that only season two results are presented, due to logistical costs, as analysis could not be done for both seasons one treatments.

3.4. Vitamin C

For vitamin C, the results showed that there was a significant ($p \leq 0.05$) difference between the interaction of different water levels and soil types under the greenhouse and open-field environment. However, there was no significant ($p > 0.05$) difference between different water levels and soil types in the shade-net environment (Table 2). Under the greenhouse environment, vitamin C content ranged from 23.2 to 30.3 mg 100 g^{-1} DW. The results illustrated that treatment of severe water stress combined with sandy loam decreased vitamin C from 30.3 to 23.2 100 g^{-1} DW, whereas moderate water stress treatment combined with sandy loam increased it from 23.2 to 30.3 100 g^{-1} DW. For the shade-net environment, vitamin C content ranged from 22.6 to 33.1 100 g^{-1} DW. Our results revealed that severe water stress treatment combined with sandy loam decreased vitamin C from 33.1 to 22.6 100 g^{-1} DW, whereas no-water-stress (control) treatment combined with loamy soil increased it from 22.6 to 33.1 100 g^{-1} DW.

Regarding the open-field environment, vitamin C content ranged from 15.5 to 27.5 100 g^{-1} DW. The results of the study indicated that the treatment of severe water stress combined with sandy loam decreased vitamin C content from 27.5 to 15.5 100 g^{-1} DW, whereas moderate water stress treatment and sandy loam increased it from 15.5 to 27 100 g^{-1} DW. It is worth to note that the treatment of severe water stress combined sandy loam soil under the open-field environment indicated the lowest vitamin C content (15.5 100 g^{-1} DW), whereas the no-water-stress (control) treatment combined with loamy soil under the shade-net environment obtained the highest vitamin C content (33.1 100 g^{-1} DW).

3.5. Vitamin E

The results of the study revealed that there was no significant ($p > 0.05$) difference for vitamin E content from the interaction between different water levels and soil types under all growing environments (greenhouse, shade net, and open field). For the greenhouse environment, vitamin E content ranged from 9.3 to 35.1 100 g^{-1} DW. In addition, the results demonstrated that no-water-stress (control) treatment combined with sandy loam decreased vitamin E content from 35.1 to 9.3 100 g^{-1} DW, whereas the treatment of severe water stress and sandy loam increased it from 9.3 to 35.1 100 g^{-1} DW (Table 2). Under the shade-net environment, the no-water-stress treatment (control) combined with sandy loam decreased vitamin E content from 18.1 to 10.0 100 g^{-1} DW, whereas no water stress (control) and loamy soil increased it from 10.0 to 18.1 100 g^{-1} DW. On the other hand, open-field vitamin E content ranged from 8.3 to 13.5 100 g^{-1} DW. Results delineated that treatment of severe water stress combined with loamy soil decreased vitamin E content from 13.5 to 8.3 100 g^{-1} DW, whereas the severe water stress and sandy loam increased it from 8.3 to 13.5 100 g^{-1} DW (Table 2).

3.6. Total Flavonoids

Table 2 illustrates that there was a significant ($p \leq 0.05$) difference in total flavonoids, depending on the interaction of different water levels and soil types under varying growing environment (greenhouse, shade net, and open field). For the greenhouse environment, total flavonoids ranged from ranged from 0.21 to 0.66 CE g^{-1} DW. In addition, the results illustrated that the treatment of severe water stress combined with sandy loam reduced total flavonoids from 0.66 to 0.21 CE g^{-1} DW, whereas treatment of no water stress (control) combined with loam soil increased it from 0.21 to 0.66 CE g^{-1} DW. Under the shade-net environment, our results showed that total flavonoids ranged from 0.49 to 0.84 CE g^{-1} DW. The results indicated that severe water stress treatment combined with sandy loam reduced total flavonoids from 0.84 to 0.49 CE g^{-1} DW, whereas no-water-stress (control) treatment increased it from 0.49 to 0.84 CE g^{-1} DW. For total flavonoids content in the open-field environment, the results showed that it ranged from 0.41 to 0.85 CE g^{-1} DW. In addition, the results illustrate that water stress and sandy loam decreased total flavonoids from 0.85 to 0.41 CE g^{-1} DW, whereas severe water stress and loamy soil increased it from 0.41 to 0.85 CE g^{-1} DW (Table 2). The observed trend shows that the combination of severe water stress and loamy soil under the open-field environment obtained the highest total flavonoids content, at 0.85 CE g^{-1} DW, whereas the lowest content was observed on treatment combined.

3.7. Total Phenols

The results indicate that there was a significant ($p \leq 0.05$) difference on the total phenolic content of African horned cucumber between interaction of different water levels and soil types under varying growing environment (greenhouse, shade net, and open field). The greenhouse environment total phenols ranged from 3.1 to 5.8 GAE g^{-1} DW. Our results illustrated that the treatment of no water stress (control) combined with loamy soil decreased total phenols content from 5.8 to 3.1 GAE g^{-1} DW, whereas the severe water stress treatment combined with sandy loam increased it from 3.1 to 5.8 GAE g^{-1} DW.

For the shade-net environment, total phenols content ranged from 3.5 to 5.3 GAE g^{-1} DW. The study results showed that the combination of severe water stress treatment and sandy loam reduced total phenols content from 5.3 to 3.5 GAE g^{-1} DW. Under the open-field environment, total phenols content ranged from 3.1 to 6.4. In addition, the results of the study indicated that severe water stress treatment combined with sandy loam decreased total phenols content from 6.4 to 31 GAE g^{-1} DW, whereas treatment combination of no water stress (control) and loamy soil increased it from 3.1 to 6.4 GAE g^{-1} DW. For the open-field environment, our results showed that total phenols ranged from 3.1 to 6.1 GAE g^{-1} DW. In addition, the results showed that severe water stress treatment combined with sandy loam decreased total phenols from 6.1 to 3.1 GAE g^{-1} DW, whereas no-water-stress level (control) combined with loamy soil increased it from 3.1 to 6.4 GAE g^{-1} DW.

3.8. Micro-Nutrients

Table 3 presents the micronutrient concentration of African horned cucumber. Significant ($p \leq 0.05$) interactions were observed for manganese and zinc, under the open environment, whereas for the shade, significant interactions were observed for iron and zinc. For the open-field environment, significant interactions were found under zinc only. The greenhouse zinc content ranged from 7.7 to 12.7 µg g DW. In addition, results illustrated that treatment of no water stress (control) combined with loam soil presented a decreased zinc content from 12.7 to 7.7 µg g DW, whereas there was a double increase in zinc content from treatment combination of no water stress (control) and sandy loam, from 7.7 to 12.7 µg g DW (Table 3). For the shade-net environment, zinc content ranged from 6.4 to 8.8 µg/g DW. The results demonstrated that treatment of severe water stress combined with sandy loam decreased zinc content from 8.8 to 6.4 µg g DW, whereas no-water-stress (control) treatment combined with sandy loam increased it from 6.4 to 8.8 µg g DW. Under an open-field environment, zinc content ranged 5.1 to 7.9 µg g DW. The lowest zinc content was observed from combination of no water stress (control) and loamy soil at 5.1 µg g DW, while treatment of moderate water stress and loamy soil presented an increase at 7.9 µg g DW. Under the shade-net environment, iron ranged from 1.4 to 1.8 µg g DW. The lowest iron content was observed from treatment of no water stress and sandy loam at 1.4 µg g DW, whereas treatment combination of moderate water stress and sandy loam illustrated higher content, at 1.8 µg g DW.

Table 3. Treatment interaction effect of irrigation water regimes, soil types, and environment on micro-nutrients (µg g DW) of African horned cucumber fruit.

Treatment	Moisture (g)	Copper	Iron	Manganese	Zinc
Greenhouse					
W1S1	193(36)	0.9(0.0)	1.8(0.1)	0.8(0.0)	7.7(1.2)
W2S1	179(40)	0.7(0.1)	2.0(1.5)	0.9(0.1)	9.3(0.8)
W3S1	78(42)	0.8(0.4)	1.6(0.2)	1.0(0.1)	10.1(1.8)
W1S2	95(35)	0.7(0.4)	2.8(1.8	1.1(0.1)	12.7(1.5)
W2S2	152(5)	0.5(0.0)	3.8(0.2)	0.9(0.2)	8.6(2.0)
W3S2	129(22)	0.5(0.4)	0.5(0.1)	0.9(0.1)	10.6(0.6)

Table 3. *Cont.*

Treatment	Moisture (g)	Copper	Iron	Manganese	Zinc
Grand mean	138	0.689	2.1	0.942	9.8
LSD0.05	98.4	0.42	1.81	0.1854	2.228
p-value	0.15	0.99	0.06	0.01	0.01
Shade net					
W1S1	162(24)	0.7(0.2)	0.9(0.1)	0.8(0.0)	7.2(1.1)
W2S1	140(30)	0.8(0.3)	2.7(0.3)	0.9(0.1)	7.1(2.2)
W3S1	83(4)	0.6(0.1)	2.7(0.9)	0.9(0.1)	7.2(0.5)
W1S2	157(5)	0.6(0.1)	1.4(0.6)	0.8(0.0)	8.8(0.0)
W2S2	146(5)	0.8(0.3)	1.8(0.2)	0.8(0.1)	12.7(0.6)
W3S2	79(25)	0.6(0.1)	1.7(0.5)	0.8(0.1)	6.4(0.8)
Grand mean	127.7	0.7	1.9	0.811	8.23
LSD0.05	35.9	0.4	0.7	0.1475	2.177
p-value	0.9	0.89	0.03	0.59	0.01
Open field					
W1S1	146(50)	0.5(0.0)	2.4(0.8)	0.5(0.1)	5.1(0.9)
W2S1	220(21)	0.8(0.1)	2.6(0.3)	0.7(0.2)	7.9(0.4)
W3S1	29(16)	0.6(0.2)	1.8(1.3)	0.7(0.1)	7.7(0.2)
W1S2	162(6)	0.7(0.5)	1.3(0.2)	0.8(0.1)	6.8(0.4)
W2S2	155(4)	0.6(0.1)	2.1(0.8)	0.7(0.0)	6.9(0.1)
W3S1	80(19)	0.7(0.1)	0.6(0.2)	0.6(0.1)	7.5(1.4)
Grand mean	137	0.7	1.8	0.7	6.98
LSD0.05	40	0.341	1.211	0.231	1.153
p-value	0.03	0.20	0.61	0.181	0.03

W1 means no water stress (control); W2 means moderate stress; W3 means severe water stress. S1 means loamy soil, and S2 means sandy loam. Values are average over treatments mentioned. Numbers in brackets represent the standard deviations of the mean. $LSD_{0.05}$ is the least significant difference of means. The *p*-values in bold are lower than 0.05. Note that only season two results are presented, due to logistical costs, as analysis could not be done for season-one treatments.

4. Discussion

This study investigated the effect of different water stress levels and varying substrates on the nutrient concentration of African horned cucumber fruit grown in three different environments (greenhouse, shade net, and open field). Previous studies conducted by [11,12] have evaluated the nutrient concentration of leafy vegetables grown under different water stress levels. In addition, studies conducted by [1,2] focused on iron and zinc. However, these studies did not evaluate biochemical constituents, such as crude protein, total soluble sugars, total flavonoids, total phenols, and vitamins. Therefore, the findings of this research study serve as a benchmark for the biochemical constituents of African horned cucumber fruit.

4.1. Bio-Chemical Constituents

Ref. [4] determined the mineral constituents and phytochemicals of crops harvested from different locations. Ref. [13] recommended that it is crucial to note the effect of water, irrigation and rainfall received by crops on the mineral constituents, such as β-carotene, total phenols, vitamins, total flavonoids, and micro- and micro-nutrients, so that growers can make an informed decision, to ensure that quality produce is supplied to their target market.

4.2. Total Soluble Sugars

Fruit sugar content is affected by a number of factors, including climate, water supply, and soil type. Total soluble sugars in fruits have a variety of health benefits, including provision of glucose, preventing colorectal cancer, and variety of diseases [14]. Fruit intake is currently recommended by most dietary practitioners for improvement of health and disease prevention. The findings of this study demonstrated that the treatment affects the total soluble sugars of African horned cucumber fruit grown under varying environment. When plants were subjected to severe water stress under shade-net conditions, total soluble sugars increased, but they decreased under no-water-stress treatment. This implies that, when plants are exposed to different water levels, there is variation in fruit nutrient content.

Refs. [10,15] reported a significant difference in total soluble sugars of kiwi fruit harvested from different sites, due to variation in temperatures and rainfall patterns.

High total soluble sugar content was expected from open-field fruit under moderate water stress, as reported by [14], on pomegranate trees. These authors concluded that active osmoregulation caused by water stress was responsible for sugar variation in fruits, since there is imbalanced fluid movement within plant cells. Similarly, this study's findings unveiled that fruits harvested from water stress treatment had a higher total soluble sugar content, when compared to the other treatments. Therefore, a relatively high total soluble sugar level in fruit is crucial for human nutrition, especially when the °Brix level is above 5. However, the values obtained from this study are slightly higher, making it an important fruit for the fresh and juice market. This suggests that the fruit is valuable and should be considered for commercialization, as the fruit shows potential benefits for human nutrition.

4.3. Crude Proteins

Crude proteins are important in human nutrition because they aid in cell formation, nutrient storage, pH balance, and immune system improvement, and they serve as a messenger [9]. Previous studies have often reached conflicting findings regarding crude protein content of crops harvested from different treatments and growing conditions. For example, [16] presented their findings on crude protein of potatoes harvested under different regions that experience varying weather conditions and treated with varying level of fertilizers. They concluded that potatoes harvested from regions with moderate temperatures subjected to moderate nitrogen fertilizers resulted in higher significant crude protein content, when compared to other treatments, due to high enzyme activities within cells, caused by different nitrogen content. For this study, shade-net conditions expressed high crude protein content, compared to the other growing environment. Perhaps the growing environment of the shade net favored higher crude proteins in moderate and no-water-stress treatments, as compared to the water stressed treatment.

When the surrounding conditions (adequate sunlight and water) are favorable, cells can carry out chemical reaction at an optimum rate, but at a lower rate under stress environment, such as excessive radiation and water stress. These results agree with the fact that excessive temperatures negatively affect protein activities (denature) and have other general destructive effects on plant cells, as reported by [17], who found higher crude protein content in fruits harvested from protected structures, but low in those harvested from open-field conditions. This advocates that African horned cucumber, if grown under optimum environmental conditions may have several health benefits in human nutrition and may also be a potential solution for a hunger and health issues globally.

4.4. β-Carotene

β-*carotene*, famously known as a major source for Vitamin A, has been reported by [18] as an important compound for human health. It is (i) responsible for the formation and maintenance of teeth, (ii) formation of muscle tissues, and (iii) improvement of eyesight. The grand mean showed that β-*carotene* content was higher in severe water stress treatment under greenhouse environment, but significantly decreased by the same water stress treatment under open-field environment. In addition, loamy soil seemed to increase β-*carotene*, whereas sandy loam reduced it.

The fact that carotene is responsible for radiation interception in plants could have been the cause for variation, since there is control of light intensity in the greenhouse, due to cladding material used for protection, as compared to an open-space area. [19] found that there was variation in *β-carotene* among some plant varieties subjected to reduced water supply. Their findings are in harmony with those of the current study, whereby varying water levels under different growing conditions significantly altered the *β-carotene* content of African horned cucumber fruit. *β-carotene* promotes cell and tissue development, strengthens the immune system, and slows the aging process. Furthermore, it effectively enhances eye vision, skin, nail, and hair function. African horned cucumber fruit contains reasonable amount of *β-carotene*, which can be converted to vitamin A in the body, to complement it. Therefore, optimum growing environment could serve as strong evidence for mass production and commercialization globally.

4.5. Vitamin C

In the present study, the grand mean showed that vitamin C was higher on fruit grown under greenhouse environment (25.3 mg 100 g−1 DW), as compared to the other growing environments. In addition, vitamin C increased in plants subjected to no water stress (control) under the shade-net environment, but it decreased under severe water stress under the open-field environment. Higher fluctuation in the vitamin C content could be the result of unbalanced turgor pressure in plants, caused by varying irrigation water levels, water holding capacity by a specific substrate, and different growing environments, as reported by [20,21], who mentioned that water and fertilizers stimulate the vitamin C content of cucumber and citrus fruit grown in an open field and semi-protected structure. This was authenticated by [22], when they reported that plants respond to harsh environmental conditions, such as excessive sunlight, heat, and water stress, by producing vitamin C as a defensive mechanism to protect themselves [23].

The mean results also showed that the vitamin C reduction was more on plants subjected to adverse conditions, such as water stress level and open space, as compared to plants that were grown under a protected environment (greenhouse and shade net). The current study findings agree with findings by [24], who reported that plants can tolerate moderate water stress. However, such alteration has a negative impact on the fruit vitamin C content of various fruit crops. Even though vitamin C deficiency is uncommon in today's world, dieticians prescribe vitamin C because it plays a critical role in the production of collagen, iron absorption, wound healing, and bone and tooth health. Determination of optimal conditions that increase African horned cucumber vitamin C content could fill the void in human nutrition, globally, and increase its consumption.

4.6. Vitamin E

For vitamin E, the means illustrated that vitamin E content was greater in the greenhouse environment (23.7 mg 100 g^{-1} DW), as compared to other growing environments. In addition, the current study findings exhibited that the treatment imposed (water levels and soils types) did not caused significant variation in vitamin E content. However, there was a slight increase on severe water stress treatment under greenhouse conditions, but there was a significant decrease under severe water stress in the open-field environment. Perhaps the evapotranspiration rate, which regulates the osmoregulation, played an important role in the vitamin E variation, since there was a change in stomatal opening and closure, due to alteration in turgor pressure within the guard cells. Carbon dioxide interception is higher when there is balance of solutes movement within the open guard cells, but they close when there is imbalance concentration due to high evapotranspiration rate caused by excessive conditions, such as high wind and radiation, subsequently limiting the ability to synthesis vitamin E due to limited activities in the chloroplast caused by stomatal closure. Closing of stoma not only prevents water loss, but also prevents the plant's ability to synthesize vitamins and other biochemical compounds. The study findings affirm that water stress levels under varying growing environment were the critical contributors of vitamin E content of African horned cucumber fruit, as compared to other factors. These findings agree with [18,21,25], who found significant differences in vitamin E content of fruit such as chilies and peppers subjected to varying water stress, due to the balanced osmotic flow within plant organs. Vitamin E has a variety of functions in the human body, including preventing free radical damage and acting as an antioxidant. In addition, the vitamin deficiency is associated with stunted development. The values in this study serve as benchmarks required by policymakers for commercialization of this crop, since it has nutritional benefit to humans.

4.7. Total Flavonoids

The shade-net grand mean showed higher total flavonoids (0.67 CE g^{-1} DW), relative to greenhouse at (0.4 CE g^{-1} DW). The study findings also remarked that the reduction was more on the severe water stress treatment, relative to moderate and no-water-stress treatment (control). The alteration in total flavonoids could have been caused by turgor pressure within the plant's cells, which subsequently allows the plant to access surrounding atmospheric elements through the epidermal cells, thus allowing the plant to absorb atmospheric elements needed by plants for cellular activities. When the stomata close, plant cellular activities get negatively affected, but they function normally when there is good movement of water within plant organs. However, contradictory findings were noticed by [13,26] on opuntia and red grapes. They determined that total flavonoids significantly increased in fruit harvested from regions with a low rainfall pattern, but decreased in fruit harvested from regions experiencing higher rainfall patterns, due to varying active osmoregulation within plant organs, since plants were trying to cope with stress caused by the environmental conditions. Their findings are consistent with observations made in this current study, whereby severe-water-stress fruit demonstrated a significant increase in total flavonoids when compared to stressed-free fruit. Total flavonoids are well-known in human health for their function in controlling cellular activity, as well as fighting free radicals that cause oxidative stress. The total flavonoids values of African horned cucumber fruit serve as benchmark information required by policymakers; therefore, the crop can be recommended for commercialization, if grown under optimal conditions.

4.8. Total Phenols

In terms of total phenolic content, the study findings outline that open space grand mean exhibited higher total phenols (4.8 GAE g^{-1} DW), relative to the greenhouse (4.5 GAE g^{-1} DW) and shade-net environment (4.2 GAE g^{-1} DW). The study findings showed increased total phenolic content under normal watering on loamy soil from the open-field environment but decreased when subjected to water stress under a similar growing environment. Perhaps variation in water stress and soil types under different growing conditions could have been the major cause in variation of total phenolic content of African horned cucumber fruit since xylem and phloem functions effectively under active-osmoregulation, but solutes uptake decrease when the is lower water movement within the cells cause by higher temperature. For example, [27] found significant differences in several edible fruits such as blackberry and cherry harvested from different locations experiencing varying rainfall patterns. [10] also found a significant difference in total phenols of walnuts' green husks harvested during different periods. They found that fruits harvested earlier have a higher total phenolic content than those which were harvested late, after ripening, due to different metabolites released by plants at different stages of growth.

The current study affirmed that different water stress levels are major triggers of metabolites responsible for this compound, since the plant has to adapt to variation in water levels, as reported by [28], who found a significant total phenolic content in strawberries exposed to different environmental conditions such as water stress and growing conditions. Total phenols are known in human health for their antioxidant properties, which stop free radicals from reacting with other molecules in the body and prevent DNA damage, which is usually caused by a variety of health effects. Therefore, values in this study serve as a concrete evidence needed by policymakers in order to consider this crop for commercialization.

4.9. Micro-Nutrients

Micronutrients deficiency, including of iron, copper, and zinc, may lead to decreased intellectual ability, development, bone mineralization, and immune response, whereas deficiency in zinc may lead to poor digestion, metabolism, reproduction, and wound healing. According to WHO, the recommended daily nutrients intake of zinc for children between four and six years should range from a minimum of 9.6 µg g DW and above. The study findings showed that zinc grand mean of greenhouse grown fruit was higher, at 9.8 µg g DW, relative to shade net at 8.2 µg g DW and open field (7.0 µg g DW). This micronutrient is vital for metabolism and reproduction. Its deficiency may lead to poor digestion and bone diseases. The zinc content of the African horned cucumber fruit serves as a benchmark for commercialization of this fruit, since it has the potential to meet human nutritional needs. In addition, the study findings have shown that moderate water stress and sandy loam increase zinc content under greenhouse environment, and this has added to the information needed by potential growers, since they will be able to create suitable growing environment in order to increase vital micronutrients content for the African horned cucumber fruit.

Iron is another micronutrient that is vital for blood health, bone development, and immune system. Shortage of iron may reduce intellectual capacity, slow growth and poor bone development. According to WHO, recommended daily nutrient intake (RNI) of iron by children between the age of one and three should be 5.8. A range of (0.5 to 3.5 µg g DW) was observed on the African horned cucumber fruit, which is slightly lesser that the recommended daily intake (RNI) by WHO [29]. However, the study findings showed that the fruit has a high potential of meeting the recommended nutrient intake (RNI) if grown from treatment of moderate water stress level combined with sandy loam soil under greenhouse environment. The study's findings serve as a benchmark on the potential nutritional benefit of African horned cucumber fruit. Other researchers, such as [21], reported that growing environment and temperature as growth factors are able to cause variation in nutrient content of crops.

They have shown that a higher evapotranspiration rate, caused by extreme temperatures, could cause a significant variation in fruit nutrient content, as osmotic balance is directly affected, subsequently causing an abnormal flow rate of water and other soluble nutrients within xylem and phloem. Similar findings were observed in the current study, whereby interaction between irrigation water levels and soil types under different growing environments affected the micro-nutrient content (Zn and Fe) African horned cucumber fruit. Several authors remarked that water levels and soil types affect micro-nutrient content of fruits. For example, [30] found that variation in nutrient content may occur when plants are subjected to water stress. They have also demonstrated that, when the plant is subjected to water stress, stomata close, but they open under normal watering.

Their findings unveiled that when the stomatal opening reduces, there is limited carbon dioxide entry in leaves, subsequently affecting the plant's ability to synthesize its own nutrients. [31] report that less frequencies in irrigation significantly increased nutrient content in tomatoes, when compared to treatment that received more irrigation frequencies [31]. They have shown that there is a direct relationship between stomatal conductance and active osmoregulation under less frequencies, but complications occur when there is over-/under-supply of water in plants, as it negatively affects the xylem functions. It worth noting that Cu and Mn were not significantly affected by treatment imposed. [23] observe that a nutrient element such as Mn depends on environmental factors such as adequate water supply, temperature, and plant genotype. However, in this study, irrigation water levels and soil type under different growing environments did not significantly cause variation in some of African horned cucumber fruit, but they significantly affected the Zn and Fe content.

5. Conclusions and Future Research

Quantification of quality parameters such total soluble sugars and macro- and micronutrients contribute to the factors required by policymakers before commercializing a specific crop. Therefore, the outcome of this study has shown that African horned cucumber fruit contain vital biochemical constituents required by humans in both larger and smaller quantities. In addition, this research has provided evidence that the African horned cucumber fruit quality content is significantly affected by treatments. This is useful information to farmers, as quality has become more significant to most consumers worldwide. When grown in the open field, total soluble sugars increased; this is important for the juice-manufacturing industry and for fresh markets, where many fruits are required to meet the demand. Quality parameters such as total flavonoids, total phenols, micro-nutrients and vitamins metabolites seem to be treatment-imposed. This is an important finding, as these factors influence the flavor of fruits. Where the market is geared towards organoleptic quality—in expensive markets, for example—it may be best to grow this crop under a specific growing environment, depending on your target market. The other advantage is that the crop can grow well under protected structures, which eliminate potential damage caused by higher rainfall, hail, and extreme heat in summer.

Author Contributions: M.K.M. was responsible in study designing, drafting, data collection, analysis and interpretation. S.J.M., M.N. and D.M.M. were involved in interpretation of data and critical revision of the draft. All authors have read and agreed to the published version of the manuscript.

Funding: This research was partially funded by the University of South Africa.

Review Board Statement: Not applicable.

Informed Consent Statement: Not applicable.

Data Availability Statement: Data generated for this study is available from the corresponding author on formal request.

Conflicts of Interest: Authors state that there have no conflict of interest.

References

1. Ali, N.A.; Abdelatif, A.H. Chemical Measurement of Total Soluble Sugars as a parameter for cotton lint stickiness grading: Seminar of the project 'Improvement of the marketability of cotton produced in zones affected by stickiness'. In Proceedings of the Seminar, Lille, France, 4–7 July 2001.
2. Backeberg, G.R.; Water, A.J.S. Underutilised indigenous and traditional crops: Why is research on water use important for South Africa? *South Afr. J. Plant Soil* **2013**, *1862*, 291–292. [CrossRef]
3. Barrett, D.M.; Beaulieu, J.C.; Shewfelt, R. Color, flavor, texture, and nutritional quality of fresh-cut fruits and vegetables: Desirable levels, instrumental and sensory measurement, and the effects of processing. *Food Sci. Nutr.* **2010**, *50*, 369–389. [CrossRef]
4. Bartova, V.; Barta, J.; Divis, J.; Svajner, J.; Peterka, J. Crude protein content in tubers of starch processing potato cultivars in dependence on different agroecological conditions. *J. Cent. Eur. Agric.* **2009**, *10*, 57–66.
5. Bernaert, N.; De Clercq, H.; Van Bockstaele, E.; De Loose, M.; Van Droogenbroeck, B. Antioxidant changes during postharvest processing and storage of leek (Allium ampeloprasum var. porrum). *Postharvest Biol. Technol.* **2013**, *86*, 8–16.
6. Dold, A.P.; Cocks, M.L. The trade in medicinal plants in the Eastern Cape Province, South Africa. *South Afr. J. Sci.* **2002**, *98*, 589–597.
7. Esch, J.R.; Friend, J.R.; Kariuki, J.K. Determination of the Vitamin C content of conventionally and organically grown fruits by cyclic voltammetry. *Int. J. Electrochem. Sci.* **2010**, *5*, 1464–1474.
8. Fenech, M.; Amaya, I.; Valpuesta, V.; Botella, M.A. Vitamin C Content in Fruits: Biosynthesis and Regulation. *Front. Plant Sci.* **2019**, *9*, 2006. [CrossRef] [PubMed]
9. Hurr, B.M.; Huber, D.J.; Vallejos, C.E.; Talcott, S.T. Developmentally dependent responses of detached cucumber (*Cucumis sativus* L.) fruit to exogenous ethylene. *Postharvest Biol. Technol.* **2009**, *52*, 207–215. [CrossRef]
10. Iglesias-Carres, L.; Mas-Capdevila, A.; Bravo, F.I.; Aragonès, G.; Arola-Arnal, A.; Muguerza, B. A comparative study on the bioavailability of phenolic compounds from organic and nonorganic red grapes. *Food Chem.* **2019**, *299*, 125092. [CrossRef]
11. Ity, Q.; Canino, O.F. Effect of spraying magnesium, boron, ascorbic acid and vitamin B complex on yield and fruit quality of "Canino" apricot. *J. Agric. Sci.* **2006**, *14*, 337–347.
12. Khattab, M.M.; Shaban, A.E.; El-Shrief, A.H.; Mohamed, A.S.E. Growth and productivity of pomegranate trees under different irrigation levels II: Fruit quality. *J. Hortic. Sci. Ornam. Plants* **2011**, *3*, 259–264.

13. Legwaila, G.M.; Mojeremane, W.; Madisa, M.E.; Mmolotsi, R.M.; Rampart, M. Potential of traditional food plants in rural household food security in Botswana. *J. Hortic. For.* **2011**, *3*, 171–177.
14. Lombardi, N.; Salzano, A.M.; Troise, A.D.; Scaloni, A.; Vitaglione, P.; Vinale, F.; Marra, R.; Caira, S.; Lorito, M.; D'Errico, G.; et al. Effect of Trichoderma Bioactive Metabolite Treatments on the Production, Quality, and Protein Profile of Strawberry Fruits. *J. Agric. Food Chem.* **2020**, *68*, 7246–7258. [CrossRef]
15. López, A.; Arazuri, S.; Jarén, C.; Mangado, J.; Arnal, P.; De Galarreta, J.I.R.; Riga, P.; López, R. Crude Protein Content Determination of Potatoes by NIRS Technology. *Procedia Technol.* **2013**, *8*, 488–492. [CrossRef]
16. Manthey, J.A.; Perkins-Veasie, P. Influences of harvest date and location on the levels of β-carotene, ascorbic acid, total phenols the *in vitro* antioxidant capacity, and phenolic profiles of five commercial varieties of mango (*Mangifera Indica* L.). *J. Agric. Food Chem.* **2009**, *57*, 10825–10830. [CrossRef] [PubMed]
17. Marta, B.; Szafrańska, K.; Posmyk, M.M. Exogenous melatonin improves antioxidant defense in Cucumber Seeds (*Cucumis sativus* L.) germinated under chilling stress. *Front. Plant Sci.* **2016**, *7*, 1–12. [CrossRef] [PubMed]
18. Maseko, I.; Mabhaudhi, T.; Tesfay, S.; Araya, H.T.; Fezzehazion, M.; Du Plooy, C.P. African Leafy Vegetables: A Review of Status, Production and Utilization in South Africa. *Sustain. J. Rec.* **2017**, *10*, 16. [CrossRef]
19. Moyo, M.; Amoo, S.O.; Aremu, A.O.; Grúz, J.; Šubrtová, M.; Jarošová, M.; Tarkowski, P.; Dolezal, K. Determination of Mineral Constituents, Phytochemicals and Antioxidant Qualities of Cleome gynandra, Compared to Brassica oleracea and Beta vulgaris. *Front. Chem.* **2018**, *5*, 128. [CrossRef] [PubMed]
20. Nyathi, M.; Mabhaudhi, T.; Van Halsema, G.; Annandale, J.; Struik, P. Benchmarking nutritional water productivity of twenty vegetables—A review. *Agric. Water Manag.* **2019**, *221*, 248–259. [CrossRef]
21. Rahil, M.; Qanadillo, A. Effects of different irrigation regimes on yield and water use efficiency of cucumber crop. *Agric. Water Manag.* **2015**, *148*, 10–15. [CrossRef]
22. Reinten, E.; Coetzee, J.; Van Wyk, B.-E. The potential of South African indigenous plants for the international cut flower trade. *South Afr. J. Bot.* **2011**, *77*, 934–946. [CrossRef]
23. Rimpapa, Z.; Toromanovic, J.; Tahirovic, I.; Sofic, E. Total content of phenols and anthocyanins in edible fruits from Bosnia. *J. Basic Med. Sci.* **2007**, *7*, 119–122. [CrossRef] [PubMed]
24. Santos-Zea, L.; Guti, J.A.; Serna-Saldivar, S.O.; Monterrey, D.; Eugenio, A.; Sada, G. Comparative analyses of total phenols, antioxidant activity, and flavanol glycoside profile of cladode flours from different varieties of *Opuntia* spp. *J. Agric. Food Chem.* **2011**, *59*, 7054–7061. [CrossRef]
25. Sezen, S.M.; Celikel, G.; Yazar, A.; Tekin, S.; Kapur, B. Effect of irrigation management on yield and quality of tomatoes grown in different soilless media in a glasshouse. *Sci. Res. Essay* **2010**, *5*, 41–48.
26. Sharma, S.; Rao, R. Nutritional quality characteristics of pumpkin fruit as revealed by its biochemical analysis. *Int. Food Res. J.* **2013**, *20*, 2309–2316.
27. Tavarini, S.; Degl'Innocenti, E.; Remorini, D.; Massai, R.; Guidi, L. Antioxidant capacity, ascorbic acid, total phenols and carotenoids changes during harvest and after storage of Hayward kiwifruit. *Food Chem.* **2008**, *107*, 282–288. [CrossRef]
28. Uusiku, N.P.; Oelofse, A.; Duodu, K.G.; Bester, M.J.; Faber, M. Nutritional value of leafy vegetables of sub-Saharan Africa and their potential contribution to human health: A review. *J. Food Compos. Anal.* **2010**, *23*, 499–509. [CrossRef]
29. Van Wyk, B.-E. *Food Plants of the World*; Timber Press: Portland, OR, USA, 2005.
30. Wang, C.; Guo, L.; Li, Y.; Wang, Z.; Yang, X.; Wang, X.; Pretorius, B. Systematic comparison of C3 and C4 plants based on metabolic network analysis. *BMC Syst. Biol.* **2017**, *65*, 1–14. [CrossRef]
31. Zhang, J.; Wang, X.; Yu, O.; Tang, J.; Gu, X.; Wan, X.; Fang, C. Metabolic profiling of strawberry (*Fragaria x ananassa* Duch.) during fruit development and maturation. *J. Exp. Bot.* **2011**, *62*, 1103–1118. [CrossRef]

Article

Growth and Competitive Infection Behaviors of *Bradyrhizobium japonicum* and *Bradyrhizobium elkanii* at Different Temperatures

Md Hafizur Rahman Hafiz [1,2], **Ahsanul Salehin** [3] **and Kazuhito Itoh** [1,3,*]

[1] Faculty of Life and Environmental Science, Shimane University, 1060 Nishikawatsu, Matsue 690-8504, Japan; hafizhstu@hotmail.com

[2] Department of Crop Physiology and Ecology, Hajee Mohammad Danesh Science and Technology University, Dinajpur 5200, Bangladesh

[3] The United Graduate School of Agricultural Sciences, Tottori University, 4-101 Koyama-Minami, Tottori 680-8553, Japan; ujanrijvi224@gmail.com

[*] Correspondence: itohkz@life.shimane-u.ac.jp; Tel.: +81-852-32-6521

Abstract: Growth and competitive infection behaviors of two sets of *Bradyrhizobium* spp. strains were examined at different temperatures to explain strain-specific soybean nodulation under local climate conditions. Each set consisted of three strains—*B. japonicum* Hh 16-9 (Bj11-1), *B. japonicum* Hh 16-25 (Bj11-2), and *B. elkanii* Hk 16-7 (BeL7); and *B. japonicum* Kh 16-43 (Bj10J-2), *B. japonicum* Kh 16-64 (Bj10J-4), and *B. elkanii* Kh 16-7 (BeL7)—which were isolated from the soybean nodules cultivated in Fukagawa and Miyazaki soils, respectively. The growth of each strain was evaluated in Yeast Mannitol (YM) liquid medium at 15, 20, 25, 30, and 35 °C with shaking at 125 rpm for one week while measuring their OD_{660} daily. In the competitive infection experiment, each set of the strains was inoculated in sterilized vermiculite followed by sowing surface-sterilized soybean seeds, and they were cultivated at 20/18 °C and 30/28 °C in a 16/8 h (day/night) cycle in a phytotron for three weeks, then nodule compositions were determined based on the partial 16S-23R rRNA internal transcribes spacer (ITS) gene sequence of DNA extracted from the nodules. The optimum growth temperatures were at 15–20 °C for all *B. japonicum* strains, while they were at 25–35 °C for all *B. elkanii* strains. In the competitive experiment with the Fukagawa strains, Bj11-1 and BeL7 dominated in the nodules at the low and high temperatures, respectively. In the Miyazaki strains, BjS10J-2 and BeL7 dominated at the low and high temperatures, respectively. It can be assumed that temperature of soil affects rhizobia growth in rhizospheres and could be a reason for the different competitive properties of *B. japonicum* and *B. elkanii* strains at different temperatures. In addition, competitive infection was suggested between the *B. japonicum* strains.

Keywords: *Bradyrhizobium japonicum*; *Bradyrhizobium elkanii*; temperature effects; growth; competitive infection; nodule composition

Citation: Hafiz, M.H.R.; Salehin, A.; Itoh, K. Growth and Competitive Infection Behaviors of *Bradyrhizobium japonicum* and *Bradyrhizobium elkanii* at Different Temperatures. *Horticulturae* **2021**, *7*, 41. https://doi.org/10.3390/horticulturae7030041

Academic Editor: Stefano Marino

Received: 12 February 2021
Accepted: 25 February 2021
Published: 28 February 2021

Publisher's Note: MDPI stays neutral with regard to jurisdictional claims in published maps and institutional affiliations.

Copyright: © 2021 by the authors. Licensee MDPI, Basel, Switzerland. This article is an open access article distributed under the terms and conditions of the Creative Commons Attribution (CC BY) license (https://creativecommons.org/licenses/by/4.0/).

1. Introduction

Soybean-nodulating bacteria have distributed worldwide [1,2] and established important symbiotic relationships with host plants to fix atmospheric nitrogen [3]. *Bradyrhizobium japonicum* and *B. elkanii* are reported as the major soybean nodulating rhizobia [4,5] and their nodulation behaviors in the field need to be clarified in relation to environmental conditions because their nodulation and nitrogen fixation are known to be highly dependent on environmental conditions [6]. In previous studies, latitudinal-characteristic nodulation of *B. japonicum* and *B. elkanii* has been reported in Japan [7,8], the United States [9], and Nepal [10], in which *B. japonicum* and *B. elkanii* dominate in soybean nodules in northern and southern regions, respectively. These results suggest that the temperature of the soybean-growing location contributes to the nodule composition of *B. japonicum* and *B. elkanii*.

To elucidate the possible reason, laboratory competitive inoculation experiments have been conducted at different temperatures. Kluson et al. [11] reported that *B. japonicum* strains dominated in nodules at lower temperatures, while *B. elkanii* strains dominated at higher temperature. Suzuki et al. [12] examined the relative population of *B. japonicum* and *B. elkanii* strains in the rhizospheres of soybeans and their nodule compositions at different temperatures and revealed that the *B. japonicum* strain dominated in nodules at lower temperature even though the relative populations of both strains were similar in the rhizosphere, while at higher temperature, the *B. elkanii* strains dominated in nodules due to their larger relative population in the rhizosphere. Shiro et al. [13] reported that the nodule occupancy of *B. elkanii* increased at higher temperatures, whereas that of *B. japonicum* increased at lower temperatures, corresponding to their temperature-dependent *nodC* gene expressions. These results suggest that the temperature-dependent infections and proliferations in soils are possible reasons for the temperature-dependent nodule compositions of rhizobia in the field. However, it has been uncertain which factor, namely, temperature-dependent infection or proliferation in soil, contributes to the temperature-dependent distribution of rhizobia in nodules.

For elucidating which factor is more involved in the soybean nodule composition under local climatic conditions, Hafiz et al. [7] examined the changes in the nodule composition when soil samples were used for soybean cultivation under the different climatic conditions from the original locations, and found that the *B. japonicum* strains nodulated dominantly in the Fukagawa location (temperate continental climate) and the dominancy of *B. japonicum* did not change when soybean was cultivated in the Matsue and Miyazaki locations (humid sub-tropical climate) using the Fukagawa soil. The results suggest that the *B. japonicum* strains proliferated dominantly in the Fukagawa soil leading to their nodule dominancy because *B. elkanii* did not appear in the Matsue and Miyazaki locations. On the other hand, the *B. elkanii* strains dominated in the Miyazaki soil and location while the *B. japonicum* strains dominated when soybean was cultivated in the Fukagawa location using the Miyazaki soil, suggesting that temperature-dependent infection would lead to nodule dominancy of the *B. elkanii* and *B. japonicum* strains in the Miyazaki and Fukagawa locations, respectively.

In addition, in the Fukagawa soil and location, phylogenetic sub-group *B. japonicum* Bj11-1, which was characterized as a slow grower, dominated the nodules compared to another sub-group *B. japonicum* Bj11-2, which was characterized as a fast grower [7], suggesting that infection preference might determine the nodule composition among the *B. japonicum* strains rather than their growth properties. In the Miyazaki soil and location, it was suggested that both *B. japonicum* and *B. elkanii* strains proliferated, and that the species-specific nodule compositions under the different local climatic conditions might be due to the temperature-dependent growth and infection properties of the *Bradyrhizobium* strains [7].

These hypotheses presented in the previous study [7] should be confirmed by in vitro growth and inoculation experiments under the controlled temperatures using the *B. japonicum* and *B. elkanii* strains isolated from the corresponding soils and locations. In this study, we compared growth and infection behaviors at different temperatures of the *B. japonicum* and *B. elkanii* strains isolated from the soybean nodules cultivated in the Fukagawa and Miyazaki soils, and elucidated the reason why the species-specific nodule compositions are present in the Fukagawa and Miyazaki soils and locations.

2. Materials and Methods

2.1. Effect of Temperature on Growth of Bradyrhizobium spp. in Liquid Culture

The strains used are listed in Table 1. They were isolated from nodules of soybean cultivated in the Fukagawa and Miyazaki soils and study locations in 2016, and selected based on their phylogenetic characteristics based on the 16S rRNA and 16S-23S rRNA internal transcribes spacer (ITS) gene sequences [7].

Table 1. *Bradyrhizobium* strains used in this study.

Strain [a]	Closest 16 rDNA	ITS Group [b]	Accession Number [c]
Hh 16-9	*B. japonicum* Bj11	Bj11-1	LC582854, LC579849
Hh-16-25	*B. japonicum* Bj11	Bj11-2	LC582860, LC579855
Hk 16-7	*B. elkanii* L7	BeL7	LC582891, LC579886
Kh 16-43	*B. japonicum* S10J	BjS10J-2	LC582874, LC579869
Kh 16-64	*B. japonicum* S10J	BjS10J-4	LC582887, LC579882
Kh 16-7	*B. elkanii* L7	BeL7	LC582901, LC579896

[a] The strains were isolated from nodules of soybean cultivated using Fukagawa (H) and Miyazaki (K) soils at Fukagawa (h) and Miyazaki (k) study locations in 2016. The isolates were designated by soil, location, year, and strain number. [b] Group based on gene sequence of 16S-23S rRNA internal transcribes spacer (ITS) region. [c] Gene accession number of 16S rRNA and ITS sequences.

Considering the temperature ranges during the soybean cultivation period in the study locations (Table 2), the temperatures were set at 15 °C (around average daily minimum temperature in the Fukagawa location), 20 °C (around average daily temperature in the Fukagawa location), 25 °C (around average daily maximum and minimum temperatures in the Fukagawa and Miyazaki locations, respectively), 30 °C (around average daily temperature in the Miyazaki location), and 35 °C (around average daily maximum temperature in the Miyazaki location).

Table 2. Geographical and climatic characteristics of the study locations in Japan [7].

Location	Latitude (°N)	Longitude (°E)	Temperature (°C) [a]	Rainfall (mm) [a]
Fukagawa	43.71	142.01	16–26/16–26 (14–24/17–27) [b]	432/243
Miyazaki	31.82	131.41	24–32/24–31 (25–32/25–33)	240/860

[a] Average daily minimum and maximum temperatures and total rainfall during the cultivation period in 2016/2017. [b] Figures in parenthesis indicate those during one month after sowing. (https://www.jma.go.jp, accessed on 28 February 2021).

Each strain was pre-incubated on Yeast Mannitol (YM) [14] agar medium at 26 °C for 5–10 days, and a part of the colony was taken into 3 mL of YM liquid medium to adjust OD_{660} at 0.03, then incubated with shaking at 125 rpm for seven days while measuring their OD_{660} at 24-h intervals. All the experiments were done in triplicate.

2.2. Effect of Temperature on Competitive Infection of Bradyrhizobium spp. in Soybean

For the competition experiment, each set consisting of three strains from each soil was used as follows: *B. japonicum* Hh 16-9 (Bj11-1), *B. japonicum* Hh 16-25 (Bj11-2), and *B. elkanii* Hk 16-7 (BeL7) from the Fukagawa soil; *B. japonicum* Kh 16-43 (Bj10J-2), *B. japonicum* Kh 16-64 (Bj10J-4), and *B. elkanii* Kh 16-7 (BeL7) from the Miyazaki soil.

The strains were cultured in YM liquid medium with shaking at 25 °C for seven days, then each cell density was adjusted to 10^9 colony forming unit (CFU)/mL with sterilized distilled water based on OD-CFU/mL correlated linear equations prepared for each strain. Each one milliliter aliquot of the culture was added onto sterilized vermiculite in a 400 mL Leonard jar [15], which was supplemented with sterilized N-free nutrient solution [16]. Three jars were prepared for each treatment. After mixing the inoculated vermiculite thoroughly, three soybean seeds, cv. Orihime (non-Rj) were sown in each Leonard jar and cultivated in a phytotron (LH-220S, NK system, Osaka, Japan) at 20/18 °C and 30/28 °C in 16/8 h (day/night) cycle with an occasional supply of the N-free nutrient solution. The soybean seeds were surface-sterilized prior to sowing with 70% ethanol for 30 s and then with 2.5% NaOCl solution for 3 min [17]. Seedlings were thinned to one plant per jar one week after germination. At three weeks after sowing, the length and weight of the shoot and root were measured, and the number of nodules was counted. Then, nodule composition of the inoculated strains was examined using ten randomly-selected nodules per plant. Control plants without inoculation were prepared to check contamination, and the experiment was conducted in triplicate. Each nodule was surface sterilized with 70% ethanol for 30 s followed by washing six times with sterilized distilled water, then each nodule was crushed with 200 µL of sterilized MilliQ water for extraction of DNA [18]. The inoculated strain in each nodule was specified by PCR and nucleotide sequence of the 16S-23S rRNA internal transcribed spacer (ITS) region, according to the procedures described previously [7].

2.3. Statistical Analysis

Statistical analysis of the soybean growth and nodule compositions of *Bradyrhizobium* spp. were performed using the MSTAT-C 6.1.4 software package [19]. The data were subjected to Duncan's multiple range test after one-way ANOVA.

3. Results

3.1. Effect of Temperature on Growth of Bradyrhizobium spp. Strains in Liquid Culture

The effects of temperature on the proliferation of the *Bradyrhizobium* spp. strains are presented in Figure 1. The responses to different temperatures varied among the strains. At 15–20 °C, the growth rates of *B. japonicum* Bj11-1 and Bj11-2 were similar and higher than those of *B. elkanii* BeL7 in the Fukagawa strains, and similar growth patterns were observed in *B. japonicum* BjS10J-2 and BjS10J-4, and *B. elkanii* BeL7 in the Miyazaki strains. At 25–35 °C, *B. elkanii* BeL7 proliferated better than the *B. japonicum* strains in the Fukagawa strains, and *B. japonicum* Bj11-1 did not proliferate at 35 °C. Similarly, in the Miyazaki strains, the growth rate of *B. elkanii* BeL7 increased at high temperatures, while those of the *B. japonicum* strains decreased at 30–35 °C, and *B. japonicum* BJS10J-2 did not proliferate at 35 °C.

For each strain, OD_{660} at 5 days of incubation is shown in Figure 2 and normalized as a relative % of OD_{660} to the maximum value in the range of temperatures examined. In the Fukagawa strains, the relative % of Bj11-1 and Bj11-2 were 93–100% at 15–20 °C, while those of BeL7 were 11–13%. The relative % of all strains were more than 80% at 25 °C. At higher temperatures, those of Bj11-1 and Bj11-2 decreased significantly at above 25 and 30 °C, respectively, while those of Bel7 were similar at 25–35 °C. In the Miyawaki strains, BjS10J-2 showed a larger relative % than BjS10j-4 at lower temperatures, and those of BeL7 were less than 20%. At higher temperature, those of BjS10J-2, BjS10j-4, and Bel7 decreased significantly at above 20, 25, and 30 °C, respectively.

3.2. Effect of Temperature on Growth and Nodule Number of Soybean Inoculated with a Set of Bradyrhizobium spp. Strains

Effect of temperature on the growth and nodule number of soybean is presented in Figure 3. The shoot and root lengths, and the shoot and root weights were significantly higher at 30/28 °C than 20/18 °C in all treatments except for the root lengths of the soybeans inoculated with Miyazaki strains. While the nodule numbers were not significantly different between the different temperature conditions. The inoculation of the *Bradyrhizobium* spp. strains significantly affected the shoot length and the root weight of soybean at 30/28 °C while the effects were not observed at 20/18 °C. Significant difference in these effects was not present between the Fukagawa and Miyazaki strains. No nodule was recorded in the control plants, indicating that there was no contamination in the experimental procedure.

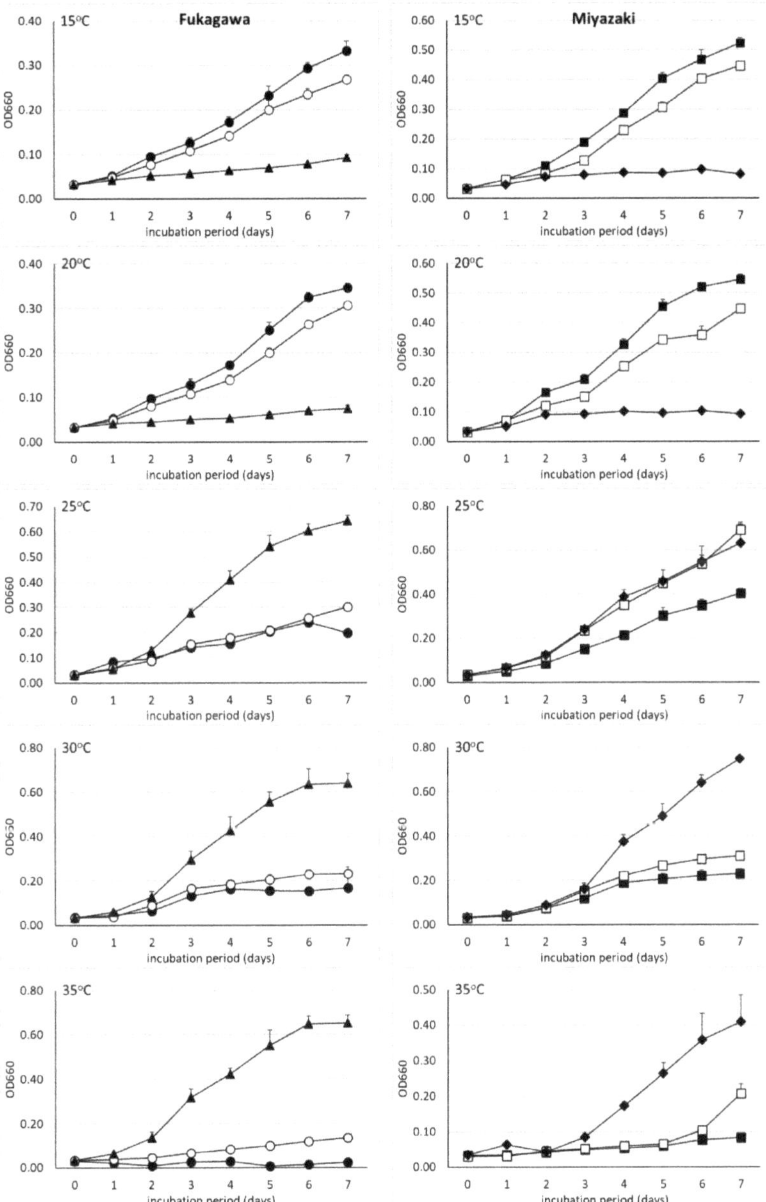

Figure 1. Effects of temperature on growth of *Bradyrhizobium* spp. strains in liquid culture. Fukagawa strains: *B. japonicum* Hh 16-9 (Bj11-1) (●), *B. japonicum* Hh 16-25 (Bj11-2) (○), and B. *elkanii* Hk 16-7 (BeL7) (▲); Miyazaki strains: *B. japonicum* Kh 16-43 (Bj10J-2) (■), *B. japonicum* Kh 16-64 (Bj10J-4) (□), and *B. elkanii* Kh 16-7 (BeL7) (♦). The bars represent the standard deviation (*n* = 3).

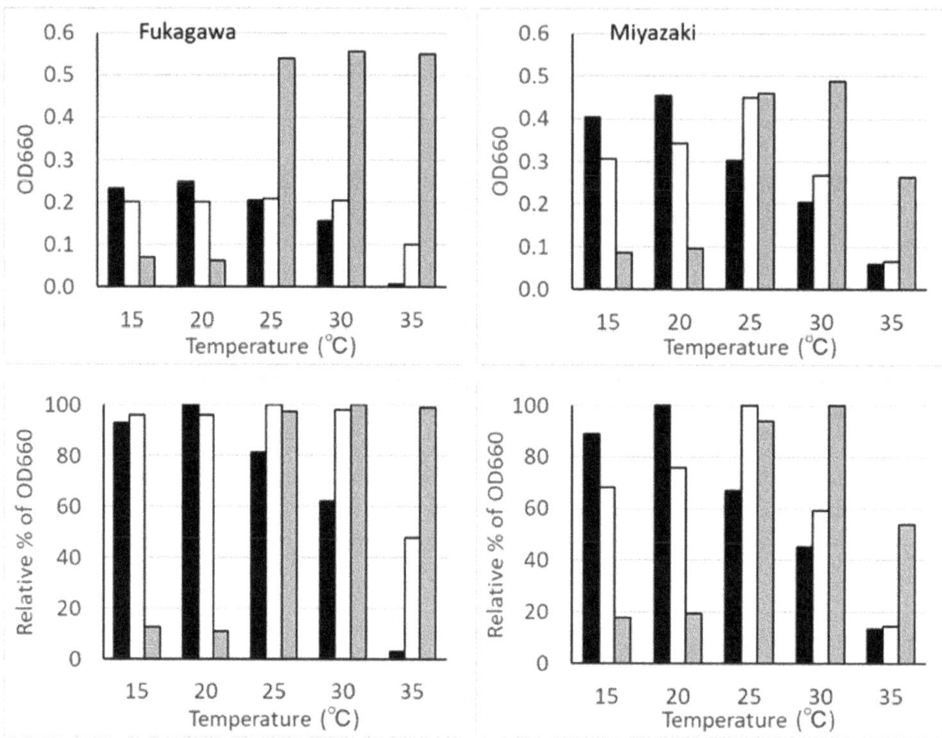

Figure 2. Effects of temperature on growth of *Bradyrhizobium* spp. strains in liquid culture. Upper: OD$_{660}$ at 5 days, lower: relative percentage of OD$_{660}$ to maximum for each strain. Fukagawa strains: *B. japonicum* Hh 16-9 (Bj11-1) (■), *B. japonicum* Hh 16-25 (Bj11-2) (□), and B. *elkanii* Hk 16-7 (BeL7) (▨); Miyazaki strains: *B. japonicum* Kh 16-43 (Bj10J-2) (■), *B. japonicum* Kh 16-64 (Bj10J-4) (□), and *B. elkanii* Kh 16-7 (BeL7) (▨).

3.3. Effect of Temperature on Soybean Nodule Composition of Inoculated Bradyrhizobium spp. Stains

The relative nodule composition of the inoculated *Bradyrhizobium* spp. strains is presented in Figure 4. Under the competitive conditions for the Fukagawa strains, only Bj11-1 formed the nodules at 20/18 °C, while only BeL7 did at 30/28 °C. For the Miyazaki strains, BjS10J-2 was dominant in the nodules at 20/18 °C with the minor presence of BjS10J-4. At high temperature (30/28 °C) BeL7 was dominant and BjS10J-4 was minor in the nodules. Mixed colonization of nodules with two or three strains in the same nodule would be possible, but minor signals were not visibly observed in the nucleotide chromatogram.

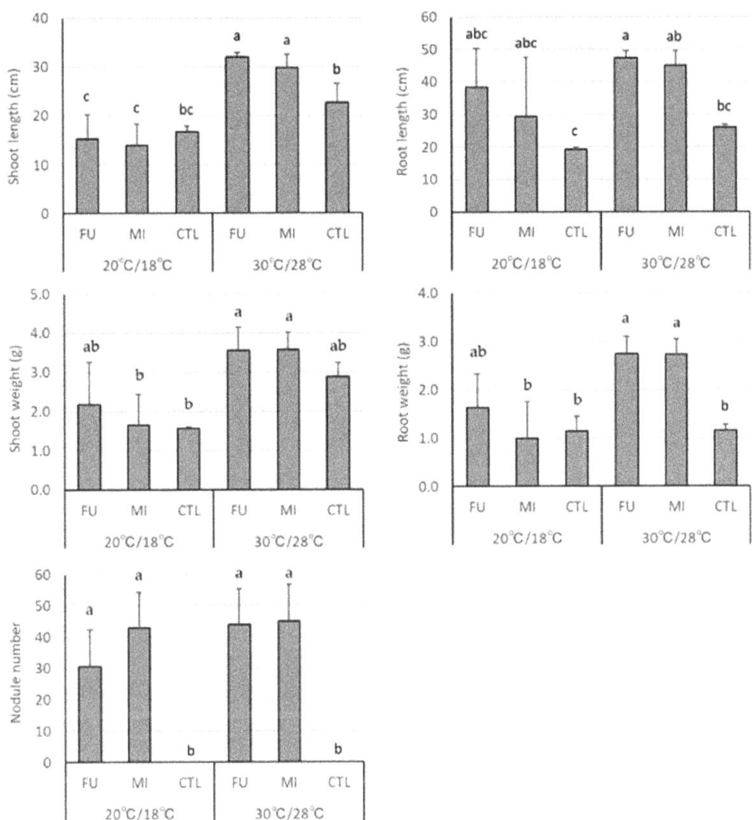

Figure 3. Effects of temperature on growth and nodule number of soybean inoculated with a mixture of the *Bradyrhizobium* spp. strains. Soybean was cultivated in a phytotron at 20/18 °C (day/night) and 30/28 °C at 16/8 h cycle. Fukagawa strains: *B. japonicum* Hh 16-9 (Bj11-1), *B. japonicum* Hh 16-25 (Bj11-2), and *B. elkanii* Hk 16-7 (BeL7); Miyazaki strains: *B. japonicum* Kh 16-43 (Bj10J-2), *B. japonicum* Kh 16-64 (Bj10J-4), and *B. elkanii* Kh 16-7 (BeL7); control: no inoculation. The bars represent the standard deviation (*n* = 3) and different letters indicate significant differences at *p* < 0.05 by Duncan's test.

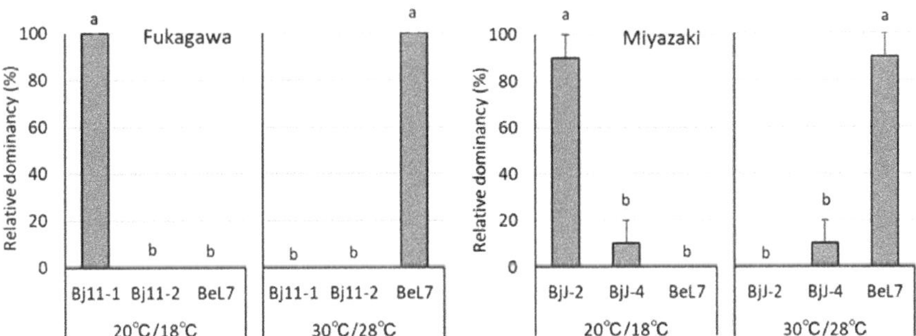

Figure 4. Effects of temperature on relative abundancy of inoculated *Bradyrhizobium* spp. strains in soybean. Soybean was cultivated in a phytotron at 20/18 °C (day/night) and 30/28 °C at 16/8 h cycle. Fukagawa strains: *B. japonicum* Hh 16-9 (Bj11-1), *B. japonicum* Hh 16-25 (Bj11-2), and *B. elkanii* Hk 16-7 (BeL7); Miyazaki strains: *B. japonicum* Kh 16-43 (Bj10J-2), *B. japonicum* Kh 16-64 (Bj10J-4), and *B. elkanii* Kh 16-7 (BeL7). The bars represent the standard deviation (*n* = 3) and different letters indicate significant differences at *p* < 0.05 by Duncan's test.

4. Discussion

Although the number is limited, a similar temperature-dependent growth tendency in liquid media of the two *Bradyrhizobium* species has been reported previously. Three *B. japonicum* strains grew better at 15 °C than 25 °C, and could not grow at 35 °C, while one *B. elkanii* strain grew better at 25–35 °C than at 15 °C [20]. Kluson et al. [11] also reported that optimum growth of two *B. elkanii* strains was around 25 °C while two *B. japonicum* strains grew best at 20 °C in the range of 20–35 °C. These results suggest that *B. japonicum* and *B. elkanii* have species-specific temperature preference in their proliferations. The tendencies are consistent with the previous results on the latitudinal characteristic nodulation of *B. japonicum* and *B. elkanii* in Japan [7,8], the United States [9], and Nepal [10].

In the infection experiment, we used sterilized vermiculite to simplify the experimental conditions—the same population of the inoculants and elimination of the effects of indigenous soil microorganisms on the competition. Sterilization of soil samples by autoclaving could change its physicochemical conditions. Actually, the population of the inoculated rhizobia decreased in the sterilized Fukagawa soil due to unknown reasons in a preliminary experiment (data not shown). Therefore, we could not use the soil samples in this study.

The better growth of soybean at higher temperature has been reported previously in the similar range of temperatures [11,21,22]. The number of nodules was temperature-independent in this study (Figure 3), while temperature-dependent nodule formation, that is, in this study, the higher temperature, the larger nodule number in the similar temperature range, has been reported when *B. japonicum* strains were inoculated in laboratory experiments [21,23]. In this study, the *B. japonicum* and *B. elkanii* strains were co-inoculated and a different strain was dominant among the inoculated strains in the nodules depending on the temperature (Figure 4), therefore, the nodule number would be dependent on the nodulating properties of the dominant strains in the nodules rather than the temperature.

The high nodule dominancy of *B. elkanii* BeL7 (Hk 16-7 and Kh 16-7) at high temperature (30/28 °C) is presumed to be due to the difference in temperature sensitivity between the *B. japonicum* and *B. elkanii* strains (Figure 2), in addition to the up-regulated expression of *nodC* in *B. elkanii* at high temperature, compared with *B. japonicum* [13]. The temperature-dependent growth properties of the *Bradyrhizobium* spp. strains suggests high nodule dominancy of the *B. japonicum* strains at low temperature (20/18 °C). However, one of the two *B. japonicum* strains for each soil was dominant in the nodules even though their growth properties were similar (Figure 1). Differences in expression levels of nodulation genes and in responses to isoflavones secreted from soybean roots might determine the nodule composition between them. The same temperature-dependent nodule composition; dominancy of B. *japonicum* and *B. elkanii* at low and high temperatures, respectively, has been reported in the other laboratory competitive studies [11–13].

Generally, the composition of soybean rhizobia in field soil has been estimated by nodule composition. Regarding the latitudinal characteristic nodule composition of soybean rhizobia [7–10], competitive inoculation experiments have revealed that the nodule composition is affected by species-specific, temperature-dependent infection and proliferation in soils [11–13]. However, it is uncertain which factor contributes to the temperature-dependent nodule composition.

In our previous study [7], we selected three study locations of different local climatic conditions in Japan, and each soil sample of the study locations was used for soybean cultivation at all the study locations to examine the changes in the nodule compositions under the different local climatic conditions. As a result, we assumed that *B. japonicum* dominantly proliferate in the Fukagawa soil, leading to their dominant nodule composition, because the nodule composition was not affected under warmer climatic conditions in Miyazaki location. To confirm our assumption, the competitive inoculation experiment was conducted using the rhizobial strains isolated from soybean nodules cultivated in Fukagawa soil, and the results showed that *B. japonicum* dominated nodules at lower temperature while *B. elkanii* dominated at higher temperature (Figure 4), supporting our assumption that *B. japonicum* dominantly proliferate in the Fukagawa soil because the dominancy of *B. elkanii* did not increase at higher temperature in the Miyazaki location.

We also assumed that both *B. japonicum* and *B. elkanii* exist in the Miyazaki soil and the dominant nodule composition of *B. elkanii* is due to their preferred infection because the nodule composition was affected under cooler climatic conditions in Fukagawa location. In the competitive inoculation experiment using the Miyazaki rhizobial strains, *B. japonicum* and *B. elkanii* dominated nodules at lower and higher temperatures, respectively (Figure 4), also supporting our assumption that both *B. japonicum* and *B. elkanii* exist in the Miyazaki soil and their preferred infection determined the nodule composition.

5. Conclusions

The experiments performed in the liquid cultures revealed better growth of *B. japonicum* at lower temperatures and *B. elkanii* at higher temperatures, and therefore it can be assumed that the temperature of soil affects rhizobia growth in the rhizosphere and could be a reason for the different competitive properties of *B. japonicum* and *B. elkanii* strains at different temperatures. In addition, competitive infection was suggested between the *B. japonicum* strains.

Author Contributions: M.H.R.H. and K.I. conceptualized the study and designed the experiments; M.H.R.H. performed the experiments; A.S. helped to conduct the experiment and in the data analysis; and M.H.R.H. wrote the article, with a substantial contribution from K.I. All authors have read and agreed to the published version of the manuscript.

Funding: This research received no external funding.

Conflicts of Interest: The authors declare no conflict of interest.

References

1. Hymowitz, T. On the domestication of the soybean. *Econ. Bot.* **1970**, *24*, 408–421. [CrossRef]
2. Hymowitz, T.; Singh, R.J. Taxonomy and speciation. In *Soybeans, Improvement, Production and Uses, 2nd. ed.*; Wilcox, J.R., Ed.; American Society of Agronomy: Madison, WI, USA, 1987; pp. 23–48.
3. Hungria, M.; Mendes, I.C.; de Bruijn, F. Nitrogen fixation with soybean: The perfect symbiosis? In *Biological Nitrogen Fixation*; De Bruijn, F., Ed.; Wiley Publisher: Hoboken, NJ, USA, 2015; Volume 2, pp. 1005–1019.
4. Jordan, D.C. NOTES: Transfer of Rhizobium japonicum Buchanan 1980 to Bradyrhizobium gen. nov., a Genus of Slow-Growing, Root Nodule Bacteria from Leguminous Plants. *Int. J. Syst. Bacteriol.* **1982**, *32*, 136–139. [CrossRef]
5. Kuykendall, L.D.; Saxena, B.N.; Devine, T.E.; Udell, S.E. Genetic diversity in Bradyrhizobium japonicum Jordan 1982 and a proposal for Bradyrhizobium elkanii sp.nov. *Can. J. Microbiol.* **1992**, *38*, 501–505. [CrossRef]
6. Hungria, M.; Vargas, M.A.T. Environmental factors impacting N2 fixation in legumes grown in the tropics, with an emphasis on Brazil. *Field Crop. Res.* **2000**, *65*, 151–164. [CrossRef]
7. Hafiz, H.R.; Salehin, A.; Adachi, F.; Omichi, M.; Saeki, Y.; Yamamoto, A.; Hayashi, S.; Itoh, K. Latitudinal Characteristic Nodule Composition of Soybean-Nodulating Bradyrhizobia: Temperature-Dependent Proliferation in Soil or Infection? *Horticulturae* **2021**, *7*, 22. [CrossRef]
8. Saeki, Y.; Aimi, N.; Tsukamoto, S.; Yamakawa, T.; Nagatomo, Y.; Akao, S. Diversity and geographical distribution of indigenous soybean-nodulating bradyrhizobia in Japan. *Soil Sci. Plant Nutr.* **2006**, *52*, 418–426. [CrossRef]
9. Shiro, S.; Matsuura, S.; Saiki, R.; Sigua, G.C.; Yamamoto, A.; Umehara, Y.; Hayashi, M.; Saeki, Y. Genetic Diversity and Geographical Distribution of Indigenous Soybean-Nodulating Bradyrhizobia in the United States. *Appl. Environ. Microbiol.* **2013**, *79*, 3610–3618. [CrossRef] [PubMed]
10. Adhikari, D.; Kaneto, M.; Itoh, K.; Suyama, K.; Pokharel, B.B.; Gaihre, Y.K. Genetic diversity of soybean-nodulating rhizobia in Nepal in relation to climate and soil properties. *Plant Soil* **2012**, *357*, 131–145. [CrossRef]
11. Kluson, R.A.; Kenworthy, W.J.; Weber, D.F. Soil temperature effects on competitiveness and growth of Rhizobium japonicum and on Rhizobium-induced chlorosis of soybeans. *Plant Soil* **1986**, *95*, 201–207. [CrossRef]
12. Suzuki, Y.; Adhikari, D.; Itoh, K.; Suyama, K. Effects of temperature on competition and relative dominance of Bradyrhizobium japonicum and Bradyrhizobium elkanii in the process of soybean nodulation. *Plant Soil* **2013**, *374*, 915–924. [CrossRef]
13. Shiro, S.; Kuranaga, C.; Yamamoto, A.; Sameshima-Saito, R.; Saeki, Y. Temperature-Dependent Expression of NodC and Community Structure of Soybean-Nodulating Bradyrhizobia. *Microbes Environ.* **2016**, *31*, 27–32. [CrossRef] [PubMed]
14. Callow, J.A.; Vincent, J.M. A Manual for the Practical Study of Root-Nodule Bacteria. In *IBP Handbook No. 15*; Blackwell Scientific Publishers: Oxford, UK, 1970; p. 164.

15. Leonard, L.T. A Simple Assembly for Use in the Testing of Cultures of Rhizobia. *J. Bacteriol.* **1943**, *45*, 523–527. [CrossRef] [PubMed]
16. Murashige, T.; Skoog, F. A revised medium for rapid growth and bioassays with tobacco tissue cultures. *J. Plant. Physiol.* **1962**, *15*, 473–497. [CrossRef]
17. Shiro, S.; Yamamoto, A.; Umehara, Y.; Hayashi, M.; Yoshida, N.; Nishiwaki, A.; Yamakawa, T.; Saeki, Y. Effect ofRjGenotype and Cultivation Temperature on the Community Structure of Soybean-Nodulating Bradyrhizobia. *Appl. Environ. Microbiol.* **2011**, *78*, 1243–1250. [CrossRef] [PubMed]
18. Saeki, Y.; Kaneko, A.; Hara, T.; Suzuki, K.; Yamakawa, T.; Nguyen, M.T.; Nagatomo, Y.; Akao, S. Phylogenetic Analysis of Soybean-Nodulating Rhizobia Isolated from Alkaline Soils in Vietnam. *Soil Sci. Plant. Nutr.* **2005**, *51*, 1043–1052. [CrossRef]
19. Freed, R. MSTAT: A software program for plant breeder. In *Principles of Plant Genetics and Breeding*, 2nd ed.; Acquaah, G., Ed.; Blackwell Publishing: Malden, MA, USA, 2007; Volume 1, pp. 426–431.
20. Saeki, Y.; Ozumi, S.; Yamamoto, A.; Umehara, Y.; Hayashi, M.; Sigua, G.C. Changes in Population Occupancy of Bradyrhizobia under Different Temperature Regimes. *Microbes Environ.* **2010**, *25*, 309–312. [CrossRef] [PubMed]
21. Stoyanova, J. Growth, nodulation and nitrogen fixation in soybean as affected by air humidity and root temperature. *Biol. Plant.* **1996**, *38*, 537–544. [CrossRef]
22. Zhang, F.; Smith, D.L. Effects of low root zone temperatures on the early stages of symbiosis establishment between soybean [Glycine max (L.) Merr.] and Bradyrhizobium japonicum. *J. Exp. Bot.* **1994**, *45*, 1467–1473. [CrossRef]
23. Montañez, A.; Danso, S.; Hardarson, G. The effect of temperature on nodulation and nitrogen fixation by five Bradyrhizobium japonicum strains. *Appl. Soil Ecol.* **1995**, *2*, 165–174. [CrossRef]

Article

Effects of Fertigation Management on the Quality of Organic Legumes Grown in Protected Cultivation

María del Carmen García-García [1,*], Rafael Font [2], Pedro Gómez [1], Juan Luis Valenzuela [3], Juan A. Fernández [4] and Mercedes Del Río-Celestino [2]

[1] Department of Agro-Food Engineering and Technology, IFAPA Centro La Mojonera, CAGPDS, 04745 Almería, Spain; pedro.gomez.j@juntadeandalucia.es
[2] Agri-Food Laboratory, CAGPDS, Avda, Menéndez Pidal, s/n, 14080 Córdoba, Spain; rafaelm.font@juntadeandalucia.es (R.F.); mercedes.rio.celestino@juntadeandalucia.es (M.D.R.-C.)
[3] Department of Biology and Geology, Higher Engineering School, University of Almería, 04120 Almería, Spain; jvalenzu@ual.es
[4] Department of Agronomical Engineering, Technical University of Cartagena, 30203 Murcia, Spain; juan.fernandez@upct.es
* Correspondence: mariac.garcia.g@juntadeandalucia.es

Abstract: Appropriate fertigation management plays an important role in increasing crop quality and economizing water. The objective of the study was to determine the effects of two fertigation treatments, normal (T100) and 50% sustained deficit (T50), on the physico-chemical quality of legumes. The determinations were performed on the edible parts of peas, French beans and *mangetout*. The trials were conducted in a protected cultivation certified organic farm. The response of legumes to the treatments varied between the cultivars tested. The fertigation treatments had a significant effect on the morphometric traits (width for *mangetout* and French bean; fresh weight for French bean; seed height for Pea cv. Lincoln). The total soluble solids and citric acid content have been shown to be increased by low soil water availability (T50) for *mangetout*. Fertigation treatments did not significantly affect the antioxidant compounds (total phenolic and ascorbic acid), minerals and protein fraction contents of legumes studied. Regarding legume health benefits, the most prominent cultivars were BC-033620 pea and French bean because of their high total phenolic (65 mg gallic acid equivalent 100 g $^{-1}$ fresh weight) and ascorbic acid content (55 mg ascorbic acid 100 g^{-1} fresh weight), respectively. The results expand our knowledge concerning the nutraceutical quality and appropriate cultivation methods of legumes in order to make the system more sustainable and to encourage their consumption.

Keywords: French bean; *mangetout*; peas; antioxidant; ascorbic acid; total phenolic content; mineral composition

Citation: García-García, M.d.C.; Font, R.; Gómez, P.; Valenzuela, J.L.; Fernández, J.A.; Del Río-Celestino, M. Effects of Fertigation Management on the Quality of Organic Legumes Grown in Protected Cultivation. *Horticulturae* **2021**, *7*, 28. https://doi.org/10.3390/horticulturae7020028

Academic Editor: Stefano Marino
Received: 24 December 2020
Accepted: 4 February 2021
Published: 7 February 2021

Publisher's Note: MDPI stays neutral with regard to jurisdictional claims in published maps and institutional affiliations.

Copyright: © 2021 by the authors. Licensee MDPI, Basel, Switzerland. This article is an open access article distributed under the terms and conditions of the Creative Commons Attribution (CC BY) license (https://creativecommons.org/licenses/by/4.0/).

1. Introduction

The production of greenhouse crops in Almería (Southeast Spain) accounts for 3.3 million tons, with a surface of 30,000 ha and a value of €1782.4 million. In terms of organic production, Almería ranks first among all Spanish provinces, with more than 3000 ha of greenhouses [1]. Since legumes for consumption of pods and fresh grain are not major crops in the greenhouses of Spain, they can be considered an important alternative, providing diversification in the organic cultivation of legumes under greenhouse conditions. The most economically important legumes consumed as vegetables are green pods of cowpea, snow pea (*mangetout*), common bean, faba bean, and green pea seeds.

Legumes constitute one of the most important botanical families (*Papilionaceae* or *Fabaceae*) from a socioeconomic point of view, with significant implications for agriculture, the environment and food. They are a valuable source of proteins for both animal and human food [2], with known health benefits [3,4], being one of the basic pillars of the

Mediterranean diet. Numerous studies show the beneficial role of the consumption of legumes for health, including blood pressure and cholesterol levels [5]; high soluble fibre and oligosaccharide content has been associated with an improvement in gastrointestinal health [6,7], and these can also play an important role in diabetes prevention and treatment [8].

Among the different legumes, the most widely eaten in the Mediterranean diet are French beans (*Phaseolus vulgaris*) and peas (*Pisum* ssp.) (Figure 1). Peas can be cultivated with the aim of obtaining dry and fresh peas (*P. sativum* L.) but also as *P. sativum* L. ssp. *arvense*, which are absent parchment pods, sweet, crisp and colloquially known as *tirabeque* or *mangetout*.

Figure 1. Plant material tested (from top to bottom): French bean "Helda" (*Phaseolus vulgaris* L.), *mangetout* "Tirabí" (*Pisum sativum* L. ssp. *arvense*), pea cv. Lincoln (*Pisum sativum* L.) and pea cv. BGE*-033620 (*Pisum sativum* L.). * BGE: Spanish germplasm bank.

In recent decades, many studies have been conducted to optimize the nutrient and water supply for maximizing crop yield and quality as well as minimizing leaching below the rooting volume according to crop requirements e.g., [9,10]. However, the scarcity of water in some intensive horticultural areas like Almería has resulted in the implementation of new sustainable technological adaptations based on improvements in water use efficiency through automated fertigation, localized irrigation systems and the use of tensiometers [11]. Nowadays, most greenhouses in the region have automated fertigation systems, allowing farmers to have greater control of irrigation parameters [12].

Nevertheless, published research results regarding the impact of fertigation on quality characteristics of green pods are scarce. Thus, previous studies have found significantly different effects on quality parameters such as length, width, number of seeds per pod, fresh fruit weight and pod colour (L* and a* parameters) in French bean (*Phaseolus vulgaris* L.) pods under both fertigation levels and frequencies [13]. Indeed, as reported by the above-mentioned authors, large fertigation intervals reduced the colour brightness (L* parameter) and increased the pod greenness (a* parameter) in French bean pods. Moreover, irrigation management could influence green pod quality. Thus, the application volume of irrigation water based on replacing 80% of evapotranspiration improved the pod parameters and nutritional composition of green beans [14]. Furthermore, inadequate supply of irrigation water to French beans may also increase the fibre content in pods, as indicated by the results of Singer et al. [15] obtained after a reduction in the water supply from 100% to 75% or 50% of the field capacity.

Since the fundamental principles of organic farming are the preservation of natural resources and the increase in biodiversity, the objective of this work was to study the effect of reducing the dose of fertigation and, consequently, the dose of irrigation on the physico-chemical quality of different legume cultivars (French beans, *mangetout* and peas) under protected organic conditions.

2. Materials and Methods

2.1. Experimental Framework, Plant Species and Applied Treatments

Four legume cultivars were used: French bean "Helda" (*Phaseolus vulgaris* L.), *mangetout* "Tirabí" (*Pisum sativum* L. ssp. *arvense*), pea cv. BGE-033620 (*Pisum sativum* L.) from: Spanish germplasm bank of the Centro de Recursos Fitogenéticos-INIA (Instituto Nacional de Investigación y Tecnología Agraria y Alimentaria and a commercial pea cv. Lincoln (*Pisum sativum* L.).

The experimental trial was carried out in the Ifapa Center La Mojonera, in the province of Almería, South-East Spain. The crops were grown in a greenhouse of 1600 m^2. The type of greenhouse was a symmetrical multi-tunnel, without active climate control, although equipped with temperature and humidity meters, fertigation system and officially certified as organic by an accredited company within the previous 15 years. The minimum, average and maximum relative humidity were 35.8, 74.3, and 98.1%, respectively. The minimum, average and maximum air temperature were 12.2, 17.2, and 26.2 °C, respectively. Pests and diseases were monitored weekly and biological control was applied as the main control method.

Sheep manure was applied to the soil at a dose of 0.7 kg m^{-2} and with a percentage composition of dry matter corresponding to 45.6%; 17.7 g kg^{-1} total nitrogen; 520.0 mg kg^{-1} nitrate; 2.2 g kg^{-1} phosphorus; 16.5 g kg^{-1} potassium; 889.0 mg kg^{-1} ammonium; 100.9 g kg^{-1} calcium. All fertilizers are listed in Annex I of the EU Regulation (Commission Regulation (EC) No. 889/2008). The legumes were transplanted in October 2016, with a density of 2 plants m^{-2}.

Fertigation was applied by a controller, via drip fertigation and a nutrient solution prepared with groundwater pH 7.3, regulated to 6.5 by acetic acid. The fertigation controller was composed of a programmer with venturis injectors and four fertilizer tanks with the following solutions: tank A: chelated calcium (MgO 0.5% + CaO 15%) and microelements; tank B: humic and fulvic acids (total humic extract 26% *w/w*; humic acids 10% *w/w* + fulvic acids 16% *w/w*); tank C: potassium sulphate (K$_2$O 52% + SO$_3$ 45%); tank D (injectors): amino acids (free amino acids 24% *w/w* + total nitrogen 3.3% *w/w*, organic nitrogen 3% *w/w* and ammonia nitrogen 0.3% *w/w*). A final electric conductivity (EC) of 2.4 dS m^{-1} was reached.

Two treatments, T100 and T50, were arranged in a randomized complete design. T100 consisted of water and fertilizer provided according to fertigation management. This programming was carried out through the use of 3 tensiometers installed at a depth of 15 centimetres, randomly allocated in the 100% treatment plots. The command used was to fertigate when the average of the tensiometers located only in the plot of T100 was 22 cb, matric potential usually applied in horticultural greenhouses in Almeria [16]. T50 consisted of the water supply corresponding to half the fertigation time compared to T100. The concentration of the nutrient solution provided in each treatment was the same, therefore, T100 consists of double the amount of water and fertilizer as T50. Fertigation times of the two treatments were varied throughout the cultivation. During the period of maximum crop growth, it was fertigated for 30 min for T100 and half of that time (15 min) for T50. At the end of the trial, the total water volume applied in T100 was 60 L m^{-2} and 30 L m^{-2} for T50.

The soil texture was sandy clay loam, determined by Bouyoucos-hydrometer analysis [17]; pH and electrical conductivity (EC) were determined in the saturated extract by pHmeter (model MicropH 2002 Crison) and conductivity meter (model GLP31 Crison); Chemical elements were determined in the water extract from saturated soil paste. Organic matter was determined using the Walkley–Black method [18]. A soil analysis was carried out before planting and after harvesting of T100 and T50 treatments.

A total of 35 legume plants were grown per cultivar (BGE-033620, Lincoln, Helda, Tirabí), per replicate and for each treatment (T100 and T50). Legume-pod samples were taken from 10 randomly distributed plants in each replicate during the maximum production period. The samples for chemical analysis consisted of pods of French-bean and *mangetout* and pea grains, consistent with the edible format for each legume. Then, vegetal material was packaged in polypropylene plastic containers and sent to the laboratory for analysis.

2.2. Physical Traits

The morphometry was determined by measuring the length, width and height of the pods (French bean and *mangetout*) and maximum and minimum diameter in peas with digital calibre Laser 4263.

The determination of fresh weight was measured with a precision digital balance Mettler Toledo XPE1203S. For the moisture determination, dry weight of the samples was calculated drying in a Memmert UF110 stove at 45 °C for 72 h. All determinations were carried out on pods. To determine fruit firmness, a texturometer was used on pods (Texture Analyzer TA.XT Plus, Stable Micro Systems Texture Analyzer, Surrey, UK) equipped with a fine-cut probe at a speed of 1 mm s^{-1}, for 5 s. The colour was determined by a CM-700d portable colorimeter (Konica Minolta Sensing Americas, Inc. Ramsey, NJ, USA). The determinations of the Hue and Chroma colour parameters were made in two different external points of the pod's equatorial plane with a colorimeter (model CR-200, Minolta, Ahrensburg, Germany).

2.3. Chemical Traits

All parameters were measured for the edible part of each legume: seeds for peas and pods for *mangetout* and French bean.

The juice was extracted from the fruits for the determination of the soluble solid content by means of a digital refractometer (Smart-1, Atago, Japan). Titratable acidity was measured by titrating 10 mL of juice with NaOH 0.1 N up to pH 8.2 using an automatic titrator (Metrohm 862 Compact. Titrosampler, Herisau, Suiza). Legume fruit acidity was reported as the percentage of citric acid. The pH value of the sample was determined using a digital pH meter (WTW pH 330; WTW; Weilheim, Germany) equipped with an electrode (Sen Tix 41; WTW, Weilheim, Germany).

Legumes (10 g) were mixed in a blender a stirred with 10 mL methanol. The mixture was homogenised (Polytron PT3100; Kinematica AG, Littau, Switzerland) and centrifuged at 4 °C (Beckman J2-21M/E; Beckman Instruments Inc., Fullerton, CA, USA) for 10 min. The supernatant was decanted into a 25 mL measuring flask. The pellet was resuspended in 10 mL 70% methanol in water (v/v), followed by centrifugation. The combined supernatants were diluted to 25 mL with 70% methanol. The extracts were frozen in small tubes at −80 °C until further analysis. The solution was diluted to volume (25 mL) with distilled deionised water. The solution was incubated at room temperature in the dark for 90 min, and the absorbance was read at 750 nm against a blank solution. Finally, results were reported in gallic acid equivalents (mg g^{-1} DW).

Total polyphenol content was determined according to the Folin–Ciocalteu procedure [19]. To the diluted methanol extract (200 μL), in a cuvette, 1 mL of Folin–Ciocalteu solution (diluted 1:10 in water) was added. After 2 min, 800 μL Na$_2$CO$_3$ (7.5%) was added, mixed for 5 s on a whirl mixer and incubated in the dark at room temperature for 60 min. The absorbance was measured at 765 nm with a ThermoSpectronic UV–visible Spectrometer (Thermo Fisher Scientific, Waltham, MA, USA). Gallic acid was used as standard and total phenolics were expressed as mg gallic acid equivalent (GAE) 100 g^{-1} fresh weight.

The reference values for Ascorbic Acid (AA) were obtained using an automatic titration (Metrohm, 862 Compact Titrosampler, Metrohm, Riverview, FL, USA) by the iodine titration method with minor modification [20]. Thus, 5 g of sample juice was mixed with distilled deioniser water until final weight of 50 g and treated with 2 mL glyoxal solution (40%), stirred briefly and allowed to stand for 5 min. After the addition of 5 mL sulphuric acid (25%), it was titrated with iodine (0.01 mol L^{-1}) up to the endpoint (EP1). The linearity of the method was determined using AA as an external standard. Finally, the ascorbic acid content was expressed as mg g^{-1} fresh weight (FW).

For mineral composition determination of the legume cultivars, the dry mineralization method was used [21]. Dried samples in a furnace at 100 °C to constant weight were homogenized and then weighed into porcelain crucibles. Later, they were incinerated in a muffle furnace at 460 °C for 15 h. The ash was bleached after cooling by adding 2 mL of 2 mol L^{-1} nitric acid, then drying it on thermostatic hotplates and finally maintaining it in a muffle furnace at 460 °C for 1 h. Ash recovery was performed with 5 mL of 2 mol L^{-1} Suprapur nitric acid, making up to 15 mL with 0.1 mol L^{-1} Suprapur nitric acid. The determinations were carried out by flame atomic absorption spectrophotometry, except for Na and K, which were analysed by flame atomic emission. Elemental analyses were performed with a PerkinElmer (Waltham, MA, USA) model 2100 atomic absorption spectrophotometer equipped with a PerkinElmer AS-50 autosampler, standard air–acetylene flame and single-element hollow cathode lamps and background correction with deuterium lamp for Mn. The nitrogen (N) content was determined according to the Kjeldahl method [22] and the protein content was calculated (N × 6.25).

The quantification of the protein fraction content was determined in fresh pea grain (Lincoln and BGE-033620 cultivars). To obtain soluble protein 10 mg of pea flour was weighed and 1 M NaOH was added. The protein fractions were obtained based on the methods postulated by Hu and Esen [23] and Knabe et al. [24]. The protein fractionation of the legumes was performed with 4 different solvents: H_2O, NaCl 0.5 M, 2-propanol (IPA) 70% and glacial acetic acid 50% for determining the albumin, globulin, prolamin and glutelin content, respectively. The supernatant absorbances were measured at 595 nm with a ThermoSpectronic UV–visible Spectrometer (Thermo Fisher Scientific, Waltham, MA, USA). A bovine serum albumin (BSA) dilution curve was used as standard [25]. The results were expressed in mg protein fraction per 100 g dry weight (DW).

2.4. Statistical Analyses

Physico-chemical data were subject to one-way analysis of variance (ANOVA), and Tukey's multiple range test was used in cases where significance at $p < 0.05$ variance was found among treatments (T50 and T100) per each cultivar. Statistical analyses were performed using SPSS 13.0 (SPSS Inc., Chicago, IL, USA).

3. Results and Discussion

3.1. Soil Analysis

The saturated extract soil indicates what plants can take from soil when irrigation water and solubilized and minerals are applied. The EC of the samples collected from the T50 (2.79 dS·m^{-1}) after harvesting were higher compared to the T100 treatment (4.45 dS·m^{-1}) (Table 1). Soil EC is affected by cropping, irrigation, land use, and application of fertilizer. The higher EC in T100 could be due to a higher salt accumulation in the soil since irrigation by means of tensiometers at the above-mentioned matric potential considerably reduces the quantities of leaching. The activation of fertigation with a threshold value of soil matric tension −22 kPa allows no drainage, as shown by the research carried out in our cultivation area [26,27]. EC of saturated extract of soil obtained in T50 (2.79 dS·m^{-1}), being similar to the electric conductivity (EC) of the nutritive solution used (2.40 dS·m^{-1}), which indicates an optimal use of the nutrients provided in T50.

Table 1. Saturated extract soil analysis.

Soil	EC (dS m^{-1})	pH	Organic Matter (%)	Total N (%)	NO$_3{}^-$ (meq·L^{-1})	P (meq·L^{-1})	K (meq·L^{-1})
Before planting	1.31	9.2	1.6	0.051	1.3	11	1.5
After harvesting (T100)	4.45	9.3	1.3	0.051	9.3	12.4	8.7
After harvesting (T50)	2.79	9.1	1.2	0.05	3.7	12	4.6

The increase in NO$_3{}^-$ and K in the analysis after harvest is due to the continuous supply of N (via aminoacids and humic and fulvic acids) and K (potassium sulphate) through the nutrient solution. Higher supply in the case of T100 treatment implied higher NO$_3{}^-$ and K content in the soil. Regarding nitrate, the used N fertilizers are normally retained in the soil and slowly mineralized. Thus, the nitrate is slowly released, and, asfar as nitrate leaching is reduced, its content in the soil is increased. Furthermore, the soil sampling was carried out after a few days of irrigation events, which would explain the high amount of nitrate released by mineralization at the end of the cycle. In addition, plant root uptake of NO$_3{}^-$ could be negatively influenced by the external supply of amino acids [28,29]; in particular, the supplying of amino acids to roots could reduce NO$_3{}^-$ uptake [30], resulting in hardly any differences in produced biomass and tissues analysed between treatments. In relation to K, its concentration tended to increase in soil saturation extract with increasing salt concentration because of the release of absorbed K [31]. Additionally, at the pH of 9 in soil (Table 1), potassium phosphates would not be produced, since phosphorus would precipitate as calcium and ferric phosphates, among others. In any case, if potassium phosphates were formed, this compound would be soluble, and, in the saturation extract, it would be redissolved (which would not occur in calcium and ferric phosphates, which are insoluble). Furthermore, the availability of this element in the soil is affected by moisture levels; more moisture makes more K available, this fact being more frequent in T100 than in T50, and in Almería greenhouse cultivation conditions [32,33].

3.2. Physical Parameters

The most important physical traits used to assess the external quality of green pods consumed as legumes are, among others: length, width and height of the pod, the individual pod weight, the firmness of the pod, and the colour of the pod when harvested. With respect to green seeds, the criteria used to assess their external quality are mainly the texture, the shape (round, oval, etc.) and the individual seed weight. The texture of pea green seeds is considered one of the most important quality attributes for consumers [34].

Table 2 shows the physical parameters for the different cultivars tested grown under two fertigation regimes. In relation to length, width and height of the pod, significant differences were found between treatments for fresh pod width in *mangetout* and French beans and also for grain height in peas cv. Lincoln (17 mm and 15 mm in T50 and T100, respectively).

Table 2. Physical parameters of pea, French beans and mangetout grown under normal (T100) and sustained deficit (T50) fertigation tested under protected cultivation conditions. Data are displayed as mean ± standard deviation.

Cultivar	Treatment	Length (mm) [a]	Width (mm) [a]	Height (mm) [a]	FW (g) [b]	Moisture (%) [b]	Firmness (N) [b]	C* [b]	Hue [b]
Mangetout	T50	102.3 ± 0.4	23 ± 1.0	5.8 ± 0.5	5.8 ± 0.3	85.2 ± 0.3	56.0 ± 3.0	33.0 ± 0.9	107.0 ± 0.6
	T100	104.7 ± 0.7	21.0 ± 1.0 *c	5.7 ± 0.5	6.3 ± 0.8	84.2 ± 0.7	63.0 ± 7.3	32.0 ± 1.9	108.0 ± 0.2
French bean	T50	210.5 ± 0.2	18.6 ± 0.4	6.7 ± 0.3	15.0 ± 0.8	91.1 ± 0.7	55.0 ± 3.5	27.0 ± 0.3	111.0 ± 0.1
	T100	229.7 ± 0.5	16.0 ± 1.8 *c	5.9 ± 0.2	20.1 ± 1.8 *c	91.2 ± 1.7	56.0 ± 11.1	29.0 ± 2.8	111.0 ± 1.7
Pea Lincoln	T50		10.9 ± 0.4	17.0 ± 0.8	4.7 ± 0.2	24.4 ± 2.0	75.7 ± 2.1	33.0 ± 1.0	109.0 ± 0.5
	T100		12.0 ± 0.8	15.0 ± 0.6 *c	4.7 ± 0.2	24.2 ± 1.4	75.5 ± 1.5	34.0 ± 0.7 *c	108.0 ± 0.4
BGE-033620	T50		9.9 ± 0.5	12 ± 0.5	2.4 ± 0.3	20.2 ± 0.9	77.1 ± 0.8	39.0 ± 1.3	105.0 ± 0.5
	T100		9.9 ± 0.4	12 ± 0.4	2.4 ± 0.4	22.9 ± 1.1	79.7 ± 1.9 *c	39.0 ± 1.9	105.0 ± 0.2

[a] Length, width and height measured in mangetout and French bean pods, and in pea seeds. [b] Fresh weight, moisture, C* and Hue parameters measured in mangetout, French bean and pea pods. [c] * $p < 0.05$ denote a statistically significant difference between treatments (T50 and T100) for each cultivar (ANOVA followed by Tukey's multiple range test).

When analyzing the impact of fertigation on the fresh weight in the legume plants, significant differences between treatments were found for French bean pods with 15.0 ± 0.8 g and 20.1 ± 1.8 g fruit^{-1} in T50 and T100, respectively.

The comparison of the yield of the diverse species and varieties, in response to fertigation deficit, evidenced that only French bean plants significantly decreased in their productivity under 50% fertigation conditions. In this regard, the commercial production of French beans under control conditions (full fertigation), 1.60 kg m^{-2}, decreased by 32.5%, until reaching 1.08 kg m^{-2}, when exposed to 50% fertigation. The sensitivity of French beans to water shortage, especially during flowering initiation and development, is well demonstrated; this directly affects the final yield [35]. Our results are in agreement with those of Martelo-Nuñez et al. [36] who demonstrated that the fertigation regime was especially relevant for the production of French beans, which were more sensitive than other legume crops to water stress.

In addition, the fertilizer requirements in legumes are low, particularly nitrogen fertilization, which is not generally required or is required in small amounts, although the application of "starter" nitrogen fertilization at a low dose rate seems to enhance the nodulation process and onset of nitrogen fixation in most of the legume crops [37], including peas [38] and French beans [39]. Thus, normally, peas grown in fertile soils are not very dependent on fertilization, particularly on N doses, except during the initial stage of development [40], while in French bean an increase in fertilization, up to a threshold, significantly augmented green pod yield [41]. Therefore, the differences in yield responses between species to different fertilizer amounts applied could be due to higher fertilizer requirements of French beans with respect to pea varieties.

In relation to moisture (%), the analysis of variance indicated no significant differences between treatments for any legume cultivars (Table 2). The percentages of dry matter were lower in grain (75–79.7%) than in pods (85–91.1%).

Regarding the firmness parameters, significant differences were not found between treatments (T100 and T50) except for the BGE cultivar with 21 N and 16 N, respectively (Table 2). Currently, published research results relevant to the impact of fertigation on quality traits like texture legumes are scarce.

Color is one of the main external characteristics that determine the acceptance of the product by the consumer. The mean values for the C* chromatic parameter varied from 27 to 59 (French bean and pea cv. BGE-033620, respectively) and the h* parameter varying from 105 to 111 (BGE-033620 and French bean, respectively), which exhibited green surface. Significant differences were not found between the fertigation treatments tested for any chromatic parameter. Previous work reported by Sezen et al. (2008) [42] showed that large fertigation intervals reduced the color brightness (L* parameter) and increased the pod greenness (a* parameter) in snap bean pods.

The total soluble solids (TSS) content ranged from 5.82 (French bean) to 11.29 °Brix (*mangetout*) (Figure 2). Previous results are in agreement with those obtained in this study for °Brix of French bean (5.0–5.7 °Brix) grown under protected crop conditions [43].

Mangetout pods are rich in TSS content in comparison with other legume pods, thus cowpea accessions from Spain, Greece and Portugal have shown low TSS contents (5.07–7.57) in previous studies [44]. Pea cultivars also displayed a high TSS content (8.54–10.83) in consonance with values obtained by Mera et al. [45], for open air spring pea crops (11 °Brix). Green peas of high quality should be tender enough, but with a high sugar content [34]. Significant differences were found between fertigation treatments for *mangetout*, reaching higher TSS contents at T50 (11.3 °Brix) compared to T100 (9.1 °Brix). No significant differences were found between treatments for French beans or for the two pea cultivars.

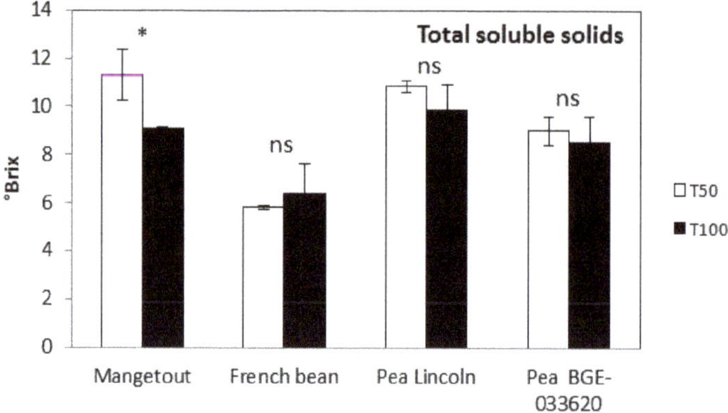

Figure 2. Total soluble solids (expressed as °Brix) in edible parts of *mangetout*, French beans and peas grown under normal (T100) and sustained deficit (T50) fertigation tested under protected cultivation conditions. * and "ns" indicate significant differences at $p < 0.05$, and non-significant differences between treatments (T50 and T100) for each cultivar, respectively (ANOVA followed by Tukey's multiple range test).

Figure 3 shows the pH of the different legume cultivars tested under protected cultivation conditions. The pH ranged from 6.1 to 6.9, which is in agreement with previous studies on legumes [43,46–48]. Significant differences were not found between fertigation treatments.

The titratable acidity of the different legume cultivars tested, expressed as percentage of citric acid (major organic acid in legumes) is shown in Figure 3. The citric acid content varied from 0.11 (French bean) to 0.29% (pea cv. Lincoln). Previous studies have found citric acid values similar for French beans (0.10–0.18% citric acid) [49]. The data on citric acid content of the pods/seeds indicated a significant increment under drought by *mangetout* (0.23) and Lincoln pea (0.32) compared to T100 treatment (0.19 and 0.14, respectively).

Although numerous studies are available on the effects of either salinity or drought/deficit fertigation on plant growth and yield of grain and vegetable legumes, only a few of them also address pod and/or immature seed quality parameters.

Soils from T100 treatment had greater salt content (4.45 dS m^{-1}) compared to T50 treatment, as indicated the highest EC value observed (Table 1). Further studies are required to confirm these results by molecular evidence. The tolerant genotypes could be utilized for further breeding programmes to evolve new legume genotypes for better salt stress tolerance with higher quality.

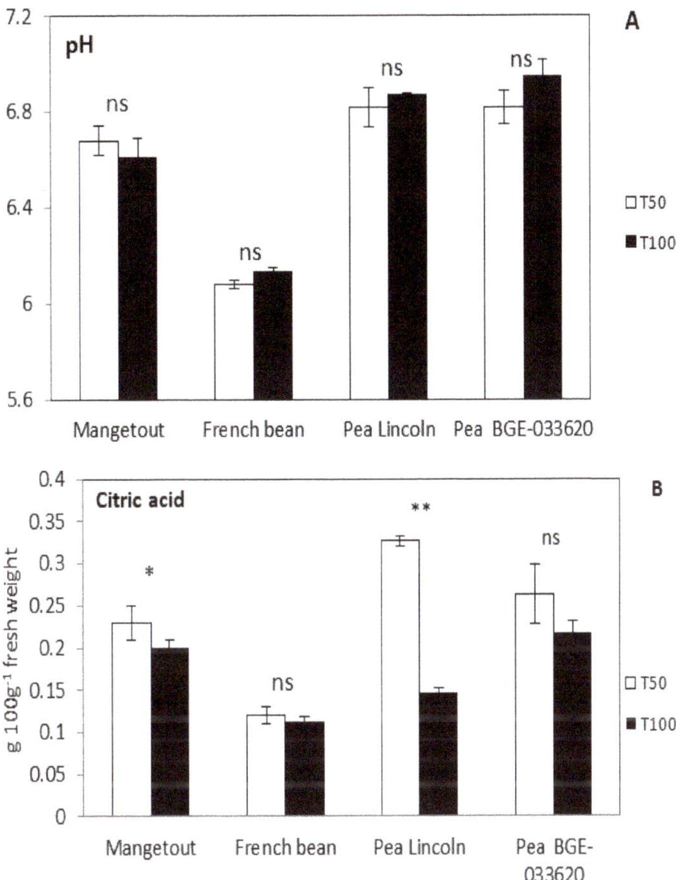

Figure 3. pH (**A**) and titratable acidity (**B**) (expressed as g of citric acid $100 \, g^{-1}$ fresh weight) in edible parts of *mangetout*, French beans and peas grown under normal (T100) and sustained deficit (T50) fertigation tested under protected cultivation conditions. *, ** and "ns" indicate significant differences at $p < 0.05$, $p < 0.01$ and non-significant differences between treatments (T50 and T100) for each cultivar, respectively (ANOVA followed by Tukey's multiple range test).

Previous studies on cucumber, melon, tomato and pepper fruits displayed similar results; thus, the salinity of the nutrient solution did not increase the TSS levels [50]. On the other hand, reduced watering application in faba beans increases the carbohydrate concentrations in seeds [51].

3.3. Nutritional Parameters

Vegetable legumes contain less proteins and more water than those consumed as dry pulses. In addition, they are richer sources of antioxidants, such as phenolics and vitamin C among other compounds [52]. Therefore, their consumption is intended to provide a balanced nutritional source full of healthy promoting compounds rather than to serve as a primary protein source.

Figure 4 shows the mean total phenol content of the different cultivars of legumes tested under organic farming, expressed as mg GAE $100 \, g^{-1}$ of fresh weight. Significant differences were not observed between the normal and restricted fertigation treatments; therefore, a deficit at 50% fertigation did not affect the total phenol content. The lower values were found in French bean and pea cv. Lincoln (20–30 mg GAE 100 g^{-1} FW). The BGE-033620 pea cultivar showed the highest total phenolic content, with 60–65 mg GAE 100 g^{-1} FW. These results are in consonance with those reported for the seeds of cultivars of *Lupinus albus*, *L. luteus*, and *L. angustifolius* with total phenol contents varying from 212 to 317 mg 100 g^{-1} DM (approximately 43–64 mg 100 g^{-1} FW) [53].

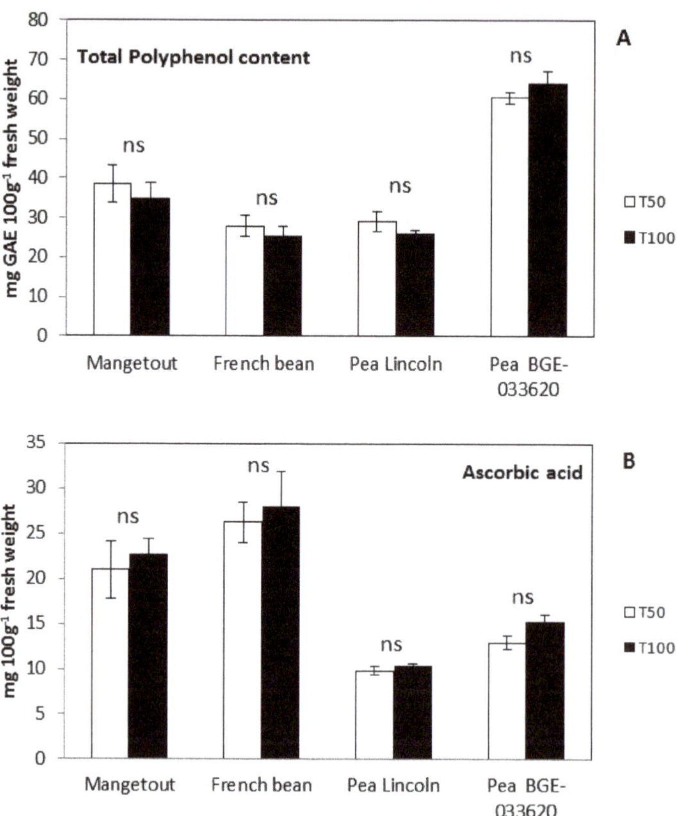

Figure 4. Antioxidant compound content (Total polyphenol (**A**) and ascorbic acid content (**B**)) in edible parts of *mangetout*, French beans and peas grown under normal (T100) and sustained deficit (T50) fertigation tested under protected cultivation conditions. "ns" indicates non-significant differences between treatments for each cultivar (ANOVA followed by Tukey's multiple range test).

According to the European Food Information Council [54], there is no official dietary recommendation for the consumption of phenolic compounds. However, some studies are being carried out to determine consumption recommendations for different adult population groups, such as that of Ovaskainen et al. [55], which proposed Recommended Dietary Allowances (RDA) for the intake of total phenolic content (461–1377 mg day^{-1} for men and 449–1185 mg day^{-1} for women). According to our data, the intake of a 200 g ration of Fresh bean, *mangetout* and pea cv. Lincoln would provide 5–13% of the RDA for adults. Pea cv. BGE-033620 would provide up to 9–26% of the RDA for adults.

Figure 4 shows the mean ascorbic acid content of the different legumes tested under greenhouse cultivation. No significant differences were observed between the normal and restricted fertigation treatments. The ascorbic acid content ranged from 10–15 mg 100 g^{-1} FW (pea cultivars) to 26.32 mg 100 g^{-1} FW for French beans (Figure 4). Previous studies reported lower concentrations for French beans with 16.7 mg 100 g^{-1} FW [56].

Vitamin C is found naturally as L-ascorbic acid, being widely distributed in fresh plant foods, among which, citrus fruits, kiwi, strawberry and melon are the largest source of vitamin C [57]. Other vegetables like tomato also contain high vitamin C content, varying from 2 to 21 mg 100 g^{-1} FW [58]. According our data, Fresh beans and *mangetout* could be considered rich sources of vitamin C. According to the RDA (Recommended Dietary Allowances) for the Spanish population [59], the daily intake of vitamin C for adult men and women is 60 mg day^{-1}. According to our data, a portion size of approximately 200 g FW of French beans and *mangetout* would provide 83 and 70%, respectively, of the recommended daily intake.

Table 3 shows the mean protein and mineral contents (dry weight) for each legume cultivar. The analyses statistically showed non-significant differences for the protein and mineral contents except for Mn in the BGE-033620 cultivar.

Table 3. Protein and mineralcontents (dry weight) in legume cultivars grown under normal (T100) and sustained deficit (T50) fertigation tested under protected cultivation conditions. Data are displayed as mean ± standard deviation.

Cultivar	Treatment	g 100g^{-1}						mg kg^{-1}			
		Protein	N	K	Ca	Mg	Na	Fe	Zn	Cu	Mn
Mangetout	T50	20.0 ± 2.7	3.2 ± 0.4	1.9 ± 0.2	0.4 ± 0.0	0.2 ± 0.1	0.05 ± 0.0	44.6 ± 8.7	44.8 ± 6.5	3.7 ± 2.4	9.1 ± 1.6
	T100	21.1 ± 2.4	3.4 ± 0.4	1.9 ± 0.2	0.4 ± 0.1	0.2 ± 0.0	0.05 ± 0.0	44.5 ± 7.0	38.8 ± 1.9	3.2 ± 2.4	9.8 ± 0.8
French bean	T50	15.5 ± 1.3	2.5 ± 0.2	2.3 ± 0.2	0.5 ± 0.0	0.3 ± 0.0	0.01 ± 0.0	47.3 ± 6.3	26.8 ± 2.0	4.7 ± 1.3	17.8 ± 0.6
	T100	18.6 ± 1.6	3.0 ± 0.3	2.2 ± 0.1	0.4 ± 0.1	0.3 ± 0.0	0.01 ± 0.0	44.0 ± 2,8	26.8 ± 1.2	5.0 ± 1.0	18.3 ± 2.5
Pea Lincoln	T50	25.6 ± 2.4	4.1 ± 0.8	1.4 ± 0.2	0.1 ± 0.0	0.1 ± 0.0	0.01 ± 0.0	52.5 ± 5.3	45.0 ± 3.0	5.0 ± 1.1	8.5 ± 0.9
	T100	27.8 ± 2.5	4.5 ± 0.4	1.6 ± 0.1	0.1 ± 0.0	0.1 ± 0.0	0.01 ± 0.0	54.0 ± 2.5	49.0 ± 2.4	5.5 ± 0.8	7.5 ± 0.9
BGE-033620	T50	24.7 ± 4.4	4.0 ± 0.7	1.4 ± 0.1	0.2 ± 0.1	0.2 ± 0.1	0.04 ± 0.0	54.5 ± 15.7	39.0 ± 10.8	2.8 ± 2.0	14.0 ± 0.5 [*a]
	T100	23.8 ± 3.7	3.8 ± 0.6	1.4 ± 0.1	0.2 ± 0.2	0.2 ± 0.0	0.03 ± 0.0	47.1 ± 6.8	39.1 ± 8.8	2.0 ± 0.3	12.5 ± 0.4

[a] $* p < 0.05$ denote a statistically significant difference between treatments within the same cultivar (ANOVA followed by Tukey's multiple range test).

The minimum and maximum mean mineral content (dry weight) in each cultivar were: 2.5 (French bean) and 4.5 g 100 g^{-1} (pea cv. Lincoln) of N; 1.4 (peas) and 2.3 g 100 g^{-1} of (French bean) K; 0.1 (pea cv. Lincoln) and 0.5 g 100 g^{-1} (French bean) of Ca; 0.1 (pea cv. Lincoln) and 0.3 g 100 g^{-1} (French bean) of Mg; 0.01 (French bean and pea cv. Lincoln) and 0.05 g 100 g^{-1} (*mangetout*) of Na; 44 (French bean) and 54 (pea cv. Lincoln) mg kg^{-1} of Fe; 26.8 (French bean) and 49 (pea cv. Lincoln) mg kg^{-1} of Zn; 2 (pea cv. BGE-033620) and 5.5 (pea cv. Lincoln) mg kg^{-1} of Cu; 7.5 (pea cv. Lincoln) and 17.8 mg kg^{-1} (French bean) of Mn. Minerals in diets are required for metabolic reactions, transmission of nerve impulses, rigid bone formation and regulation of water and salt balance [60]. The daily requirements of an adult person are as follows (mg d^{-1}): 9–18 Fe, 1.1 Cu, 3100 K, 7–9.5 Zn, 1300–1500 Na, 300–350 Mg, 1.8–2.3 Mn, 900–1000 Ca and 41–56 g protein [61]. According to our data, supposing that a person consumes a course of legumes of approximately 200 g d^{-1} (and taking into account a mean moisture content of 80%), the calculated content for all the minerals is below the recommended values (2.16 Fe, 0.22 Cu, 920 K, 1.96 Zn, 16 Na, 120 Mg, 0.72 Mn, 200 Ca, expressed in mg and 11.2 g protein). The highest contents were observed for K, Mg and protein. Therefore, consumption of 200 g of legumes can provide 24% Fe, 20% Cu, 1.22% Na, 22.22% Ca, 40.66% Mn, 28% Zn, 29.66% K, 40% Mg and 27.30% protein of the recommended intake. The low Na content (<2%) and the high K concentration recommend the use of these legumes in an antihypertensive diet. Thus, K from vegetables and fruits can reduce blood pressure [62].

Figure 5 shows the mean protein fraction content in edible parts of peas, French beans and *mangetout* grown under two different fertigation treatments.

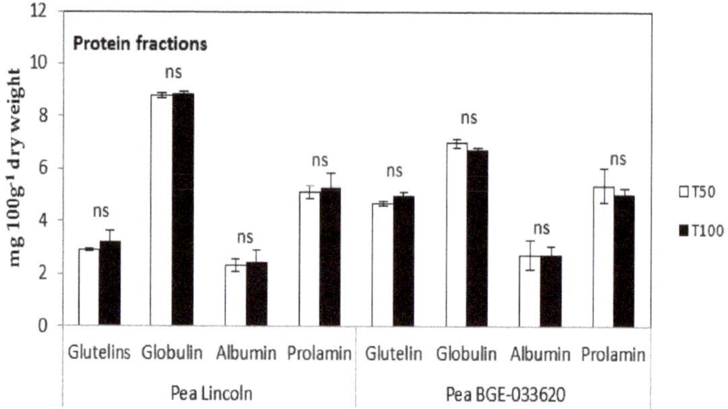

Figure 5. Protein fractions in edible parts of Lincoln and BGE-033620 peas grown under normal (T100) and sustained deficit (T50) fertigation tested under protected cultivation conditions. "ns" indicates non-significant differences between treatments (T50 and T100) for each cultivar (ANOVA followed by Tukey's multiple range test).

The protein fractions analysed were: glutelins which are only produced in plant material, and are mostly found in cereals and legume grains; globulins are present in numerous seeds and in legumes; albumins are proteins that are found in blood plasma and are necessary for the correct distribution of body fluids; and prolamins, contained in legumes and that neutralize the anticoagulant effect of heparin [63].

Significant differences were not found between T50 and T100 for protein fractions (glutelins, globulins, albumins and prolamines). The highest protein fraction content was globulin (Lincoln: 8.9 and 8.8 mg 100 g^{-1} dry weight for T50 and T100, respectively, BGE: 6.8 and 7.0 mg·100 g^{-1} dry weight for T50 and T100, respectively). Previous studies [64,65], indicated that most of the protein fractions of legume seeds contained globulins and albumins and, in some cases, prolamins and glutelins. Therefore, it should be noted that legume seeds do not show a specific soluble fraction profile, in comparison to cereals [23].

Paredes et al. [66] agree with our findings; in fact, they indicated that the globulin fraction in legumes was the highest (60–90%), followed by prolamins and glutelins (Figure 5). The peas studied can be considered as a rich source of proteins, an alternative to the protein from cereals and animals.

In summary, the results, under the controlled system of the presented study using two fertigation treatments, supported the hypothesis that physical and chemical traits varied between varieties and species in different ways. Our results are in accordance with the findings of Saleh et al., 2018 [14], El-Noemani et al., 2010 [67], and Shalaby et al., 2016 [68], having showed a wide variation among legume cultivars in terms of their performance and response to water stress.

4. Conclusions

This work describes, firstly, the different responses to abiotic stress (fertigation deficit) in terms of the physico-chemical quality of peas, *mangetout* and French beans; this information contributes to rational decisions regarding the agronomical management of such crops.

The response of legumes to the treatments varied between the cultivars tested. The fertigation treatments had significant effects on the morphometric traits (width for *mangetout* and French bean; fresh weight for French bean; seed height for Pea cv. Lincoln). Furthermore, only French bean plants significantly lowered productivity under 50% fertigation conditions

Interestingly, from the present study, *mangetout* came out as the highest source of total soluble solids, reaching higher content at 50% fertigation treatment. It was found that fertigation treatments did not significantly affect the antioxidant compounds (total polyphenols and ascorbic acid), minerals and protein fraction contents of the legumes studied. According to our data, French beans and *mangetout* are a rich source of vitamin C.

This study also reflects the importance of legumes in the contribution of mineral content, especially in the contribution of K, Mg and protein to the human diet.

On the other hand, the landraces seemed to be an interesting genetic material. Thus, the BGE-033620 landrace showed the highest total polyphenol content.

This study has shown that these legume species have relevant interest and benefits, at both the agronomic and nutritional levels, and open good perspectives for the improvement of cropping systems and the creation of innovative food products. The available biodiversity and the identified large variation in quality between cultivars and species of legumes is an underutilized resource, still requiring further studies to expand our knowledge about the quality and uses of landraces for consumption, either as fresh vegetables, or after canning or freezing.

Author Contributions: M.d.C.G.-G. wrote this manuscript and performed the crop assay; M.d.C.G.-G., R.F. and P.G., performed nutritional analysis; M.d.C.G.-G. and J.L.V. performed physical and chemical analysis; M.d.C.G.-G. and J.A.F. reviewed the manuscript; M.D.R.-C. designed this study and revised the manuscript. All authors have read and agreed to the published version of the manuscript.

Funding: The authors wish to express their thanks to the Projects (PP.AVA.AVA201601.7 and PP.TRA.TRA201600.9) and FEDER for the funding of this research.

Institutional Review Board Statement: Not applicable.

Informed Consent Statement: Not applicable.

Conflicts of Interest: The authors declare no conflict of interest.

References

1. CAPDER. *Informe de Estadísticas de La Producción Ecológica a 31/12/16*; Junta de Andalucía: Sevilla, Spain, 2017.
2. Rubio, L.A.; Molina, E. Las leguminosas en alimentación animal. *Arbor* **2016**, *192*, 315 [CrossRef]
3. Arnoldi, A.; Zanoni, C.; Lammi, C.; Boschin, G. The role of grain legumes in the prevention of hypercholesterolemia and hypertension. *Crit. Rev. Plant Sci.* **2015**, *34*, 144–168. [CrossRef]
4. Delgado-Andrade, C.; Olías, R.; Jiménez-López, J.; Clemente, A. Aspectos de las legumbres nutricionales y beneficiosos para la salud humana. *Arbor* **2016**, *192*, 313.
5. Bazzano, L.A.; Thompson, A.M.; Tees, M.T.; Nguyen, C.H.; Winham, D.M. Non-soy legume consumption lowers cholesterol level: A meta-analysis of randomized controlled trials. *Nutr. Metab. Cardiovasc. Dis.* **2011**, *21*, 94–103. [CrossRef]
6. Fernández, J.; Redondo-Blanco, S.; Villar, C.; Clemente, A.; Lombó, F. Healthy effects of prebiotics and their metabolites against intestinal diseases and colorectal cancer. *AIMS Microbiol.* **2015**, *1*, 48–71. [CrossRef]
7. Simpson, H.L.; Campbell, B.J. Dietary fibre-microbiota interactions. *Aliment. Pharmacol. Ther.* **2015**, *42*, 158–179. [CrossRef]
8. Jenkins, D.J.A.; Kendall, C.W.C.; Augustin, L.S.A.; Mitchell, S.; Sahye-Pudaruth, S.; Blanco Mejia, S.; Chiavaroli, L.; Mirrahimi, A.; Ireland, C.; Bashyam, B.; et al. Effect of legumes as part of a low glycemic index diet on glycemic control and cardiovascular risk factors in type 2 diabetes mellitus: A randomized controlled trial. *Arch. Intern. Med.* **2012**, *172*, 1653–1660. [CrossRef]
9. Kumar, M.; Rajput, T.B.S.; Kumar, R.; Patel, N. Water and nitrate dynamics in baby corn (*Zea mays* L.) under different fertigation frequencies and operating pressures in semi-arid region of India. *Agric. Water Manag.* **2016**, *163*, 263–274. [CrossRef]
10. Granados, M.R.; Thompson, R.B.; Fernandez, M.D.; Martinez-Gaitan, C.; Gallardo, M. Prescriptive corrective nitrogen and irrigation management of fertigated and drip-irrigated vegetable crops using modelling and monitoring approaches. *Agric. Water Manag.* **2013**, *119*, 121–134. [CrossRef]
11. Garcia-Caparros, P.; Contreras, J.I.; Baeza, R.; Segura, M.L.; Lao, M.T. Integral Management of Irrigation Water in Intensive Horticultural Systems of Almería. *Sustainability* **2017**, *9*, 2271. [CrossRef]
12. García-García, M.C.; Céspedes, A.J.; Lorenzo, P.; Pérez-Parra, J.J.; Escudero, M.C.; Sánchez-Guerrero, M.C.; Medrano, E.; Baeza, E.; López, J.C.; Magán, J.J.; et al. *El Sistema de Producción Hortícola de La Provincia de Almería*; IFAPA (Instituto de Formación Agraria y Pesquera de Andalucía): Huelva, Spain, 2016; p. 179.
13. Sezen, S.M.; Yazar, A.; Canbolat, M.; Eker, S.; Celikel, G. Effect of drip irrigation management on yield and quality of field grown green beans. *Agric. Water Manag.* **2005**, *71*, 243–255. [CrossRef]

14. Saleh, S.; Liu, G.; Liu, M.; Ji, Y.; He, H.; Gruda, N. Effect of irrigation on growth, yield, and chemical composition of two green bean cultivars. *Horticulturae* **2018**, *4*, 3. [CrossRef]
15. Singer, S.M.; Helmy, Y.I.; Karas, A.N.; Abou-Hadid, A.F. Influences of different water-stress treatments on growth, development and production of snap bean (*Phaseolus vulgaris* L.). *Acta Hort.* **2003**, *614*, 605–611. [CrossRef]
16. Bonachela, S.; González, A.M.; Fernández, M.D. Irrigation scheduling of plastic greenhouse vegetable crops based on historical weather data. *Irrig. Sci.* **2006**, *25*, 53–62. [CrossRef]
17. Bouyoucos, G.J. Hydrometer method improved for making particle size analyses of soils. *Agron. J.* **1962**, *54*, 464–465. [CrossRef]
18. Black, C.A. *Methods of Soils Analysis, Part II: Chemical and Microbiological Properties*; American Society of Agronomy: Madison, WI, USA, 1965.
19. Kähkönen, M.P.; Hopia, A.I.; Vuorela, H.J.; Rauha, J.P.; Pihlaja, K.; Kujala, T.S. Antioxidant activity of plant extracts containing phenolic compounds. *J. Agric. Food Chem.* **1999**, *47*, 3954–3962. [CrossRef] [PubMed]
20. Suntornsuk, L.; Gritsanapun, W.; Nilkamhank, S.; Paochom, A. Quantitation of vitamin C content in herbal juice using direct titration. *J. Pharm. Biomed. Anal.* **2002**, *28*, 849–855. [CrossRef]
21. Moreno-Rojas, R.; Sánchez-Segarra, P.J.; García-Martínez, M.; Gordillo-Otero, M.J.; Amaro López, M.A. Mineral composition of skimmed milk fruit-added yoghurts, nutritional assessment. *Milchwissenschaft* **2000**, *55*, 510–512.
22. AOAC. *Official Methods of Analysis*; Association of Official Analytical Chemists: Arlington, VA, USA, 1990.
23. Hu, B.; Esen, B. Heterogeneity of soybean and proteins: One dimensional electrophoretic profiles of six different solubility fractions. *J. Agric. Food Chem.* **1981**, *29*, 297–501. [CrossRef]
24. Knabe, D.; Laure, D.; Gregg, E.; Martínez, G.; Tankley, T. Apparent digestibility of nitrogen and amino acids in protein feed stuffs by growing pigs. *J. Anim. Sci.* **1989**, *67*, 441–458. [CrossRef]
25. Bradford, M. Arapid and sensitive method for the quantification of microgram quantities of protein utilizing the principle of protein-dye banding. *Anal. Biochem.* **1976**, *72*, 248–259. [CrossRef]
26. Contreras, J.I.; Baeza, R.; López, J.G.; Cánovas, G.; Alonso, F. Management of Fertigation in Horticultural Crops through Automation with Electrotensiometers: Effect on the Productivity of Water and Nutrients. *Sensors* **2021**, *21*, 190. [CrossRef]
27. Contreras, J.I.; Alonso, F.; Cánovas, G.; Baeza, R. Irrigation management of greenhouse zucchini with different soil matric potential level. Agronomic and environmental effects. *Agric. Water Manag.* **2017**, *183*, 26–34. [CrossRef]
28. Lee, R.B.; Purves, J.V.; Ratcliffe, R.G.; Saker, L.R. Nitrogen assimilation and the control of ammonium and nitrate absorption by maize roots. *J. Exp. Bot.* **1992**, *43*, 1385–1396. [CrossRef]
29. Muller, B.; Tilliard, P.; Touraine, B. Nitrate fluxes in soybean seedling roots and their response to amino acids: An approach using 15N. *Plant Cell Environ.* **1995**, *18*, 1267–1279. [CrossRef]
30. Dluzniewska, P.; Gessler, A.; Kopriva, S.; Strand, M.; Novak, O.; Dietrich, H.; Rennenberg, H. Exogenous supply of glutamine and active cytokinin to the roots reduces NO_3 uptake rates in poplar. *Plant Cell Environ.* **2006**, *29*, 1284–1297. [CrossRef] [PubMed]
31. Oanzen, H.H.; Chang, C. Cation concentrations in the saturation extract and soil solution extract of soil salinized with various sulfate salts. *Soil Sci. Plant Anal.* **1988**, *19*, 405–430. [CrossRef]
32. Contreras, J.I. Optimización de Las Estrategias de Fertirrigación de Cultivos Hortícolas en Invernadero Utilizando Aguas de Baja Calidad (Agua Salina y Agua Regenerada) en Condiciones Del Litoral de Andalucía. Ph.D. Thesis, Universidad de Almería, Almería, Spain, 2014.
33. Segura, M.L.; Contreras París, J.I.; Plaza, B.M.; Lao, M.T. Assessment of the Nitrogen and Potassium Fertilizer in Green Bean Irrigated with Disinfected Urban Wastewater. *Soil Sci. Plant Anal.* **2012**, *43*, 426–433. [CrossRef]
34. Edelenbos, M.; Thybo, A.; Errichsen, L.; Wienberg, L.; Andersen, L. Relevant measurements of green pea texture. *J. Food Qual.* **2001**, *24*, 91–110. [CrossRef]
35. Lizana, C.; Wentworth, M.; Martinez, J.P.; Villegas, D.; Meneses, R.; Murchie, E.H.; Pastenes, C.; Lercari, B.; Vernieri, P.; Horton, P.; et al. Differential adaptation of two varieties of common bean to abiotic stress: I. Effects of drought on yield and photosynthesis. *J. Exp. Bot.* **2006**, *57*, 685. [CrossRef]
36. Martelo-Nuñez, J.M.; Ruiz-Nogueira, B.; Sau-Sau, F. Prácticas de cultivo de la judía grano. *Agricultura* **1996**, *767*, 493–495.
37. Droga, R.C.; Dudeja, S.S. Fertilizer N and Nitrogen Fixation in legume—Rhizobium Symbiosis. *Ann. Biol.* **1993**, *9*, 149–164.
38. Clayton, G.; Rice, W.; Blade, S.; Grant, C.; Harker, N.; Johnston, A.; Lafond, G.; Lupwayi, N. Minimizing Risk and Increasing Yield Stability in Field Pea Production. In Proceedings of the 10th Annual Meeting, Conference and Trade Show of the Saskatchewan Soil Conservation Association, Regina, SK, Canada, 11–12 February 1998.
39. Kontopoulou, C.K.; Giagkou, S.; Stathi, E.; Savvas, D.; Iannetta, P.P.M. Responses of hydroponically grown common bean fed with nitrogen-free nutrient solution to root inoculation with N-2-fixing bacteria. *Hortscience* **2015**, *50*, 597–602. [CrossRef]
40. Macák, M.; Candráková, E.; Đalović, I.; Prasad, P.V.V.; Farooq, M.; Korczyk-Szabó, J.; Kováčik, P.; Šimanský, V. The Influence of Different Fertilization Strategies on the Grain Yield of Field Peas (*Pisum sativum* L.) under Conventional and Conservation Tillage. *Agronomy* **2020**, *10*, 1728. [CrossRef]
41. Sen, R.; Rahman, M.A.; Hoque, A.K.M.S.; Zaman, S.; Noor, S. Response of Different Levels of Nitrogen and Phosphorus on the Growth and Yield of French Bean. *Bangladesh J. Sci. Ind. Res.* **2010**, *45*, 169–172. [CrossRef]
42. Sezen, S.M.; Yazar, A.; Akyildiz, A.; Dasgan, H.Y.; Gencel, B. Yield and quality response of drip irrigated green beans under full and deficit irrigation. *Sci. Hortic.* **2008**, *117*, 95–102. [CrossRef]

43. Segura, M.L.; Contreras, J.I.; García, I.I.; García, M.C.; Cuadrado, I.M. Fertilización nitrogenada de judía verde bajo invernadero con criterios agroecológicos. In *VII Congreso SEAE Zaragoza*; Sociedad Española de Agricultura Ecológica: Catarroja, Spain, 2006.
44. Ntatsi, G.; Gutiérrez-Cortines, M.E.; Karapanos, I.; Barros, A.; Weiss, J.; Balliu, A.; Savvas, D. The quality of leguminous vegetables as influenced by preharvest factors. *Sci. Hortic.* **2018**, *232*, 191–205. [CrossRef]
45. Mera, M.; Kehr, E.; Mejías, J.; Ihl, M.; Bifani, V. Arvejas (*Pisum sativum*) de Vaina Comestible Sugar Snap: Antecedentes y Comportamiento en el Sur de Chile. *Agric. Tec.* **2007**, *67*, 343–352. [CrossRef]
46. Fernández, N.E.P.; López, G.P.; Domínguez, C.R.; Peñuelas, V.M.L.; Izaguirre, S.C.O. Composición química, características funcionales y capacidad antioxidante de formulaciones de garbanzo ("*Cicer arietinum*" L.) Blanco Sinaloa 92. *Agrociencia* **2019**, *53*, 35–44.
47. Khah, M.E.; Arvanitoyannis, S.I. Yield, nutrient content and physico-chemical and organoleptic properties en green bean are affected by N:K ratios. *Food Agric. Environ.* **2003**, *1*, 17–26.
48. Dawo, M.; Wilkindon, J.M.; Sanders, F.; Pilbean, D.J. The yield and quality of fresh and ensiled plant material from intercropped maize (Zea mays) and beans (*Phaseolus vulgaris*). *J. Sci. Food Agric.* **2007**, *87*, 1391–1399. [CrossRef]
49. Martínez, C.; Ros, G.; Periago, M.J.; López, G.; Ortuño, J. Physico-chemical and sensory quality criteria of green beans. *Lebensm. Wiss. Technol.* **1995**, *28*, 515–520. [CrossRef]
50. Navarro, J.M.; Garrido, C.; Flores, P.; Martínez, V. The effect of salinity on yield and fruit quality of pepper grown in perlite. *Span. J. Agric. Res.* **2010**, *8*, 142–150. [CrossRef]
51. Al-Suhaibani, N.A. Influence of early water deficit on seed yield and quality of faba bean under arid environment of Saudi Arabia. *Am. Eurasian J. Agric. Environ. Sci.* **2009**, *5*, 649–654.
52. Bhattacharya, S.; Malleshi, N.G. Physical, chemical and nutritional characteristic of premature-processed and matured green legumes. *J. Food Sci. Technol.* **2012**, *49*, 459–466. [CrossRef] [PubMed]
53. Siger, A.L.; Czubinski, J.; Kachlicki, P.; Dwiecki, K.; Lampart-Szczapa, E.; Nogala-Kalucka, M. Antioxidant activity and phenolic content in three Lupin species. *J. Food Compos. Anal.* **2012**, *25*, 190–197. [CrossRef]
54. European Food Information Council. Nutrition information & food labelling-results of the EUFIC consumer research conducted in May-June 2004. *EUFIC Forum* **2005**, *2*, 1–6.
55. Ovaskainen, M.L.; Torronen, R.; Koponen, J.M.; Sinkko, H.; Hellstrom, J.; Reinivuo, H.; Mattila, P. Dietary intake and major food sources of polyphenols in finnish adults. *J. Nutr.* **2008**, *138*, 562–566. [CrossRef]
56. Tee, E.S.; Young, S.I.; Ho, S.K.; Mizura, S.S. Determination of vitamin C in fresh fruits and vegetables using the dye-titration and microfluorometric methods. *Pertanika* **1988**, *11*, 39–44.
57. Rekha, C.; Poornima, G.; Manasa, M.; Abhipsa, V.; Devi, J.P.; Kumar, V.H.T.; Kekuda, T.R.P. Ascorbic acid, total phenol content and antioxidant activity of fresh juices of four ripe and unripe citrus fruits. *Chem. Sci. Trans.* **2012**, *1*, 303–310. [CrossRef]
58. Frusciante, L.; Carli, P.; Ercolano, M.R.; Pernice, R.; Di Mateo, A.; Fogliano, V.; Pellegrini, N. Antioxidant nutritional quality of tomato. *Mol. Nutr. Food Res.* **2007**, *51*, 609–617. [CrossRef]
59. Moreiras, O.; Carbajal, A.; Cabrera, L.; Cuadrado, C. Ingestas Recomendadas de energía y nutrientes (Revisadas 2002). In *Tablas de Composición de Alimentos*; Ediciones Pirámide: Madrid, Spain, 2004; pp. 127–131.
60. Kalac, P.; Svoboda, L. A review of trace element concentrations in edible mushrooms. *Food Chem.* **2000**, *69*, 273–281. [CrossRef]
61. Cuervo, M.; Abete, I.; Baladia, E.; Corbalán, M.; Manera, M.; Basulto, J. *Ingestas Dietéticas de Referencia (IDR) para la Población Española*; Ediciones Universidad de Navarra; Federación Española de Sociedades de Nutrición, Alimentación y Dietética (FESNAD): Madrid, Spain, 2010.
62. Genccelep, H.; Uzun, Y.; Tunctürk, Y.; Demirel, K. Determination of mineral contents of wild-grown edible mushrooms. *Food Chem.* **2009**, *113*, 1033–1036. [CrossRef]
63. Serratos-Arévalo, J.C. Aislamiento y Caracterización de Proteínas de Las Semillas Maduras de Enterolobium Cyclocarpium Para Su Aprovechamiento Alimenticio. Ph.D. Thesis, Universidad de Colima, Colima, Mexico, 2000.
64. Casey, R.; Domoney, C. The protein composition of legume seeds its variability and potential. In Proceedings of the Ire Conference Europeenne sur les Proteagineux, Angers, France, 1–3 June 1992; pp. 381–386.
65. Jambunathan, R.; Singh, U.; Subramanian, V. Grain quality of sorghum, pearl millet, pigeon-pea, and chick-pea. In Proceedings of the Workshop on Interfaces between Agriculture, Nutrition, and Food Science, Hyderabad, India, 10–12 November 1981.
66. Paredes, L.; Ordorica, F.; Guevara, L.; Covarrubias, A. *Las proteínas vegetales. Presente y futuro de la alimentación. In Prospectiva de la Biotecnología en México*; Quintero, R., Ed.; CONACYT: México, Mexico, 1985; pp. 331–349.
67. El-Noemani, A.; El-Zeiny, H.; El-Gindy, A.; El-Sahhar, E.A.; El-Shawadfy, M. Performance of some bean (*Phaseolus vulgaris* L.) varieties under different irrigation systems and regimes. *Aust. J. Basic Appl. Sci.* **2010**, *1*, 6185–6196.
68. Shalaby, M.A.; Ibrahim, S.K.; Zaki, E.M.; Abou-Sedera, F.A.; Abdallah, A.S. Effect of sowing dates and plant cultivar on growth, development and pod production of snap bean (*Phaseolus vulgaris* L.) during summerseason. *Int. J. Pharm. Tech. Res.* **2016**, *9*, 231–242.

Article

Latitudinal Characteristic Nodule Composition of Soybean-Nodulating Bradyrhizobia: Temperature-Dependent Proliferation in Soil or Infection?

Md Hafizur Rahman Hafiz [1,2], Ahsanul Salehin [3], Fumihiko Adachi [1], Masayuki Omichi [4], Yuichi Saeki [5], Akihiro Yamamoto [5], Shohei Hayashi [1] and Kazuhito Itoh [1,3,*]

1   Faculty of Life and Environmental Science, Shimane University, 1060 Nishikawatsu, Matsue, Shimane 690-8504, Japan; hafizhstu@hotmail.com (M.H.R.H.); fadachi@life.shimane-u.ac.jp (F.A.); shohaya@life.shimane-u.ac.jp (S.H.)
2   Department of Crop Physiology and Ecology, Hajee Mohammad Danseh Science and Technology University, Dinajpur 5200, Bangladesh
3   The United Graduate School of Agricultural Sciences, Tottori University, 4-101 Koyama-Minami, Tottori 680-8553, Japan; ujanrijvi224@gmail.com
4   Department of Agricultural Science and Business, Hokkaido College, Takushoku University, Memu 4558, Hokkaido 074-8585, Japan; omichi@takushoku-hc.ac.jp
5   Faculty of Agriculture, University of Miyazaki, Gakuen-Kibanadai-Nishi 1-1, Miyazaki 889-2192, Japan; yt-saeki@cc.miyazaki-u.ac.jp (Y.S.); ahyama@cc.miyazaki-u.ac.jp (A.Y.)
*   Correspondence: itohkz@life.shimane-u.ac.jp; Tel.: +81-852-32-6521

Citation: Hafiz, M.H.R.; Salehin, A.; Adachi, F.; Omichi, M.; Saeki, Y.; Yamamoto, A.; Hayashi, S.; Itoh, K. Latitudinal Characteristic Nodule Composition of Soybean-Nodulating Bradyrhizobia: Temperature-Dependent Proliferation in Soil or Infection? *Horticulturae* **2021**, *7*, 22. https://doi.org/10.3390/horticulturae7020022

Academic Editor: Stefano Marino
Received: 1 December 2020
Accepted: 26 January 2021
Published: 29 January 2021

Publisher's Note: MDPI stays neutral with regard to jurisdictional claims in published maps and institutional affiliations.

Copyright: © 2021 by the authors. Licensee MDPI, Basel, Switzerland. This article is an open access article distributed under the terms and conditions of the Creative Commons Attribution (CC BY) license (https://creativecommons.org/licenses/by/4.0/).

Abstract: A species-specific latitudinal distribution of soybean rhizobia has been reported; *Bradyrhizobium japonicum* and *B. elkanii* dominate in nodules in northern and southern areas, respectively. The aim of this study was to elucidate whether temperature-dependent proliferation in soil or infection is more reliable for determining the latitudinal characteristic distribution of soybean-nodulating rhizobia under local climate conditions. Three study locations, Fukagawa (temperate continental climate), Matsue and Miyazaki (humid sub-tropical climate), were selected in Japan. Each soil sample was transported to the other study locations, and soybean cv. Orihime (non-Rj) was pot-cultivated using three soils at three study locations for two successive years. Species composition of *Bradyrhizobium* in the nodules was analyzed based on the partial 16S rRNA and 16S–23S rRNA ITS gene sequences. Two *Bradyrhizobium japonicum* (Bj11 and BjS10J) clusters and one *B. elkanii* (BeL7) cluster were phylogenetically sub-grouped into two (Bj11-1-2) and four clusters (BjS10J-1-4) based on the ITS sequence. In the Fukagawa soil, Bj11-1 dominated (80–87%) in all study locations. In the Matsue soil, the composition was similar in the Matsue and Miyazaki locations, in which BeL7 dominated (70–73%), while in the Fukagawa location, BeL7 decreased to 53% and Bj11-1 and BjS10J-3 increased. In the Miyazaki soil, BeL7 dominated at 77%, and BeL7 decreased to 13% and 33% in the Fukagawa and Matsue locations, respectively, while BjS10J-2 and BjS10J-4 increased. It was supposed that the *B. japonicum* strain preferably proliferated in the Fukagawa location, leading to its nodule dominancy, while in the Miyazaki location, temperature-dependent infection would lead to the nodule dominancy of *B. elkanii*, and both factors would be involved in the Matsue location.

Keywords: *Bradyrhizobium*; temperature-dependent distribution; nodule composition; proliferation in soil; infection

1. Introduction

Soybean (*Glycine max* [L] Merr.) originated in north-eastern China and is presently cultivated around the globe under various soils and climatic conditions [1–3]. The high concentrations of protein and oil in soybean seeds indicate its significance in daily life. Soybean is an easy-to-cultivate crop belonging to the Leguminosae family that can grow in nitrogen-poor soils. Soybean-nodulating rhizobia can establish symbiosis with soybeans through effective nitrogen fixation.

Diverse soybean-nodulating rhizobia belong to the genera *Bradyrhizobium, Sinorhizobium* (*Ensifer*) and *Mesorhizobium* [4,5], among which *Bradyrhizobium* is recognized as a slow grower, while *Sinorhizobium* (*Ensifer*) is recognized as a fast grower and *Mesorhizobium* as a variable one [6]. *B. japonicum* and *B. elakanii* are the major soybean-nodulating rhizobia having a high nitrogen-fixing ability and have been used as inoculants for improving crop production. However, the inoculants could not dominate in nodules due to competition with indigenous rhizobia in the field [7]. Therefore, it is essential to evaluate the ecological behavior of the indigenous soybean-nodulating rhizobia in relation to the environmental conditions.

Saeki et al. [8] studied the geographical distribution of soybean-nodulating rhizobia in Japan using soil samples around the country as inoculants and showed the species-specific latitudinal distribution of *B. japonicum* and *B. elkanii*, in which the former dominated in nodules when the northern soils were used, while the latter did in southern soils. Shiro et al. [9] examined the genetic diversity of indigenous soybean-nodulating rhizobia in the USA using a similar method and found the same latitudinal distribution of *B. japonicum* and *B. elkanii* from north to south. Adhikari et al. [10] examined nodules from different locations in Nepal and found that *B. japonicum* dominated in temperate regions, while in subtropical locations, *B. elkanii, B. yuanmingense* and *B. liaoningense* dominated in acidic, moderately acidic and slightly alkaline soils, respectively. Li et al. [11] also reported the pH-dependent distribution of rhizobia in Chinese soils, in which *B. japonicum* and *B. elkanii* dominated in neutral soils, while *B. yuanmingense, B. liaoningense* and *Sinorhizobium* dominated in alkaline soils. These results suggest that temperature and soil pH determine the species-specific distribution of soybean-nodulating rhizobia in soils.

To examine the possible reasons for the temperature-dependent distribution of soybean-nodulating rhizobia, competitive inoculation experiments at different temperatures have been conducted. Kluson et al. [12] reported that *B. japonicum* USDA 6 and *B. diazoefficiens* USDA 110 dominated in nodules at lower temperatures, while *B. elkanii* USDA 76 and *B. elkanii* USDA 94 did at higher temperature. Suzuki et al. [13] examined the nodule occupancy as well as relative population of *B. japonicum* and *B. elkanii* strains in the rhizosphere of soybean cultivated using sterilized vermiculite. Under competitive conditions, *B. japonicum* strains dominated in nodules at lower temperature even though the relative populations of both strains were similar in the rhizosphere, while at higher temperature, *B. elkanii* strains dominated in nodules due to their larger relative population in the rhizosphere. Shiro et al. [14] examined the gene expression of *nodC* and the nodule occupancy of *Bradyrhizobia* at different temperatures. In the inoculation experiment with mixes of three strains, the nodule occupancy of *B. elkanii* USDA 31 increased at higher temperatures, whereas that of *B. japonicum* USDA 123 increased at lower temperatures, corresponding to their temperature-dependent *nodC* gene expressions. These results support the temperature-dependent distribution of soybean-nodulating rhizobia in the field and suggest that the temperature influenced their preference for infection and/or proliferation in soils. Since the species-specific distribution of rhizobia in field soils is evaluated by their distribution in nodules, it is uncertain which factor, namely, temperature-dependent infection or proliferation in soil, contributes to the temperature-dependent distribution of rhizobia in nodules.

Considering the two above-mentioned factors, we selected three study locations with different climatic conditions in Japan, and each soil sample of the sites was used for soybean cultivation at all the locations for two successive years to examine the changes in the distribution of rhizobia in the nodules after the transfer of the soil samples to the different climatic conditions and to follow the changes in the second year in the new environments. If the predominance of some rhizobia in the soil determines the nodule occupancy, changing climatic conditions would not affect the nodule occupancy; on the other hand, if temperature-dependent infection determines the nodule occupancy, it would be changed in different climatic conditions. The aim of this study is to elucidate the possible reasons for the latitudinal characteristic distribution of soybean-nodulating rhizobia in local climate conditions.

2. Materials and Methods

2.1. Study Locations

To examine the temperature-dependent nodule occupancy of soybean rhizobia, three study locations, Fukagawa (fu), Matsue (ma) and Miyazaki (mi), were selected in Japan. According to Koppen's climatic classification, Fukagawa belongs to the Dfb (temperate continental climate) region, and Matsue and Miyazaki belong to the Cfa (humid sub-tropical climate) region. Soil samples were collected from the experimental fields of Takushoku University of Hokkaido College, Shimane University and Miyazaki University and used for soybean cultivation at all study locations. There had been no history of legumes cultivation in all soils. Basic information on the site and climatic parameters are presented in Table 1. The soil properties were reported previously (Table S1, [15]).

Table 1. Geographical and climatic characteristics of the study locations in Japan.

Location	Latitude (°N)	Longitude (°E)	Temperature (°C) [a]	Rainfall (mm) [a]
Fukagawa	43.71	142.01	16–26/16–26 (14–24/17–27) [b]	432/243
Matsue	35.48	133.06	23–30/23–30 (22–29/25–32)	177/481
Miyazaki	31.82	131.41	24–32/24–31 (25–32/25–33)	240/860

[a] Average daily minimum and maximum temperatures and total rainfall during the cultivation period in 2016/2017. [b] Figures in parenthesis indicate those during one month after sowing. (https://www.jma.go.jp).

2.2. Soybean Cultivation

The soil samples with a total weight of about 25 Kg were collected from several sites of the experimental field and mixed together for each study location. Each soil sample was divided into three parts (ca. 7.5 Kg) and used for soybean cultivation at each study location. Each soil sample was put in three plastic pots (20 cm in diameter and 25 cm in height), which were placed on a plastic sheet or a wooden duck board in the open field. Seeds of soybean cv. Orihime (non-Rj) from the same lot were used at all study locations. Three healthy seedlings per pot remained at seven days after germination, then they were cultivated for ca. 2–3 months depending on the conditions of the study locations in each year until harvest without fertilization and then the fresh weight of the whole plant and number of nodules were measured. After harvesting of the soybean plants in 2016, in the case of the Matsue location, the pots with the soil were kept in the open filed until the next cultivation season. For the Miyazaki and Fukagawa locations, the triplicate soil samples were mixed and kept in a paper bag in a warehouse under the same temperature conditions as outdoors until the next cultivation in 2017. The different procedures were due to space issues at the study locations.

2.3. Nodule Sampling and Isolation of Rhizobia

After harvesting from flowering to the early fruiting period, the roots were washed carefully with tap water and the whole plant fresh weight was measured after removal of surface water with tissue towel, and the number of nodules in each plant was counted, then the nodules were preserved at low temperature in a vial containing desiccating silica gel until isolation of rhizobia.

For isolation of rhizobia, ten nodules were randomly selected from one plant for each replication and kept in sterilized distilled water overnight. When the number of nodules was less than 10, two plants were used. After surface sterilization with 95% ethanol for 30 s followed by 3% sodium hypochlorite solution for 30 s, and rinsing in sterilized distilled water at least seven times, each nodule was crushed in an Eppendorf tube with

1 mL of sterilized distilled water, then a drop of suspension was streaked onto yeast mannitol agar (YMA) medium [16] and incubated at 25 °C for 5–12 days. Two randomly selected colonies per nodule were purified, and a total of 540 isolates (3 replications, 3 soils from 3 study locations, 10 nodules per plant and 2 isolates per nodule) were further analyzed molecularly.

2.4. Phylogenetic Analysis of the Rhizobia Based on Genes of 16S rRNA and 16S–23S rRNA Internal Transcribed Spacer (ITS) Region

A small amount of the colony was directly subjected as the template for the PCR amplification of the partial 16S rRNA gene using the universal primers fD1 and rP2 [17]. The components of the PCR mixtures and the PCR running conditions are summarized in Tables S2 and S3, respectively. PCR products were purified and subjected to PCR cycle sequencing, according to the procedures described previously [10]. Taxonomic position of the isolates was determined based on the database (https://www.ncbi.nlm.nih.gov/) using a BLAST [18] search. Multiple sequence alignments were constructed using ClustalW 2.1 [19]. Alignments were manually edited and phylogenetic trees with the related reference strains were constructed using ClustalW 2.1 with the neighbor-joining method and the tree was visualized by MEGA 7 [20].

Among the isolates with the same phylogeny in the 16S rRNA gene in each Soil–Location–Year combination, two representatives were randomly selected for analysis of the ITS region. PCR amplification of the ITS region was conducted using the universal ITS primers 1512F and 23R [21]. The procedures were the same as described above.

2.5. Nucleotide Sequence Accession Numbers

The sequence data generated in this study were deposited in the DDBJ Nucleotide Submission System under the accession numbers LC582850 to LC582907 for the 16S rRNA gene, and LC579845 to LC579902 for the 16S–23S rRNA ITS region.

2.6. Statistical Analysis

Statistical analysis of the soybean cultivation data was carried out using the MSTAT-C 6.1.4 [22] software package. The data were subjected to Duncan's multiple range test after one-way ANOVA.

3. Results

3.1. Fresh Plant Weight and Number of Nodules of Soybean

At each study location, the fresh plant weight and the nodule numbers were not significantly different among soils in both years with a few exceptions (Figure 1). When all data in each study location were analyzed, the fresh plant weight showed the tendency of increasing from northern (FU) to southern (MI) sites, whereas the nodule number showed the opposite tendency of significantly decreasing from northern to southern sites (Figure 2).

3.2. Phylogenetical Characterizations of the Rhizobia

Based on the 16S rRNA gene analysis, the isolated rhizobia were most closely related to one of the three groups, *Bradyrhizobium japonicum* Bj11 (KY000638), *Bradyrhizobium japonicum* S10J (MF664374) and *Bradyrhizobium elkanii* L7 (KY412842). Similarities (%) of the sequences between the isolates and the corresponding type strains were 98–100%, 96–100% and 97–100% for *B. japonicum* Bj11, *B. japonicum* S10J and *B. elkanii* L7, respectively. The phylogenetic tree of the ITS region of the selected isolates indicated that the rhizobial strains were further grouped into sub-groups (Figure 3). The most similar sequences in the database are listed in Table 2.

B. japonicum Bj11 was grouped into Bj11-1 and Bj11-2 based on the ITS sequence, and the two groups were characterized by their physiological properties, that is, it took more than one week for Bj11-1 to form visible colonies on the YMA agar plate, compared to 5–6 days for Bj11-2. *B. japonicum* S10J was grouped from BjS10J-1 to BjS10J-4 based on the

ITS sequence. The ITS sequences of BjS10J-1 had more similarity to those of *B. japonicum* Bj11 than the other groups of *B. japonicum* S10J. BjS10J-2 and BjS10J-3 were characterized by their origin, that is, BjS10J-2 and BjS10J-3 were isolated from soybeans cultivated in Miyazaki and Matsue soils, respectively. *B. elkanii* L7 was isolated from soybeans cultivated in all soils and study locations, and its ITS sequences were not distinguished among them.

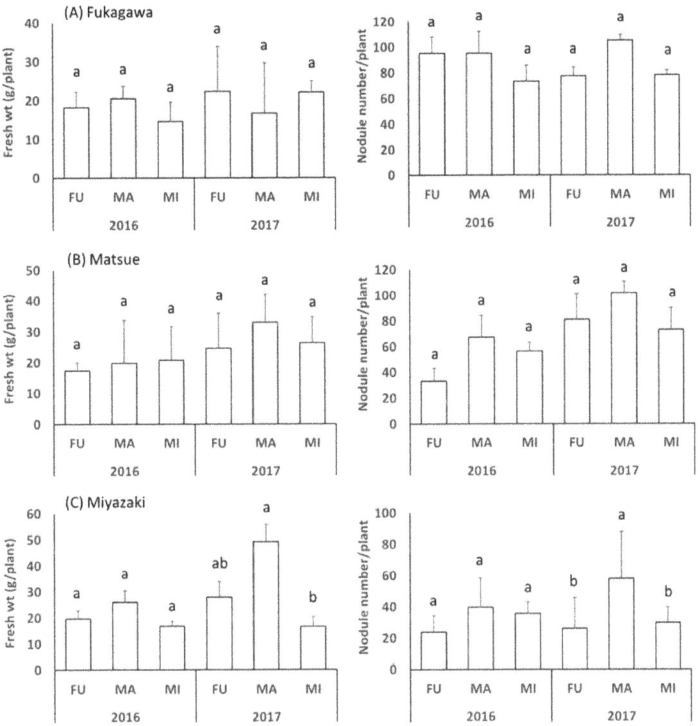

Figure 1. Fresh weight and number of nodules of soybean cultivated at Fukagawa (**A**), Matsue (**B**) and Miyazaki (**C**) locations using the soil samples (FU, MA and MI) collected from the corresponding study locations. The bars represent standard deviation (n = 3) and different letters indicate significant differences at $p < 0.05$ by Duncan's test.

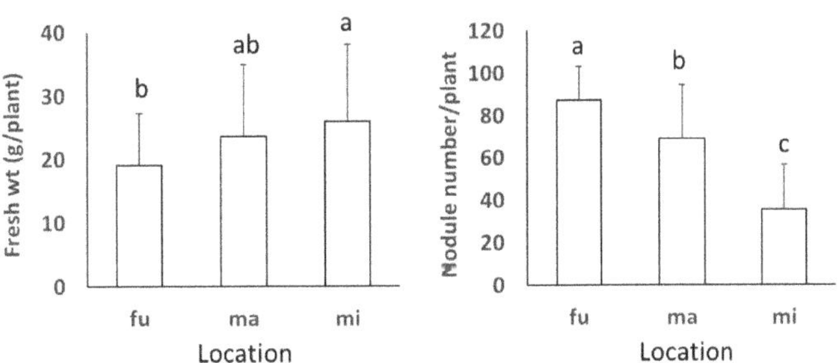

Figure 2. Fresh weight and number of nodules of soybean cultivated at Fukagawa (fu), Matsue (ma) and Miyazaki (mi) locations. The bars represent standard deviation (n = 18) and different letters indicate significant differences at $p < 0.05$ by Duncan's test.

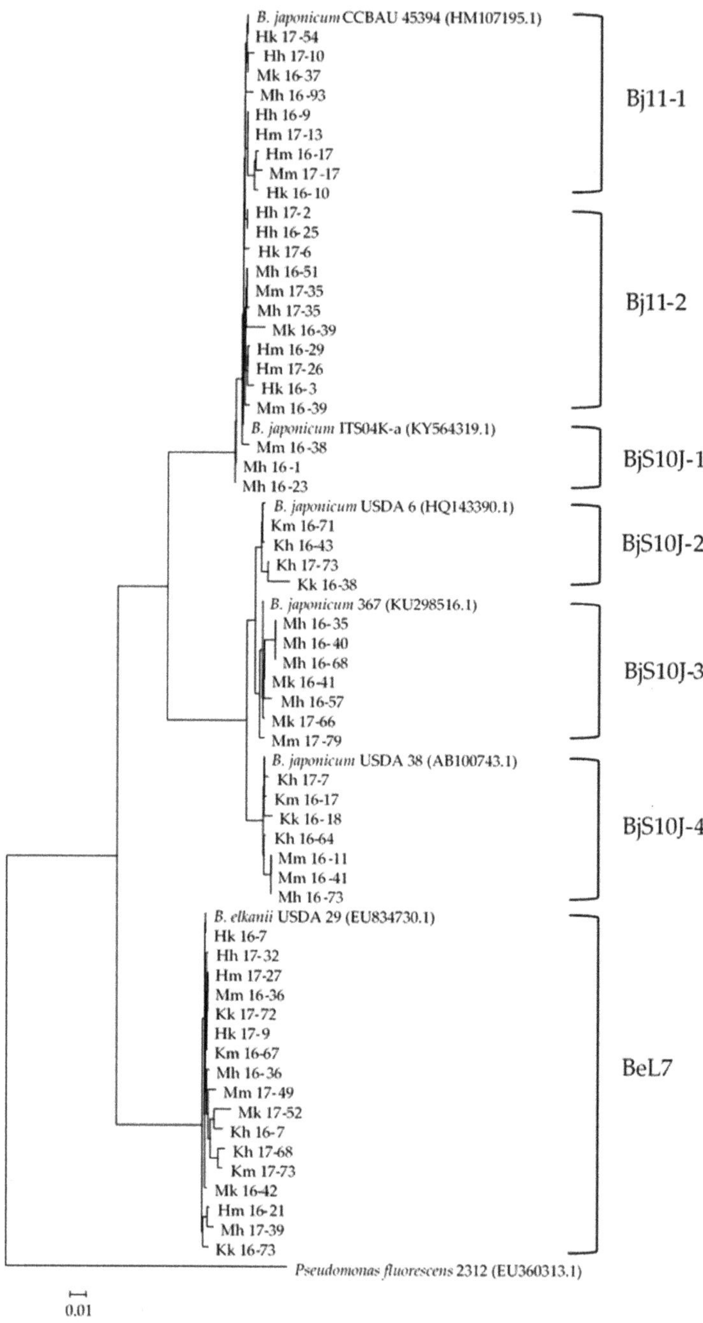

Figure 3. Phylogenetic tree of the 16S–23S rRNA ITS gene regions of the soybean rhizobial strains isolated in this study with reference strains. The isolates were designated by the soil [Fukagawa (H), Matsue (M) and Miyazaki (K)], the study location (h, m and k), year of the cultivation and the strain number. The scale bar indicates the number of substitutions per site.

Table 2. Group of soybean rhizobial strains isolated in this study based on phylogeny of 16S–23S rRNA genes' ITS region.

Closest 16S rDNA	ITS Group	Closest ITS [a]	Acc. No. [b]	Id. (%)	Remarks
B. japonicum Bj11	Bj11-1	Bj CCBAU 45394	HM107195	98–99	Slow grower Primarily Fukagawa soil
	Bj11-2	Bj CCBAU 45394	HM107195	99–100	Fast grower Fukagawa and Matsue soils
B. japonicum S10J	BjS10J-1	Bj ITS04K-a	KY564319	99–100	Matsue soil
	BjS10J-2	Bj USDA 6	HQ143390	98–99	Miyazaki soil
	BjS10J-3	Bj 367	KU298516	96–100	Matsue soil
	BjS10J-4	Bj USDA 38	AB100743	99–100	Primarily Miyazaki soil
B.elkanii L7	BeL7	Be USDA 29	EU834730	97–100	Ubiquitous in all soils

[a] Bj; *Bradyrhizobium japonicum*, Be; *Bradyrhizobium elkanii*. [b] Gene accession number in database.

3.3. Relative Composition of the Strains in Relation to Soil and Climate in 2016 and 2017

In the Fukagawa soil, the soybean rhizobia consisted of Bj11-1, Bj11-2 and BeL7 in all study locations (Table 3 and Figure 4). Bj11-1 dominated (80–87%) in all study locations in 2016. In 2017, Bj11-1 was maintained in the Fukagawa soil at 80%; however, the compositions decreased in the Matsue and Miyazaki locations at 53% and 60%, respectively, along with the increase in BeL7 to 40% and 30%, respectively. Bj11-2 was present minorly at 7–17% in all study locations and in both years.

Table 3. Relative composition (%) of rhizobial strains isolated in this study based on phylogeny of the 16S–23S rRNA ITS gene regions.

Soil/ Location	Year	Bj11		BjS10J				BeL7
		1	2	1	2	3	4	
FU/fu	2016	87	13	-	-	-	-	-
	2017	80	17	-	-	-	-	3
FU/ma	2016	80	17	-	-	-	-	3
	2017	53	7	-	-	-	-	40
FU/mi	2016	83	7	-	-	-	-	10
	2017	60	10	-	-	-	-	30
MA/fu	2016	17	7	7	-	13	3	53
	2017	-	23	-	-	-	-	77
MA/ma	2016	-	20	3	-	-	7	70
	2017	3	7	-	-	3	-	87
MA/mi	2016	7	17	-	-	3	-	73
	2017	-	-	-	-	3	-	97
MI/fu	2016	-	-	-	73	-	13	13
	2017	-	-	-	13	-	7	80
MI/ma	2016	-	-	-	53	-	13	33
	2017	-	-	-	-	-	-	100
MI/mi	2016	-	-	-	7	-	17	77
	2017	-	-	-	-	-	-	100

In the Matsue soil, Bj11, BjS10J and BeL7 were isolated in all study locations (Table 3 and Figure 4). The composition and behavior were similar between the Matsue and Miyazaki locations, with Bj11 decreasing from 20% and 24% in 2016 to 10% and 0% in 2017, respectively, while BeL7 increased from 70% and 73% in 2016 to 87% and 97% in 2017, respectively. BjS10J was present minorly at 3–10% in both years. In the Fukagawa location, Bj11 was present at 24% in 2016 and was maintained at 23% in 2017, although the major group shifted from Bj11-1 to Bj11-2. The dominant group BeL7 increased from 53%

in 2016 to 77% in 2017 as with the other study locations, while BjS10J, which was 23% in 2016, disappeared in 2017.

In the Miyazaki soil, the rhizobia consisted of BjS10J-2, BjS10J-4 and BeL7 (Table 3 and Figure 4). In the Miyazaki location, BeL7 was dominant at 77% in 2016 and completely eliminated BjS10J in 2017. In the Fukagawa and Matsue locations, BjS10J-2, which was dominant at 73% and 53% in 2016, decreased to 13% and 0% in 2017, respectively, while BeL7 increased from 13% and 33% in 2016 to 80% and 100% in 2017, respectively. BjS10J-4 also decreased from 13% in 2016 to 0–7% in 2017, respectively.

When the Fukagawa soil was moved to the Matsue and Miyazaki locations, the dominant rhizobia changed from Bj11 to BeL7 in the second year. For the Matsue and Miyazaki soils, BeL7 decreased in the Fukagawa and Matsue locations in the first year and recovered to the original level in the second year.

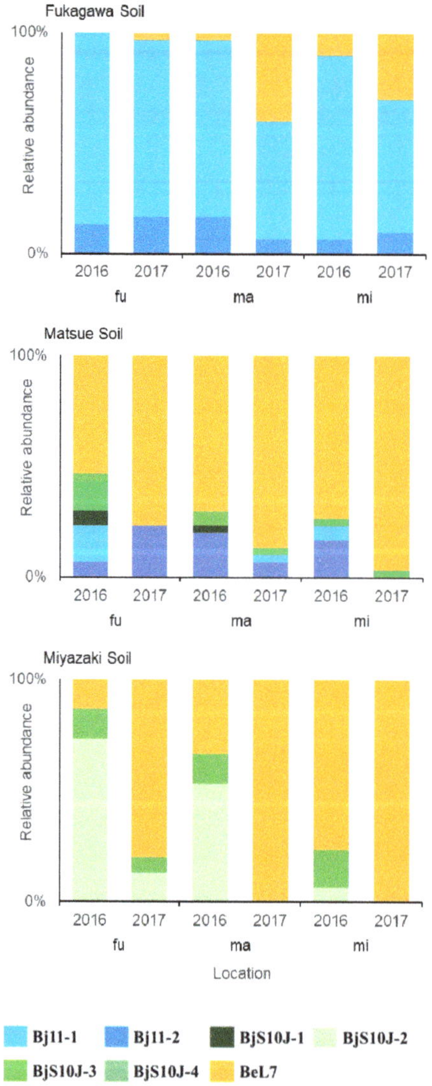

Figure 4. Relative abundance of the soybean rhizobial strains.

4. Discussion

Although the fresh weight and number of nodules were not significantly different among soils at all study locations (Figure 1), the fresh weight increased from northern to southern study locations, while the number of nodules showed the opposite tendency depending on the study location (Figure 2). The cultivation temperature might be involved in the change in the parameters (Table 1), and a similar tendency of the temperature-dependent growth of soybean has been reported [12,23–25].

In the case of the number of nodules, previous reports showed opposite temperature-dependent tendencies from ours [23–25]. Reduction in nodules at higher temperature might be due to strain-specific properties. Shiro et al. [14] reported that the nodule numbers were different by about 10 times depending on the strains, and the temperature-dependent expression of the *nodC* gene was also strain-specific but it was not related to the nodule numbers of the corresponding strains. Hungria and Vargas [26] showed an example of adverse effects of high temperature on the soil population of bradyrhizobia and the nodule number of soybean. Strain-dependent tolerance against high temperature in soil was also reported [13]. Since the bradyrhizobial community structure was changed at the different study locations, the microbial transition might be a reason for the reduction in the nodules at the southern study location.

The phylogenetic analysis of the 16S rRNA genes of the *Bradyrhizobium* spp. isolates showed three clusters, Bj11, BjS10J and BeL, which mostly corresponded to the three clusters in the phylogeny of the ITS sequences. As the pHs of the soil samples were slightly acidic (Table S1, [15]), the dominant presence of *B. japonicum* and *B. elkanii* in the nodules is reasonable [11]. The three clusters were phylogenetically comparable with the results of Saeki et al. [27] and Willems et al. [28] (data not shown). Willems et al. [28] showed that each cluster had more than 95.5% similarity in ITS sequences, whereas the similarity within each cluster in this study ranged 95–97%, and those within the subclusters (Bj11-1-2 and BjS10J-1-4) were 98–99% (data not shown). These results suggest that the *Bradyrhizobium* spp. isolates in this study were phylogenetically positioned in the same groups as previously reported, and they were further grouped by physiological property (Bj11-1 and 2), phylogeny of 16S rRNA genes (BjS10J-1) and origin of the soil (BjS10J-1, 2 and 3).

The topology of the phylogenetic trees of 16S rDNA and the ITS region was almost the same except for BjS10J-1. The variable position of a subcluster of *B. japonicum* has also been reported in the reports of Saeki et al. [8] and Adhikari et al. [10]. The ITS nucleotide sequence similarity of the BjS10J-1 strains was more than 98% with those of the Bj11 strains, while it was 88 to 90% with those of the other BjS10J strains having the same 16S rRNA gene sequences. Horizontal gene transfer among them would be one of the possible reasons for the discrepancy in their topologies.

Each cluster of BjS10J was characterized by its origin; on the other hand, BeL7 originated from all soils having undistinguishable gene sequences, and Bj11 was isolated only from Fukagawa and Matsue soils. These results suggest that the range of distribution of the strains differed among the groups. As the wide range of distributions was generally reported in previous studies [10,29], the limited range of distribution of the BjS10J strains suggests that their presence might depend on soil characteristics.

It has been well known that the species-specific distribution of soybean-nodulating rhizobia in the field is temperature-dependent [8–10] and that the temperature effect is mainly due to their infection preference [14] and/or proliferation in soil [12,13]. However, it is uncertain which factor, temperature-dependent infection or proliferation in soil, contributes to the temperature-dependent distribution of the rhizobia in nodules.

In the case of the Fukagawa soil, *B. japonicum* was dominant in the nodules in the high-latitude Fukagawa location, and the dominancy was maintained for two years (Table 3 and Figure 4). This tendency is the same as the temperature-dependent nodule dominancy of *B. japonicum* as previously reported [8–10]. When the Fukagawa soil was moved to the warmer Matsue and Miyazaki locations, the nodule composition of *B. japonicum* and

B. elkanii was not changed, suggesting an originally lower population of *B. elkanii* in the Fukagawa soil. If *B. elkanii* was present in the Fukagawa soil to a certain extent and low temperature prevented their infection, thus resulting in their lower nodule dominancy, the nodule dominancy of *B. elkanii* would increase when the temperature increased in the sub-tropical study locations. In the second year, however, the nodule dominancy of *B. elkanii* increased in the Matsue and Miyazaki locations, suggesting an increase in the soil population of *B. elkanii* in the warmer environment. Regarding *B. japonicum* Bj11 in the Fukagawa soil, the composition of Bj11-2 was maintained in both years in the Matsue and Miyazaki locations, while that of Bj11-1 decreased in the second year, suggesting a higher sensitivity of Bj11-1 to high temperature.

In the case of the Matsue soil, the dominancy of *B. elaknii* was observed in the Matsue and Miyazaki locations (Table 3 and Figure 4). Similar temperature-dependent nodule occupancy of *B. elaknii* has been reported [8–10]. The dominancy of *B. elkanii* increased in both study locations in the second year. When this soil was moved to the cooler Fukagawa location, the dominancy of *B. japonicum* increased in the first year, suggesting that *B. japonicum* was originally present in the Matsue soil and its nodule dominancy increased due to the lower temperature. The composition of *B. elkanii* increased in the second year, supposing a decrease in the population of *B. japonicum* and/or an increase in that of *B. elkanii*. Although the average minimum and maximun temperatures during the cultivation period were similar between both years, those during one month after sowing, when frequent nodulation would be expected, seemed to be a little higher in the second year (Table 1). It was supposed that the higher temperature in the second year might cause the increase in the relative dominancy of *B. elaknii*. Among the *B. japonicum* strains in the first year, Bj11-2 was dominant in the Matsue and Miyazaki locations, while Bj11-1 and BjS10J were dominant in the Fukagawa location. Along with the decrease in Bj11-1 and BjS10J in the second year, the relative dominancy of Bj11-2 increased. The transition of the *B. japonicum* strains might be due the difference in sensitivity to high temperature among them.

In the Miyazaki soil, BeL7 was dominant and BjS10J-4 was minor in the Miyazaki location in 2016, while in the Fukagawa and Matsue locations, BjS10J-2 appeared dominant (Table 3 and Figure 4), suggesting that *B. japonicum* was originally present in the Miyazaki soil and its nodulation increased due to the lower temperature in the cooler environments. In the second year, however, BeL7 recovered to 80–100%. A slightly higher temperature during one month after sowing in the second year might be the reason (Table 1), but the temperature in the Fukagawa location in 2017 was lower than that in the Matsue location in 2016; therefore, only the change in temperature could not explain the increase in BjS10J-2 and BeL7 in the Matsue (2016) and Fukagawa (2017) locations, respectively. The difference in rainfall that changed at all study locations each year might be another possible reason (Table 1). As the nodule occupancy of BeL7 increased in the Matsue and Miyazaki soils in 2017, but not in the Fukagawa soil in the Fukagawa location, BeL7 in the different soils might have different properties for a competitive relationship with the coexisting Bj strains.

5. Conclusions

Various conditions of soil storage until the next year in the study location might be differentiated in environmental conditions even in the same temperature conditions as outdoors. Potentially other environmental factors and their correlation with temperature might also affect the microorganisms. In addition, rainfall might affect the soil conditions between outdoor and indoor storage during the winter. Although the effects could not be verified, the shift of the nodule composition in the second year showed the same tendency in all soils and study locations, suggesting that the effects of the difference in the soil storage conditions did not seem to be serious on the composition of rhizobia. Fluctuating rainfall over two successive years in the study locations (Table 1) also suggests that the difference in rainfall would not significantly affect the nodule composition.

By the novel methodology used in this study, we could assume that *B. japonicum* (Bj11-1) dominantly proliferated in the Fukagawa soil and led to its dominant nodule composition and that both *B. japonicum* (BjS10J-2) and *B. elkanii* (BeL7) existed in the Miyazaki soil and the dominant nodule composition of *B. elkanii* (BeL7) was due to the temperature-dependent infection.

Supplementary Materials: The following are available online at https://www.mdpi.com/2311 -7524/7/2/22/s1, Table S1: Soil property of the study sites [15], Table S2: PCR ingredients for amplification of 16S rRNA and 16S-23S rRNA ITS region, Table S3: PCR running conditions.

Author Contributions: M.H.R.H. and K.I. conceptualized the study and designed the experiments; M.H.R.H. performed the experiments; F.A., M.O., Y.S., A.Y. performed the field experiment; S.H., A.S. helped to conduct the experiment and data analysis; M.H.R.H. wrote the article, with a substantial contribution from K.I. All authors have read and agreed to the published version of the manuscript.

Funding: This research received no external funding.

Conflicts of Interest: The authors declare no conflict of interest.

References

1. Hymowitz, T.; Harlan, J.R. Introduction of soybean to north America by Samuel Bowen in 1765. *Econ. Bot.* **1983**, *37*, 371–379. [CrossRef]
2. Dupare, B.U.; Billore, S.D.; Joshi, O.P.; Husain, S.M. Origin, domestication, introduction, and success of soybean in India. *Asian Agrihist* **2008**, *12*, 179–195.
3. Khojely, D.M.; Ibrahim, S.E.; Sapey, E.; Han, T. History, current status, and prospects of soybean production and research in sub-Saharan Africa. *Crop J.* **2018**, *6*, 226–235. [CrossRef]
4. Vinuesa, P.; Rojas-Jimenez, K.; Contreras-Moreira, B.; Mahna, S.K.; Prasad, B.N.; Moe, H.; Selvaraju, S.B.; Thierfelder, H.; Werner, D. Multilocus sequence analysis for assessment of the biogeography and evolutionary genetics of four Bradyrhizobium species that nodulate soybeans on the asiatic continent. *Appl. Environ. Microbiol.* **2008**, *74*, 6987–6996. [CrossRef] [PubMed]
5. Tan, Y.Z.; XU, X.D.; Wang, E.T.; Gao, J.I.; Martinez-Romero, E.; Chen, W.X. Phylogenetic and genetic relationships of Mesorhizobium tianshanense and related rhizobia. *Int. J. Syst. Bacteriol.* **1997**, *47*, 874–879. [CrossRef] [PubMed]
6. Chen, W.; Wang, E.; Wang, S.; Li, Y.; Chen, X.; Li, Y. Characteristics of Rhizobium tianshanense sp. nov., a moderately and slowly growing root nodule bacterium isolated from an arid saline environment in Xinjiang, People's Republic of China. *Int. J. Syst. Bacteriol.* **1995**, *45*, 153–159. [CrossRef] [PubMed]
7. Streeter, J.G. Failure of inoculant rhizobia to overcome the dominance of indigenous strains for nodule formation. *Can J. Microbiol.* **1994**, *40*, 513–522. [CrossRef]
8. Saeki, Y.; Aimi, N.; Tsukamoto, S.; Yamakawa, T.; Nagatomo, Y.; Akao, S. Diversity and geographical distribution of indigenous soybean-nodulating bradyrhizobia in Japan. *Soil Sci. Plant Nutr.* **2006**, *52*, 418–426. [CrossRef]
9. Shiro, S.; Matsuura, S.; Saiki, R.; Sigua, G.C.; Yamamoto, A.; Umehara, Y.; Hayashi, M.; Saeki, Y. Genetic diversity and geographical distribution of indigenous soybean-nodulating bradyrhizobia in the United States. *Appl. Environ. Microbiol.* **2013**, *79*, 3610–3618.
10. Adhikari, D.; Kaneto, M.; Itoh, K.; Suyama, K.; Pokharel, B.B.; Gaihre, Y.K. Genetic diversity of soybean-nodulating rhizobia in Nepal in relation to climate and soil properties. *Plant Soil* **2012**, *357*, 131–145. [CrossRef]
11. Li, Q.Q.; Wang, E.T.; Zhang, Y.Z.; Zhang, Y.M.; Tian, C.F.; Sui, X.H.; Chen, W.F.; Chen, W.X. Diversity and biogeography of rhizobia isolated from root nodules of Glycine max grown in Hebei province, China. *Microb. Ecol.* **2011**, *61*, 917–931. [CrossRef] [PubMed]
12. Kluson, R.A.; Kenworthy, W.J.; Weber, D.F. Soil temperature effects on competitiveness and growth of Rhizobium japonicum and on Rhizobium-induced chlorosis of soybeans. *Plant Soil* **1986**, *95*, 201–207. [CrossRef]
13. Suzuki, Y.; Adhikari, D.; Itoh, K.; Suyama, K. Effects of temperature on competition and relative dominance of Bradyrhizobium japonicum and Bradyrhizobium elkanii in the process of soybean nodulation. *Plant Soil* **2014**, *374*, 915–924. [CrossRef]
14. Shiro, S.; Kuranaga, C.; Yamamoto, A.; Sameshima-Saito, R.; Saeki, Y. Temperature-dependent expression of NodC and community structure of soybean-nodulating bradyrhizobia. *Microbes Environ.* **2016**, *31*, 27–32. [CrossRef]
15. Puri, R.R.; Adachi, F.; Omichi, M.; Saeki, Y.; Yamamoto, A.; Hayashi, S.; Itoh, K. Culture-dependent analysis of endophytic bacterial community of sweet potato (*Ipomoea batatas*) in different soils and climates. *J. Adv. Microbiol.* **2018**, *13*, 1–12.
16. Vincent, J.M. A manual for practical study of root nodule bacteria. In *IBP Handbook No. 15*; Blackwell Scientific Publishers: Oxford, UK, 1970; p. 164.
17. Weisburg, W.G.; Barns, S.M.; Pelletier, D.A.; Lane, D.J. 16S ribosomal DNA amplification for phylogenetic study. *J. Bacteriol.* **1991**, *173*, 697–703. [CrossRef]
18. Altschul, S.F.; Madden, T.L.; Schäffer, A.A.; Zhang, J.; Zhang, Z.; Miller, W.; Lipman, D.J. Gapped BLAST and PSI-BLAST: A new generation of protein database search programs. *Nucleic Acids Res.* **1997**, *25*, 3389–3402. [CrossRef]

19. Larkin, M.A.; Blackshields, G.; Brown, N.P.; Chenna, R.; McGettigan, P.A.; McWilliam, H.; Valentin, F.; Wallace, I.M.; Wilm, A.; Lopez, R.; et al. Clustal W and Clustal X version 2.0. *Bioinformatics* **2007**, *23*, 2947–2948. [CrossRef]
20. Kumar, S.; Stecher, G.; Tamura, K. MEGA7: Molecular evolutionary genetics analysis version 7.0 for bigger datasets. *Mol. Biol. Evol.* **2016**, *33*, 1870–1874.
21. Hiraishi, A.; Inagaki, K.; Tanimoto, Y.; Iwasaki, M.; Kishimoto, N.; Tanaka, I. Phylogenetic characterization of a new thermoacidophilic bacterium isolated from hot springs in Japan. *J. Gen. Appl. Microbiol.* **1997**, *43*, 295–304. [CrossRef]
22. Freed, R. MSTAT: A software program for plant breeder. In *Principles of Plant Genetics and Breeding*, 2nd ed.; Acquaah, G., Ed.; Blackwell Publishing: Malden, MA, USA, 2007; Volume 1, pp. 426–431.
23. Stoyanova, J. Growth, nodulation and nitrogen fixation in soybean as affected by air humidity and root temperature. *Biol. Plant* **1996**, *38*, 537–544. [CrossRef]
24. Zhang, F.; Smith, D.L. Effects of low root zone temperatures on the early stages of symbiosis establishment between soybean [*Glycine max* (L.) Merr.] and *Bradyrhizobium japonicum*. *J. Exp. Bot.* **1994**, *45*, 1467–1473.
25. Montañez, A.; Danso, S.K.A.; Hardarson, G. The effect of temperature on nodulation and nitrogen fixation by five Bradyrhizobium japonicum strains. *Appl. Soil Ecol.* **1995**, *2*, 165–174. [CrossRef]
26. Hungria, M.; Vargas, M.A.T. Environmental factors affecting N2 fixation in grain legumes in the tropics, with an emphasis on Brazil. *F. Crop Res.* **2000**, *65*, 151–164. [CrossRef]
27. Saeki, Y.; Aimi, N.; Hashimoto, M.; Tsukamoto, S.; Kaneko, A.; Yoshida, N.; Nagatomo, Y.; Akao, S. Grouping of bradyrhizobium usda strains by sequence analysis of 168 rRDNA and 168-238 rDNA internal transcribed spacer region. *Soil Sci. Plant Nutr.* **2004**, *50*, 517–525.
28. Willems, A.; Munive, A.; de Lajudie, P.; Gillis, M. In most Bradyrhizobium groups sequence comparison of 16S-23S rDNA internal transcribed spacer regions corroborates DNA-DNA hybridizations. *Syst. Appl. Microbiol.* **2003**, *26*, 203–210. [CrossRef]
29. Risal, C.P.; Yokoyama, T.; Ohkama-Ohtsu, N.; Djedidi, S.; Sekimoto, H. Genetic diversity of native soybean bradyrhizobia from different topographical regions along the southern slopes of the Himalayan Mountains in Nepal. *Syst. Appl. Microbiol.* **2010**, *33*, 416–425. [CrossRef]

 *horticulturae*

Article

A Novel Method for Estimating Nitrogen Stress in Plants Using Smartphones

Ranjeeta Adhikari and Krishna Nemali *

Department of Horticulture and Landscape Architecture, Purdue University, West Lafayette, IN 47907, USA; adhikar@purdue.edu
* Correspondence: knemali@purdue.edu; Tel.: +1765-494-8179; Fax: +1765-494-0391

Received: 1 October 2020; Accepted: 26 October 2020; Published: 29 October 2020

Abstract: For profits in crop production, it is important to ensure that plants are not subjected to nitrogen stress (NS). Methods to detect NS in plants are either time-consuming (e.g., laboratory analysis) or require expensive equipment (e.g., a chlorophyll meter). In this study, a smartphone-based index was developed for detecting NS in plants. The index can be measured in real time by capturing images and processing them on a smartphone with network connectivity. The index is calculated as the ratio of blue reflectance to the combined reflectance of blue, green, and red wavelengths. Our results indicated that the index was specific to NS and decreased with increasing stress exposure in plants. Further, the index was related to photosynthesis based on the path analysis of several physiological traits. Our results further indicate that index decreased in the NS treatment due to increase in reflectance of red and green (or yellow) wavelengths, thus it is likely related to loss of chlorophyll in plants. The index response was further validated in strawberry and hydrangea plants, with contrasting plant architecture and N requirement than petunia.

Keywords: electrical conductivity; greenhouse; image processing; nutrient stress; remote sensing

---

1. Introduction

Nitrogen (N) is one of the major elements essential for plant growth, development, and quality. Maintaining optimal N concentration in the plant tissue is essential for increasing productivity and profitability in controlled environment agriculture (CEA). In spite of supplying plants with optimal fertilizer solution concentration, plant N uptake can vary from pot to pot due to differences in substrate pH, leaching, water content, and crop growth. Therefore, monitoring N concentration of the plant tissue is more useful than measuring N concentration supplied to plants or present in the substrate, to ensure that plants are not exposed to N stress in CEA.

Nitrogen concentration in the tissue can be measured in a laboratory. However, plant sample analysis in a laboratory can be both expensive and time-consuming. Sensors recommended for indirect measurement of plant N status in CEA systems are expensive (e.g., chlorophyll meter, Soil Plant Analysis Development (SPAD), Normalized Deviation Vegetation Index (NDVI) sensor). Moreover, some sensor measurements (e.g., NDVI) can be potentially confounded by the signal from the background when the canopy is not fully closed [1], leading to errors. Other sensors for measuring crop N status including Cropscan, Greenseeker, Yara N-sensor and Fieldspec Spectroradiometer [2] are more suitable for conventional agriculture and not CEA. Therefore, regularly monitoring plant N status in CEA can be challenging with available techniques.

Plant N status can be assessed using plant images. Chlorophyll pigment in the leaves absorb red, blue, and a small proportion of green wavelengths incident on plants [3]. Because tissue N concentration affects chlorophyll synthesis in plants [4,5], a deficiency of N in the tissue can decrease the concentration of chlorophyll and increase reflectance of red, blue, and green wavelengths from plants. Therefore, an indirect assessment of N stress experienced by plants can be made by measuring the reflectance of red, blue, and green wavelengths from a canopy [2,6–8]. Images are comprised of pixels that store information on the intensity of reflected light from an object. Reflectance from plants can be measured to estimate N status of plants by processing images using image analysis software [9].

Using this technique, hyper-spectral and multi-spectral imaging platforms are being developed for N stress assessments in plants [10–12]. Several indices for N stress have been developed using reflectance in the red, blue, green, red-edge, and near infrared regions of the light spectrum [12–19]. Although these platforms and indices are available, they are not widely used in academic research and industry. Some of the reasons for this include high equipment cost, complicated hardware and software, and the selective nature of developed applications. Smartphones can capture high quality images of plants. Color images captured by a smartphone can be separated into blue, green, and red channels. Thus, the images captured by smartphones can be processed to measure reflectance in the blue, green, and red wavebands. From this, it is possible to develop indices for N stress in plants. With advancements in cloud computing, software can be developed and made accessible on smartphones with network connectivity. The images can be captured, processed, and N stress index measured in real-time using smartphones, similar to other remote sensing platforms. Nitrogen stress assessments made using smartphones can be highly valuable as these devices are universally available and simple to use. However, there is limited research that has tested or developed smartphone- based applications for assessing N stress in plants.

It is well known that carotenoids in addition to chlorophyll affect blue light absorption [20–22], while mainly chlorophyll absorbs red light [20,21,23]. Furthermore, the xanthophyll (a carotenoid) pool can increase in response to N stress in plants [24,25]. Therefore, it is possible that the reflectance of blue light is relatively less (or absorption is relatively more) than other wavelengths under N stress, as blue light can be absorbed by carotenoids in addition to chlorophyll. Based on this, we hypothesized that the ratio of blue light reflectance to that of the combined blue, green, and red wavebands will decrease under N stress in plants. The objectives of this research were to (i) test the hypothesis that the ratio of blue light reflectance to that of combined reflectance in the visible band can be used as an index for N stress, (ii) study the association between the N stress index and the physiological pathways in plants, and (iii) develop a smartphone application to measure the N stress index in species with differences in plant architecture.

2. Materials and Methods

The study comprised of "proof-of-concept" and "product development" experiments. Hypothesis testing, associating index with physiological pathways, and testing index specificity to N stress were conducted in the proof-of-concept experiment using a multispectral image station with tight control on incident light intensity, spectrum, and distance between object and camera. The purpose of the product development experiment was to test a smartphone application for measuring the N stress index in two different species with contrasting plant architecture under real-world conditions in a greenhouse. Incident light intensity, spectral composition, and distance between the plant and camera were similar but not tightly controlled in the product development experiment.

2.1. Proof-of-Concept Experiment

The experiment was conducted during July and August of 2017 in a temperature-controlled glass greenhouse at Purdue University, West Lafayette, IN using petunia (*Petunia* × *hybrida* L. var. "Easy Wave Red Velour"). It is well known that growth rate and nutrient requirements of petunia is higher than other herbaceous greenhouse crops [26]. Therefore, large effects can be observed in N stress index among treatments (see below) using petunia. Seeds were purchased from Ball Seed Company (West Chicago, IL, USA) and germinated in plug flats (72-cell, Landmark Plastics, Akron, OH, USA) filled with a propagation mix (Fafard®, germination mix, Sungro Horticulture, Agawam, MA, USA). The trays were placed in a mist until germination, after which, seedlings were transplanted into 0.45 L containers (Hummert International, Earth City, MO, USA) filled with a peat-based soilless substrate (Sunshine mix #8, Sungro Horticulture) containing 75% peat, 20% perlite, and 5% vermiculite. Plants were fertilized with a solution made by mixing 15N-2.2P-12.5K and 21N-2.2P-16.6K commercial fertilizers (Peters Excel, ICL specialty fertilizer, UK) in a 3:1 ratio every alternate day. The electrical conductivity (EC) of the fertilizer solution (a measure of total fertilizer ions dissolved in the solution) was 2.0 dS·m^{-1} and contained an N concentration of 198 mg N·L^{-1}, except in the nitrogen stress treatment (see *Treatments* section below). The pH of the substrate was maintained between 6.0 to 6.5 during the study. Greenhouse was maintained at a day/night temperature of 26/20 ± 2.4/1.1 °C, daily light integral of 20 to 25 mol·m^{-2}·d^{-1}, and relative humidity close to 50% during the study.

Plants were grown under optimal conditions for two weeks after transplanting. After this, plants were subjected to three treatments including optimal or control (C), drought stress (DS), and nitrogen stress (NS). Drought stress was applied by maintaining a low substrate volumetric water content (θ) of 0.15 m^3·m^{-3} and supplying a fertilizer solution with EC of 2.0 dS·m^{-1}. Nitrogen stress was provided by supplying a fertilizer solution with an EC of 0.75 dS·m^{-1} and maintaining a θ level of 0.48 m^3·m^{-3}. Plants in the optimal treatment were grown at an θ level of 0.47 m^3·m^{-3} using a fertilizer solution with an EC of 2.0 dS·m^{-1}. Plants were grown under different treatments for five weeks.

Solution EC, substrate EC (EC$_s$), and θ were measured weekly using a dielectric sensor (ECHO 5TE, Meter Group, Pullman, WA, USA). A line quantum sensor (SQ-326-SS, Apogee instruments, Logan, UT, USA) was used to measure photosynthetic photon flux density (PPFD) at the canopy level during the middle of the day.

A custom measuring station with three quantum sensors (LI190, LI-COR Biosciences, Lincoln, NE, USA) was used to measure the light absorption fraction (I_{abs}) of plants. A group of four plants in a tray was moved from the main experiment to the station for measurement of incident, transmitted and reflected light intensity ($PPFD_i$, $PPFD_t$ and $PPFD_r$, respectively) in different treatments. We measured $PPFD_i$ by placing a quantum sensor horizontally on a flat surface at approximately canopy height. A second quantum sensor was placed at the bottom of the canopy to measure $PPFD_t$. In addition, a third quantum sensor was placed upside down at an angle of 45° towards the canopy and 0.3 m above the plants was used to measure $PPFD_r$. The intensity of light absorbed by the plants ($PPFD_a$) was calculated as described by [27]:

$$PPFD_a = PPFD_i - (PPFD_r + PPFD_t) \tag{1}$$

Fraction of incident light absorbed by plants was calculated as follows:

$$I_{abs} = \frac{PPFD_a}{PPFD_i} \tag{2}$$

Leaf photosynthetic rate (A) and quantum efficiency in light (ϕ_{PSII}) were measured according to the procedure described by [28] using a leaf chamber fluorometer with an LED light source attached to an open-flow leaf gas exchange system (LI-COR-6400XT, LI-COR Biosciences). Measurements were taken on three separate leaves belonging to different plants within each treatment at midday prior to harvest. Fully expanded new leaves were clamped and exposed to a reference CO_2 concentration of 400 µmol·mol^{-1} and a light intensity of 400 µmol·m^{-2}·s^{-1} inside the chamber. The proportion of red and blue light was 90 and 10%, respectively. Relative humidity and temperature inside the leaf chamber were maintained at 40–70% and 25 °C, respectively.

Canopy area (CA) and reflectance at 450 (blue), 521 (green), and 660 (red) nm were measured on the 4th, 8th, 16th, 22nd, 27th, and 34th day after imposing treatments using a multi-spectral image station (TopView, Aris, Eindhoven, The Netherlands). A group of four plants from each treatment were placed inside the image station and sequentially exposed to 450, 521, 625, and 660 nm of light using strobe light-emitting diodes (OSLON SSL80, Osram, Munich, Germany). A monochromatic camera (acA3800; Basler Ace, 10 MP with MT9J003 CMOS sensor, 8-bits resolution, ON Semiconductor, AZ,

USA) inside the image station captured grayscale images (Figure 1) for each light exposure. The images were automatically stored with unique file names.

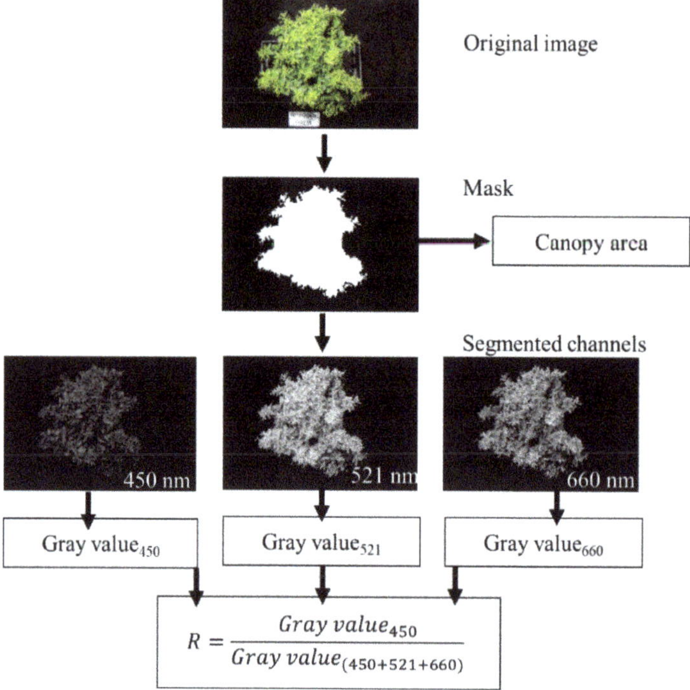

$$R = \frac{Gray\ value_{450}}{Gray\ value_{(450+521+660)}}$$

Figure 1. Procedure for estimating canopy area and N stress index (R) of petunia plants using multi-spectral image station. Original image, mask, and grayscale images captured by a monochromatic camera after sequentially exposing plants to 450 nm, 521 nm and 660 nm.

Captured images were processed automatically using built-in MultiSpec software V2.0 (Aris, The Netherlands). Image processing involved developing a mask of plant, separating plant pixels from the background by super-imposing a mask on the image, counting plant pixels, and measuring average gray value of plant pixels from each image exposed to 450, 521, and 660 nm wavelengths. The average gray value of a grayscale image is related to average reflectance of light from the objects (i.e., plants) captured in the image. As plants absorb more blue and red wavelengths in photosynthesis, images from blue (450 nm) and red (660 nm) exposures are less bright (lower gray value) than those from green (521 nm) exposure as relatively more green light is reflected by plants (Figure 1). From the gray values, N stress index (R) was calculated as the ratio of average gray value of 450 nm image to combined gray value of 450, 521, and 660 nm images.

$$R = \frac{Gray\ Value_{450}}{Gray\ Value_{(450+521+660)}} \tag{3}$$

Image-processing software automatically measured CA by counting the number of plant pixels, and multiplying the pixel number by the individual pixel area and magnification factor (specific to the camera inside the image station). Plants were harvested after five weeks of exposure to different treatments. Shoot material was dried in a forced oven maintained at 70 °C for one week. The dried samples were weighed to measure shoot dry weight (SDW).

2.2. Product Development Experiment

The experiment used strawberry (*Fragaria* × *ananassa* var. "Quinault") and hydrangea (*Hydrangea paniculate* var. "Bobo"). These species were selected due to their differences in leaf shape, growth rate, N requirement, and architecture to petunia. Strawberry runners were separated from stock plants available with researchers. Hydrangea plants were purchased from Spring Meadow Nursery Inc. (MI, USA). Strawberry runners were transplanted in plastic containers (10 cm diameter, 0.45 L, Hummert International, Earth City, MO, USA) and hydrangea plants were transplanted in nursery containers (16 cm diameter, 3.78 L, Greenhouse Megastore, Danville, IL, USA). Containers for both strawberry and hydrangea plants were filled with the same media used in the proof-of-concept experiment. A fertilizer solution containing EC of 1.0 dS·m^{-1} was supplied to strawberry plants during the establishment stage. After two weeks, strawberry plants were exposed to C and NS treatments. Plants in the C and NS treatments received fertilizer solutions containing an EC of 2.0 and 0.75 dS·m^{-1}, respectively twice a week. The hydrangea plants were grown in four N fertilizer treatments containing 9, 15, 21 and 30 g·pot^{-1} of 21N-2.2P-16.6K commercial fertilizer (Peters Excel, ICL specialty fertilizer, UK), respectively. The substrate water content and environmental conditions were similar to the proof-of-concept experiment.

Smartphone images of strawberry plants and hydrangea branches were captured after three and five weeks of exposure to treatments, respectively. In addition, hydrangea branches were imaged inside the multi-spectral image station used in the proof-of-concept experiment. This was done to compare the N stress indices measured by the smartphone (R_{sp}) and multi-spectral image station (R). Prior to capturing images, strawberry plants were placed on the greenhouse floor. The images of whole strawberry plants were captured by placing the smartphone approximately 60 cm above plants (Figure 2). A black plastic sheet (0.45 m × 0.45 m) was used as the background for hydrangea branches. Each branch, while attached to the mother plant, was inserted through a slit in the middle of the plastic sheet. The smartphone was placed approximately 30 cm above the plastic sheet for capturing images. After capturing smartphone images, the branch was cut and placed inside the image station. The images of hydrangea branches were captured inside the multispectral image station as described above for petunia plants in the proof-of-concept experiment. The time between cutting the branch from the mother plant and imaging the branch inside the image station was less than a minute.

The image-processing software for analyzing images of strawberry plants and hydrangea branches collected by the smartphone was developed using Matlab (R2017B, MathWorks, Natic, MA, USA). The image processing method used was similar to that described in other published works [19,29]. The developed software was loaded to an online drive (Matlab Drive, MathWorks) and accessed on the smartphone using an app (Matlab Mobile, MathWorks). The software controls the camera of the smartphone and displays a video of the plant on the screen to enable users to capture images from a preferred height. The software on the Matlab Drive automatically processed images after capture. Image processing involved separating the color image into red, green, and blue channels, enhancing green color and developing a mask, segmenting plant pixels by superimposing the mask on red, green, and blue channels, and measuring the average gray value of plant pixels in each channel (Figure 2). From the average gray values, software automatically calculated R_{sp} and stored the results of the analysis as a Microsoft excel file:

$$R_{sp} = \frac{Gray\ Value_{blue}}{Gray\ Value_{(blue+green+red)}} \tag{4}$$

Equation (4) is similar to, but slightly different from Equation (3) used to measure R from images captured by the multi-spectral image station. Plants were exposed to narrow wavebands of 450, 521, and 660 nm in the multi-spectral image station using strobed LED lights. Such exposure to narrow wave bands is not possible using a smartphone. The images captured by the smartphone are based on the broadband blue (400 to 499 nm), green (500 to 599 nm) and red (600 to 700 nm) wavelengths in the natural light. Therefore, average gray values of images captured by the multi-spectral image station and smartphone were based on narrow and broadband wavelengths, respectively.

$$R_{sp} = \frac{Gray\ value_{blue}}{Gray\ value_{(blue+green+red)}}$$

Figure 2. Procedure for the estimation of N stress index by a smartphone (R_{sp}). Strawberry plant images are shown in the illustration. The green channel of the original image was enhanced to make mask, blue, green, and red channels were separated, and the mask was used to segment plants in three channels. Average gray values were calculated for each segmented channel to estimate R_{sp}.

2.3. Experimental Design and Data Analyses

A randomized complete block design with four replications was used in both the proof-of-concept and product development experiments. Data were analyzed using a linear mixed model (Proc Mixed) procedure of statistical analysis software (SAS, SAS Institute, Cary, NC, USA) with repeated measures as needed. Tukey's honestly significant difference procedure was used to separate least square means. Path analyses in the main experiment was conducted using the "Proc Calis" procedure of SAS. For all analyses, a $p \leq 0.05$ was considered statistically significant.

3. Results and Discussion

3.1. Proof-of-Concept Experiment

Statistical analyses indicated that the environmental conditions were significantly different in the NS and DS compared to C treatment (Table 1). Photosynthetic photon flux density incident on plants was not significantly different among the treatments and averaged 417 μmol·m^{-2}·s^{-1} (Table 1). However, θ was significantly lower in the DS compared to the other two treatments. A θ value of 0.15 m^3·m^{-3} was maintained in the DS treatment based on a previous work [30] that showed a decline in the growth of bedding plants including petunia at this level. Substrate EC was significantly lower in both the NS and the DS compared to the C treatment. In addition, EC$_s$ was significantly lower in the DS compared to the NS treatment. The ECH_2O-5TE sensor used in our experiment measures electrical resistance to calculate EC. Electrical resistance increases or conductivity decreases when the current flow through the solution decreases. The current flow can decrease significantly when θ is low, as in the DS treatment. In addition, the sensor measures bulk EC (influenced by dielectric permittivity of water, dissolved ions, substrate particles, and air), therefore the values are lower than other commonly used sensors measuring pore–water conductivity. In previous research, pore–water EC was 1.8 times higher than bulk water EC measurements for the substrate used in this experiment [31]. Based on this, the equivalent pore–water EC_s in the C and NS treatments can be estimated as 1.3 and 0.6 dS·m^{-1}, respectively.

Table 1. Photosynthetic photon flux density (PPFD), substrate electrical conductivity (EC_s) and volumetric water content (θ) maintained in the control (C), nitrogen stress (NS), and drought stress (DS) treatments in the main experiment. Treatment means followed by the same letter are not statistically different ($p \leq 0.05$). Values in parenthesis indicate standard error of mean.

Treatment	PPFD	EC_s	θ
	μmol·m^{-2}·s^{-1}	dS·m^{-1}	m^3·m^{-3}
C	415 (13.4) a	0.72 (0.027) a	0.47 (0.016) a
NS	417 (31.0) a	0.32 (0.017) b	0.48 (0.013) a
DS	419 (18.8) a	0.08 (0.014) c	0.15 (0.01) b

Statistical analyses indicated that SDW was higher in the C compared to the other treatments. Furthermore, SDW was significantly higher in the NS than DS treatment (Table 2). This confirms that the stress treatments in our experiment decreased plant growth compared to C treatment. A significant decrease in CA, I_{abs} and A were observed in the DS whereas only A was significantly lower in the NS compared to the control (Table 2). This may suggest that nitrogen stress mainly affects plant growth by reducing A. Photosynthesis is affected by light absorption, generation of energy in the light-dependent reactions and utilization of energy in the Calvin cycle [32,33]. Nitrogen stress can reduce both light absorption (by decreasing chlorophyll concentration) [4,34] and utilization of energy in the Calvin cycle (due to decreased enzymatic activity) [35,36] in plants. There were no differences in ϕ_{PSII} among the treatments, although a numerically lower value was observed in the NS treatment. Lack of significance could be due to small effect size and/or large variability in the ϕ_{PSII} measurements. A similar decline ϕ_{PSII} (without statistical significance) of wheat plants under N stress was previously reported [37].

Table 2. Shoot dry weight (SDW), canopy area (CA), leaf photosynthesis (A), light absorption fraction (I_{abs}), N stress index (R) and quantum efficiency in light (ϕ_{PSII}) of petunia at harvest stage in the control (C), nitrogen stress (NS) and drought stress (DS) treatments. Treatment means followed by the same letter are not statistically different ($p \leq 0.05$). Values in parenthesis indicate standard error of mean.

Treatment	SDW	CA	A	I_{abs}	R	ϕ_{PSII}
	g·plant^{-1}	m^2	μmol·m^{-2}·s^{-1}			
C	12.2 (0.47) a	0.20 (0.002) a	5.2 (0.26) a	0.95 (0.008) a	0.296 (0.008) b	0.31 (0.018) a
NS	10.0 (0.69) b	0.19 (0.002) a	3.8 (0.56) b	0.95 (0.013) a	0.271 (0.021) a	0.26 (0.005) a
DS	6.0 (0.78) c	0.14 (0.008) b	1.8 (0.27) c	0.92 (0.003) b	0.292 (0.006) b	0.28 (0.022) a

Nitrogen stress index was significantly lower in the NS than C, but not different between the DS and C treatments (Table 2). Reflectance-based measurements are mostly affected by chlorophyll concentration [3,38]. Nitrogen stress can significantly reduce the chlorophyll concentration [4,34], while DS may have a relatively smaller effect on chlorophyll in plants [39]. This may be the reason for the observed R differences in the NS than the DS compared to C treatment in our experiment (Table 2). Furthermore, the result supports our hypothesis that the ratio of blue light reflectance to that of combined reflectance in the visible band can be used as an index for N stress. In addition, the index was specific to NS and was not affected by DS. Analyses of changes in R with time indicated no significant differences on any day between the DS and C treatments (Figure 3). However, a gradual decrease in R was observed with stress progression in the NS treatment. There were no differences in R on the 4th, 8th, 16th and 22nd day after imposing treatments, but the differences became gradually larger. By the 27th and 34th day of stress exposure, the decrease in R was large to significant in the NS compared to C. A significantly lower R was associated with a significantly lower A and a numerically lower ϕ_{PSII} in the NS compared to the C treatment (Table 2 and Figure 3). This may suggest that R measurements are related to photosynthetic pathways in plants.

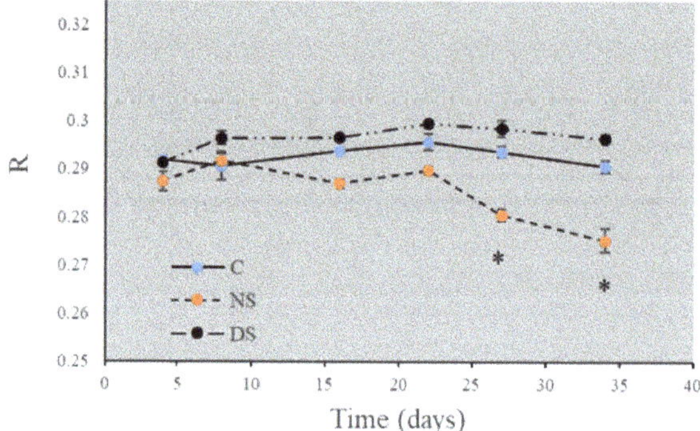

Figure 3. Reflectance index (R) of petunia plants measured on different days after exposure to optimal (C), nitrogen stress (NS) and drought stress (DS) treatments in the main experiment. The R-value was measured based on images of petunia plants captured inside a multi-spectral image station. A. * denotes statistical significance ($p \leq 0.05$) between NS and C treatments on a given day.

The path analysis tested the model where SDW was considered as a primary response affected by several lower-order responses including I_{abs}, A, CA, ϕ_{PSII} and R. When the data were pooled from the C, NS and DS treatments, path analyses indicated that R, ϕ_{PSII} and CA were exogenous (not affected by other variables) and A, I_{abs} and SDW were endogenous (affected by other variables) in nature (Figure 4). Furthermore, the model indicated that SDW was dependent on both A and CA and the reliability of the effects on SDW was high (error was 0.32; Figure 3). Both CA and A are known to influence biomass production in plants [40–42]. This supports that A and CA are secondary responses affecting SDW. Furthermore, CA affected I_{abs}. Light absorption is proportional to light interception by the canopy, which in turn is proportional to CA [43,44]. The model also indicated that ϕ_{PSII} directly affected A, which is expected. Interestingly, the model indicated that R inversely affected A. This supports our finding that R measurements are related to the photosynthetic pathway, however the effects observed in the model are opposite to those observed between R and A in the NS treatment. This could be because the model included data from the DS and C treatments in addition to the NS treatment. Furthermore, the effects of A on SDW and that of R and ϕ_{PSII} on A were not significant when NS data were removed from the model. This may suggest that NS effects on plants are primarily due to the reduction in A. In addition, the model supports that R is related to photosynthesis pathways. Both R and ϕ_{PSII} showed covariance with CA, but there were no causal relationships among them. Based on the model, R and ϕ_{PSII} can be considered as independent tertiary responses affecting A.

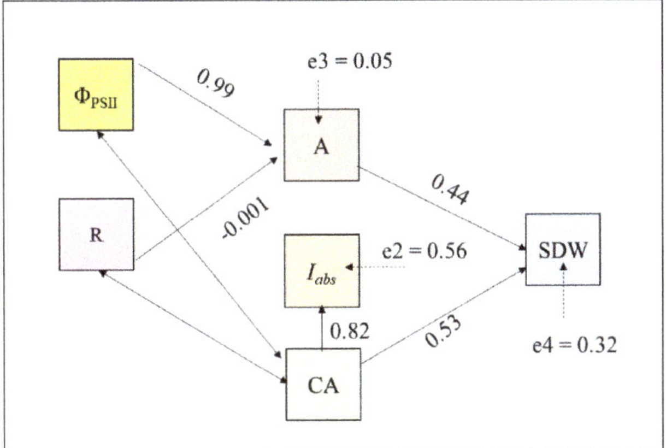

Figure 4. Path analyses of physiological measurements associated with shoot dry weight (SDW). Other variables include leaf photosynthesis (A), canopy area (CA), light absorption (I_{abs}), reflectance index (R) and quantum efficiency in light (Φ_{PSII}). Beta (or linear coefficient) and error values are shown for different effects. The model used terms that showed statistical significance ($p \leq 0.05$).

3.2. Product Development Experiment

Image analysis software effectively segmented strawberry plants from the background in the images captured by the smartphone (Figure 2). In spite of broadband wavelengths used in the smartphone method, R_{sp} of strawberry plants in the NS treatment was significantly lower than that of plants in the C treatment (Figure 5), similar to responses observed for petunia in the proof-of-concept experiment. This indicates that the broadband wavelengths used in the R_{sp} estimation were equally effective as narrowband wavelengths used in the proof-of-concept experiment.

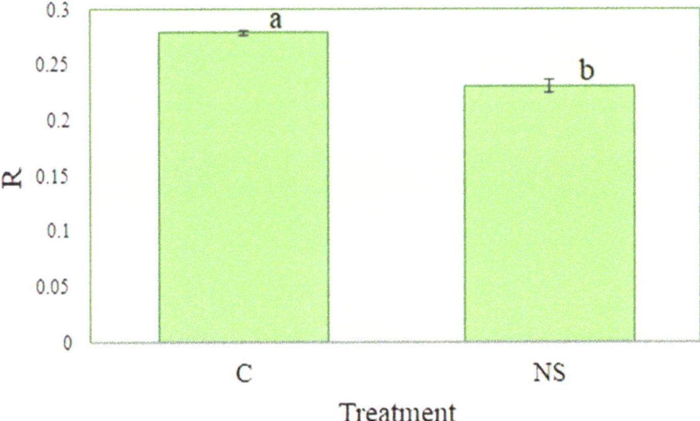

Figure 5. Nitrogen stress index (R) assessment using a smartphone. Strawberry plants were exposed to nitrogen stress (NS) and optimal (C) treatments. Letters 'a' and 'b' indicate that the means are statistically different. Error bars represent standard error of the mean.

The decrease in R or R_{sp} in the NS treatment was related to an increase in the gray value (or reflectance) of combined blue, green and red wavelengths as opposed to decreases in the gray value of blue wavelength in both petunia and strawberry (Table 3). This indicates that N stress effects were more pronounced on the reflectance of red and green wavelengths than the blue wavelength. As described before, carotenoids and xanthophylls in addition to chlorophyll can absorb blue light [20–22]. While chlorophyll synthesis is affected by N stress, the xanthophyll (a carotenoid) pool can increase in response to N stress in plants [24,25]. Thus, reflectance (or gray value) of blue wavelengths is relatively less affected than green or red wavelengths under N stress. In addition, decrease in chlorophyll can expose xanthophyll pigments, which are yellow in color [25]. As yellow is a combination of red and green colors, an increase in yellow color on the leaf surface may result in increased gray values for red and green channels. Given this, the decrease in R-value in the NS treatment is likely due to a loss of chlorophyll or increased xanthophyll absorption (appearance of yellow coloration on the leaf).

Table 3. Average gray values of blue and combined wavelengths in petunia (proof-of-concept experiment) and strawberry (product development experiment) under control (C) and nitrogen stress (NS) treatments. Treatment means followed by the same letter are not statistically different ($p \leq 0.05$). Values in parenthesis indicate standard error of mean.

Treatment	Gray Value (0 to 255 Scale)			
	Petunia		Strawberry	
	450 nm	(450 + 521 + 660) nm	blue	(blue + green + red)
C	74.2 (1.18) a	253.3 (3.87) b	77.4 (3.90) a	277.1 (12.26) b
NS	76.4 (1.22) a	267.6 (3.94) a	73.2 (1.68) a	317.0 (5.83) a

There was a linear relationship between N stress indices measured using smartphone and multi-spectral image stations in hydrangea (Figure 6). This indicates that the N stress index estimated using a smartphone is comparable to the values estimated using a multi-spectral image station. Interestingly, R_{sp} value changed approximately by 1.7-folds for one-fold change in R-value. Furthermore, statistical analysis (data not shown) indicated that R_{sp} values of hydrangea plants grown at the two highest N fertilizer treatments (30 and 21 g·pot^{-1}) were significantly higher than those in the two lowest N fertilizer treatments (3 and 9 g·pot^{-1}), whereas R-values trended lower in the two lowest N fertilizer treatments compared to those in the two higher N fertilizer treatments. This may suggest that R_{sp} is more sensitive than R in the detection of differences between the treatments. One possible reason for this could be due to the broadband wavelengths used in measuring R_{sp}. The difference can be larger when multiple wavelengths are included in the estimation of an index, especially if the effects are spread across the broadband.

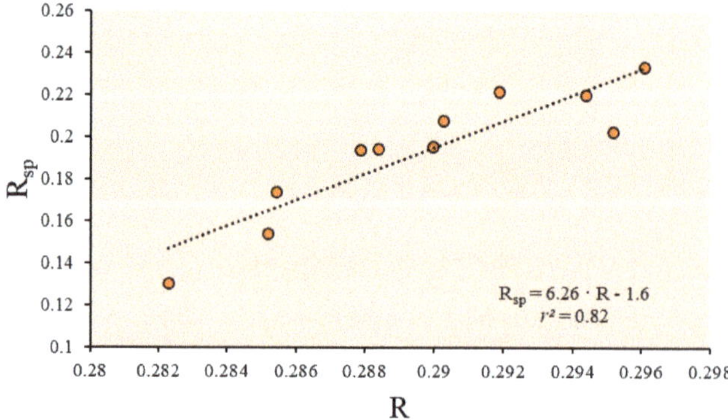

Figure 6. Linear relationship between N stress indices measured by smartphone (R_{sp}) and multi-spectral image station (R) in hydrangea.

4. Conclusions

In this study, we tested an index for N stress based on the images of plants. The index was calculated as the ratio of reflectance of blue relative to the reflectance of combined wavelengths in the visible band. The index value decreased when plants were exposed to NS relative to optimal conditions. Furthermore, the index value decreased gradually with increasing N stress in plants. Therefore, the continuous measurement of index can aid in the timely detection of N stress in plants. The index can be estimated using images captured by smartphones and image processing software loaded on network drives. The smartphone-based approach can be attractive to users in academia and industry. It is possible to make image-processing software available to users on a webserver. Using the network connectivity on smartphones, users can connect to the webserver, capture images using a smartphone, and process images on the webserver in real time to estimate N stress index.

Author Contributions: R.A. conducted experiments, analyzed data, and helped in the drafting of an early version of the manuscript; K.N. was responsible for overall project management, provided resources and helped in data interpretation and the final draft of the manuscript. All authors have read and agreed to the published version of the manuscript.

Funding: This research received funding from Fred Gloeckner Foundation, American Floral Endowment and Horticultural Research Institute.

Acknowledgments: We thank Jacob Brasseur for helping with capturing images of plants.

Conflicts of Interest: The authors declare no conflict of interest.

References

1. Xue, J.; Su, B. Significant remote sensing vegetation indices: A review of developments and applications. *J. Sens.* **2017**. [CrossRef]
2. Muñoz-Huerta, R.F.; Guevara-Gonzalez, R.G.; Contreras-Medina, L.M.; Torres-Pacheco, I.; Prado-Olivarez, J.; Ocampo-Velazquez, R.V. A review of methods for sensing the nitrogen status in plants: Advantages, disadvantages and recent advances. *Sensors* **2013**, *13*, 10823–10843. [CrossRef] [PubMed]
3. Sims, D.A.; Gamon, J.A. Estimation of vegetation water content and photosynthetic tissue area from spectral reflectance: A comparison of indices based on liquid water and chlorophyll absorption features. *Remote Sens. Environ.* **2003**, *84*, 526–537. [CrossRef]
4. Evans, J.R. Photosynthesis and nitrogen relationships in leaves of C 3 plants. *Oecologia* **1989**, *78*, 9–19. [CrossRef] [PubMed]
5. Xue, L.; Cao, W.; Luo, W.; Dai, T.; Zhu, Y. Monitoring leaf nitrogen status in rice with canopy spectral reflectance. *Agron. J.* **2004**, *96*, 135–142. [CrossRef]
6. Thomas, J.R.; Oerther, G.F. Estimating nitrogen content of sweet pepper leaves by reflectance measurements1. *Agron. J.* **1972**, *64*, 11–13. [CrossRef]
7. Meyer, G.E.; Troyer, W.W.; Fitzgerald, J.B.; Paparozzi, E.T. Leaf nitrogen analysis of poinsettia (*Euphorbia pulcherrima* Will D.) using spectral properties in natural and controlled lighting. *Appl. Eng. Agric.* **1992**, *8*, 715–722. [CrossRef]
8. Erdle, K.; Mistele, B.; Schmidhalter, U. Comparison of active and passive spectral sensors in discriminating biomass parameters and nitrogen status in wheat cultivars. *Field Crop. Res.* **2011**, *124*, 74–84. [CrossRef]
9. Tewari, V.K.; Arudra, A.K.; Kumar, S.P.; Pandey, V.; Chandel, N.S. Estimation of plant nitrogen content using digital image processing. *Agric. Eng. Int. CIGR J.* **2013**, *15*, 78–86.
10. Kim, Y.; Reid, J.F. Modeling and calibration of a multi-spectral imaging sensor for in-field crop nitrogen assessment. *Appl. Eng. Agric.* **2006**, *22*, 935–941. [CrossRef]
11. Noh, H.; Zhang, Q.; Shin, B.; Han, S.; Feng, L. A neural network model of maize crop nitrogen stress assessment for a multi-spectral imaging sensor. *Biosyst. Eng.* **2006**, *94*, 477–485. [CrossRef]
12. Leemans, V.; Marlier, G.; Destain, M.F.; Dumont, B.; Mercatoris, B. Estimation of leaf nitrogen concentration on winter wheat by multispectral imaging. In *Hyperspectral Imaging Sensors: Innovative Applications and Sensor Standards*; International Society for Optics and Photonics: Bellingham, WA, USA, 2017; Volume 10213, p. 102130I.

13. Peñuelas, J.; Filella, I. Visible and near-infrared reflectance techniques for diagnosing plant physiological status. *Trends Plant Sci.* **1998**, *3*, 151–156. [CrossRef]

14. Ma, B.L.; Morrison, M.J.; Dwyer, L.M. Canopy light reflectance and field greenness to access nitrogen fertilization and yield of maize. *Agric. J.* **1996**, *88*, 915–920.

15. Baret, F.; Fourty, T. Radiometric estimates of nitrogen status of leaves and canopies. In *Diagnosis of the Nitrogen Status in Crops*; Springer: Berlin/Heidelberg, Germany, 1997; pp. 201–227.

16. Zhu, Y.; Yao, X.; Tian, Y.; Liu, X.; Cao, W. Analysis of common canopy vegetation indices for indicating leaf nitrogen accumulations in wheat and rice. *Int. J. Appl. Earth Obs. Geoinf.* **2008**, *10*, 1–10. [CrossRef]

17. Tremblay, N.; Wang, Z.; Ma, B.L.; Belec, C.; Vigneault, P.A. Comparison of crop data measured by two commercial sensors for variable-rate nitrogen application. *Precis. Agric.* **2009**, *10*, 145. [CrossRef]

18. Corti, M.; Gallina, P.M.; Cavalli, D.; Cabassi, G. Hyperspectral imaging of spinach canopy under combined water and nitrogen stress to estimate biomass, water, and nitrogen content. *Biosyst. Eng.* **2017**, *158*, 38–50. [CrossRef]

19. Adhikari, R.; Li, C.; Kalbaugh, K.; Nemali, K. A low-cost smartphone- controlled sensor based on image analysis for estimating whole-plant tissue nitrogen (N) content in floriculture crops. *Comput. Electron. Agric.* **2020**, *169*, 105173. [CrossRef]

20. Sims, D.A.; Gamon, J.A. Relationships between leaf pigment content and spectral reflectance across a wide range of species, leaf structures and developmental stages. *Remote Sens. Environ.* **2002**, *81*, 337–354. [CrossRef]

21. Croft, H.; Chen, J.M. Leaf pigment content. In *Reference Module in Earth Systems and Environmental Sciences*; Elsevier Inc.: Oxford, UK, 2017; pp. 1–22.

22. Bartley, G.E.; Scolnik, P.A. Plant carotenoids: Pigments for photoprotection, visual attraction, and human health. *Plant Cell* **1995**, *7*, 1027.

23. Gates, D.M.; Keegan, H.J.; Schleter, J.C.; Weidner, V.R. Spectral properties of plants. *Appl. Opt.* **1965**, *4*, 11–20. [CrossRef]

24. Tóth, V.R.; Mészáros, I.; Veres, S.; Nagy, J. Effects of the available nitrogen on the photosynthetic activity and xanthophyll cycle pool of maize in field. *J. Plant Physiol.* **2002**, *159*, 627–634. [CrossRef]

25. Verhoeven, A.S.; Demmig-Adams, B.; Adams, W.W., III. Enhanced employment of the xanthophyll cycle and thermal energy dissipation in spinach exposed to high light and N stress. *Plant Physiol.* **1997**, *113*, 817–824. [CrossRef]

26. Nemali, K.; van Iersel, M.W. Light intensity and fertilizer concentration: II. Optimal fertilizer solution concentration for species differing in light requirement and growth rate. *HortScience* **2004**, *39*, 1293–1297. [CrossRef]

27. Olascoaga, B.; Mac Arthur, A.; Atherton, J.; Porcar-Castell, A. A comparison of methods to estimate photosynthetic light absorption in leaves with contrasting morphology. *Tree Physiol.* **2016**, *36*, 368–379. [CrossRef] [PubMed]

28. Long, S.P.; Bernacchi, C.J. Gas exchange measurements, what can they tell us about the underlying limitations to photosynthesis? Procedures and sources of error. *J. Exp. Bot.* **2003**, *54*, 2393–2401. [CrossRef]

29. Li, C.; Adhikari, R.; Yao, Y.; Miller, A.G.; Kalbaugha, K.; Li, D.; Nemali, K. Measuring plant growth characteristics using smartphone based image analysis technique in controlled environment agriculture. *Comput. Electron. Agric.* **2020**. [CrossRef]

30. Nemali, K.S.; van Iersel, M.W. An automated system for controlling drought stress and irrigation in potted plants. *Sci. Hortic.* **2006**, *110*, 292–297. [CrossRef]

31. Adhikari, R.; Nemali, K. *Substrate versus Fertilizer-based Electrical Conductivity Measurements*; HO-322-W; The Education Store; Purdue Extension; Purdue University: West Lafayette, IN, USA, 2020.

32. Haehnel, W. Photosynthetic electron transport in higher plants. *Ann. Rev. Plant Physiol.* **1984**, *35*, 659–693. [CrossRef]

33. Osborne, B.A.; Raven, J.A. Light absorption by plants and its implications for photosynthesis. *Biol. Rev.* **1986**, *61*, 1–6. [CrossRef]

34. Ding, L.; Wang, K.J.; Jiang, G.M.; Biswas, D.K.; Xu, H.; Li, L.F.; Li, Y.H. Effects of nitrogen deficiency on photosynthetic traits of maize hybrids released in different years. *Ann. Bot.* **2005**, *96*, 925–930. [CrossRef]

35. Evans, J.R.; Terashima, I. Effects of nitrogen nutrition on electron transport components and photosynthesis in spinach. *Funct. Plant Biol.* **1987**, *14*, 59–68. [CrossRef]

36. Fredeen, A.L.; Gamon, J.A.; Field, C.B. Responses of photosynthesis and carbohydrate-partitioning to limitations in nitrogen and water availability in field-grown sunflower. *Plant Cell Environ.* **1991**, *14*, 963–970. [CrossRef]
37. Feng, W.; He, L.; Zhang, H.Y.; Guo, B.B.; Zhu, Y.J.; Wang, C.Y.; Guo, T.C. Assessment of plant nitrogen status using chlorophyll fluorescence parameters of the upper leaves in winter wheat. *Eur. J. Agron.* **2015**, *64*, 78–87. [CrossRef]
38. Buschmann, C.; Nagel, E. In vivo spectroscopy and internal optics of leaves as basis for remote sensing of vegetation. *Int. J. Remote Sens.* **1993**, *14*, 711–722. [CrossRef]
39. Mohanty, P.; Boyer, J.S. Chloroplast Response to Low Leaf Water Potentials: IV. Quantum Yield Is Reduced. *Plant Physiol.* **1976**, *57*, 704–709. [CrossRef]
40. Beadle, C.L.; Long, S.P. Photosynthesis—Is it limiting to biomass production? *Biomass* **1985**, *8*, 119–168. [CrossRef]
41. Marcelis, L.F.; Heuvelink, E.; Goudriaan, J. Modelling biomass production and yield of horticultural crops: A review. *Sci. Hortic.* **1998**, *74*, 83–111. [CrossRef]
42. Peng, S.; Krieg, D.R.; Girma, F.S. Leaf photosynthetic rate is correlated with biomass and grain production in grain sorghum lines. *Photosynth. Res.* **1991**, *28*, 1–7. [CrossRef]
43. Wilson, J.W. Analysis of light interception by single plants. *Ann. Bot.* **1981**, *48*, 501–505. [CrossRef]
44. Rosati, A.; Badeck, F.W.; Dejong, T.M. Estimating canopy light interception and absorption using leaf mass per unit leaf area in *Solanum melongena*. *Ann. Bot.* **2001**, *88*, 101–109. [CrossRef]

Publisher's Note: MDPI stays neutral with regard to jurisdictional claims in published maps and institutional affiliations.

 © 2020 by the authors. Licensee MDPI, Basel, Switzerland. This article is an open access article distributed under the terms and conditions of the Creative Commons Attribution (CC BY) license (http://creativecommons.org/licenses/by/4.0/).

 *horticulturae*

Article

Cucurbita Rootstocks Improve Salt Tolerance of Melon Scions by Inducing Physiological, Biochemical and Nutritional Responses

Abdullah Ulas [1,*], Alim Aydin [2], Firdes Ulas [3], Halit Yetisir [3] and Tanveer Fatima Miano [3,4]

[1] Department of Soil Science and Plant Nutrition, Faculty of Agriculture, Erciyes University, 38280 Kayseri, Turkey

[2] Department of Horticulture, Faculty of Agriculture, Ahi Evran University, 40100 Kirsehir, Turkey; alim.aydin@ahievran.edu.tr

[3] Department of Horticulture, Faculty of Agriculture, Erciyes University, 38280 Kayseri, Turkey; fulas@erciyes.edu.tr (F.U.); yetisir1@erciyes.edu.tr (H.Y.); tfmiano@sau.edu.pk (T.F.M.)

[4] Department of Horticulture, Faculty of Crop Production, Sindh Agriculture University, Tandojam 70060, Pakistan

* Correspondence: agrulas@erciyes.edu.tr; Tel.: +90-352-437-17-90

Received: 15 September 2020; Accepted: 12 October 2020; Published: 14 October 2020

Abstract: A hydroponic experiment was conducted to assess whether grafting with *Cucurbita* rootstocks could improve the salt tolerance of melon scions and to determine the physiological, biochemical, and nutritional responses induced by the rootstocks under salt stress. Two melon (*Cucumis melo* L.) cultivars (Citirex and Altinbas) were grafted onto two commercial *Cucurbita* rootstocks (Kardosa and Nun9075). Plants were grown in aerated nutrient solution under deep water culture (DWC) at two electrical conductivity (EC) levels (control at 1.5 dS m^{-1} and salt at 8.0 dS m^{-1}). Hydroponic salt stress led to a significant reduction in shoot and root growths, leaf area, photosynthetic activity, and leaf chlorophyll and carotenoid contents of both grafted and nongrafted melons. Susceptible plants responded to salt stress by increasing leaf proline and malondialdehyde (MDA), ion leakage, and leaf Na$^+$ and Cl$^-$ contents. Statistically significant negative correlations existed between shoot dry biomass production and leaf proline (r = −0.89), leaf MDA (r = −0.85), leaf Na$^+$ (r = −0.90), and leaf (r = 0.63) and root (r = −0.90) ion leakages under salt stress. Nongrafted Citirex tended to be more sensitive to salt stress than Altinbas. The *Cucurbita* rootstocks (Nun9075 and Kardosa) significantly improved growth and biomass production of grafted melons (scions) by inducing physiological (high leaf area and photosynthesis), biochemical (low leaf proline and MDA), and nutritional (low leaf Na$^+$ and ion leakage and high K$^+$ and Ca^{++} contents) responses under salt stress. The highest growth performance was exhibited by the Citirex/Nun9075 and Citirex/Kardosa graft combinations. Both *Cucurbita* cultivars have high rootstock potential for melon, and their significant contributions to salt tolerance were closely associated with inducing physiological and biochemical responses of scions. These traits could be useful for the selection and breeding of salt-tolerant rootstocks for sustainable agriculture in the future.

Keywords: photosynthesis; chlorophyll; proline; ion leakage; susceptibility

---

1. Introduction

Salinity is the one of the major environmental stress factors limiting crop growth and productivity in many arid and semiarid regions, in spite of the advanced management techniques developed in recent decades [1]. Worldwide, up to 20% of arable land and up to 50% in irrigated areas is detrimentally affected by salinity, while in Turkey almost 4 million hectares of land has salinity problems [2]. As long as the current situation in salinization remains, half of the presently cultivated agricultural land may be lost by 2050 [3]. Crops that are grown under excessively saline conditions usually exhibit shorter life cycles or limited plant growth and biomass yield [4]. Internal damages and metabolic disturbances [5], ion toxicity [6], water deficiency in older leaves and carbohydrate deficiency in younger leaves [7], and reductions in root growth, nutrient uptake [8], photosynthetic activity [9,10], and protein synthesis are some of the major problems exhibited by crops grown under salt stress conditions.

As a horticultural crop, melon (*Cucumis melo* L.) has economic significance in the world due to its intensive and wide cultivation particularly in arid and semiarid regions. Global melon production was almost 31.6 million tons (Mt) in 2018 [11], and the main producing countries were China (16 Mt), Turkey (1.8 Mt), Iran (1.6 Mt), and Egypt (1.06 Mt). As melon is an arid and semiarid region crop, several studies have focused on the salt stress problems of melon and have determined that melon is a salt-sensitive or moderately tolerant crop in terms of yield and fruit quality characteristics [12,13]. To improve the salt tolerance of melon for sustainable agriculture production, integrated management strategies that take into consideration improved soil and crop management practices are necessary. Moreover, another way to avoid or reduce salt stress impacts and hinder yield losses in melon production affected by salt stress in high-yielding susceptible cultivars (as scions) would be to graft them onto resistant genotypes (as rootstocks) capable of improving the salt tolerance of the scions. Some studies [7–9] have revealed that *Cucurbita* genotypes exhibit salt tolerance and may therefore be used as rootstocks to improve the growth and yield of some horticultural crops (i.e., cucumber and melon) under salt stress. Grafting onto suitable rootstocks is an important technique in the horticultural area for the suitable cultivation of some Cucurbitaceae and Solanaceae species in Japan, Korea, China, and some other Asian and European countries [14]. Previously, other studies [15–18] were carried out to determine the contribution of grafting to several abiotic stress tolerance mechanisms of many plant species. However, no comprehensive hydroponic studies were found in the literature with regard to the salinity problem of melon plants. Therefore, the aim of the present study was to evaluate whether grafting with hybrid *Cucurbita maxima* × *Cucurbita moschata* rootstocks could improve the salt tolerance of melon scions and to determine the physiological, biochemical, and nutritional responses induced by *Cucurbita* rootstocks under hydroponic salt stress.

2. Materials and Methods

2.1. Plant Material, Treatments, and Experimental Design

A hydroponic trial was set up using an aerated deep water culture (DWC) technique in a fully automated climate room in the Plant Physiology Laboratory of Erciyes University's Faculty of Agriculture, Department of Soil Science and Plant Nutrition, in Kayseri, Turkey. For the vegetation period, the room temperatures were maintained at 25/22 °C (day/night) with a relative humidity of 65–70%. The supplied photon flux in the growth chamber was almost 350 µmol m^{-2} S^{-1} with an intensity of 16/8 h (light/dark) photoperiod. As plant materials, two melon cultivars [Galia type (Citirex F1) and standard type (Kirkagac Manisa Altinbas)] were used as scions, while two commercial *Cucurbita* hybrid (*Cucurbita maxima* × *C. moschata*) cultivars (Kardosa and Nun9075) were used as rootstocks. Maintaining homogeneity among the germinated seedlings is very crucial in a hydroponic study. Therefore, melon seeds were sown 1 week earlier than rapidly growing *Cucurbita* hybrid rootstocks' seeds in multipots containing a mixture of peat (pH: 6.0–6.5) and perlite in a 2:1 (*v/v*) ratio for 2 weeks. When the seedlings reached the stage of three or four true leaves, the melon scions were grafted by using the cleft grafting technique onto the *Cucurbita* rootstocks. As control plants, the nongrafted melon varieties were used. For the healing and acclimatization process, the grafted plants were transferred to double-layered and shaded plastic growth boxes and placed in the growth chamber for 7 days. When the healing and acclimatization process was completed, the grafted and nongrafted control plants were removed from the growth medium of the multipots. The roots were washed without root damage, and the stem of each seedling was carefully covered with a thin sponge. After that, each seedling was placed onto the cover of 8 L plastic pots filled with nutrient solution (modified

Hoagland) in the fully automated climate chamber. The sufficiently dissolved oxygen (8.0 mg/L) in nutrient solution was supplied by using a continuously working air pump.

The trial was set up in a completely randomized block design (RBD) with four replicated blocks and two plants of each ungrafted cultivar and cultivar by rootstock combination in each block treated with one of two different electrical conductivity (EC) levels (control at 1.5 dS m^{-1} and salt at 8.0 dS m^{-1}). The salt stress was created by adding NaCl in nutrient solution. The salt application was done gradually in an increasing manner (2 dS/m per day) 5 days after transplanting. The total growth period of the plants from transplant into 8 L plastic pots to final harvest was 42 days after treatment (DAT). To prepare the nutrient solution for the hydroponic experiment, analytical grade (99% pure) chemicals with distilled water were used according to the Hoagland (modified) formulation. In the solution, 2000 µM nitrogen was supplied by using 75% calcium nitrate (Ca(NO$_3$)$_2$) and 25% ammonium sulfate ((NH$_4$)$_2$SO$_4$) as the N sources. Moreover, the composition of the basic nutrient solution was as follows (µM): CaSO$_4$ (1000), K$_2$SO$_4$ (500), MgSO$_4$ (325), KH$_2$PO$_4$ (250), NaCl (50), H$_3$BO$_3$ (8.0), Fe-EDDHA (80), ZnSO$_4$ (0.4), CuSO$_4$ (0.4), MnSO$_4$ (0.4), MoNa$_2$O$_4$ (0.4). All the nutrients were replaced to prior concentrations when the N concentration in the solution fell from 2.0 mM to below 1.0 mM. Daily nitrogen concentration was checked by nitrate test strips (Merck, Darmstadt, Germany) with the aid of a NitracheckTM reflectometer. Distilled water was added every 2 days to replenish the water lost to evaporation, and the solution was changed weekly.

2.2. Harvest, Shoot, and Root Dry Weight Measurements

At the final harvest, the plants were separated into leaves, stems, and roots. To determine the dry biomass, plant tissues were dried in a forced-air oven at 70 °C for 72 h. They were then weighed on an electronic digital scale. The sum of aerial vegetative plant parts (leaves + stems) is equal to total shoot biomass. To calculate the shoot-to-root ratio, the sum of leaf and stem dry weights was divided by the total root dry weight.

2.3. Leaf Area and Photosynthetic Activity Measurements

Prior to the harvest, nondestructive measurements of the leaf-level CO$_2$ gas exchange (µmol CO$_2$ m^{-2} s^{-1}) were performed using a portable photosynthesis system (LI-6400XT; LI-COR Inc., Lincoln, NE, USA). The leaf net photosynthesis measurement (photosynthetically active radiation (PAR) = 1000 µmol m^{-2} s^{-1}, CO$_2$ at 400 µmol mol^{-1}) was performed on the youngest fully expanded leaves, using four replicate leaves per treatment in the third and fifth weeks of the growth period. Leaf area of the plants was measured destructively during the harvesting process by using a portable leaf-area meter (LI-3100, LI-COR. Inc., Lincoln, NE, USA). Total leaf area was recorded as cm^2.

2.4. Leaf Total Chlorophyll and Carotenoid Content Measurements

A day before harvesting, 100 mg of fresh leaf samples from each replication of the two treatments was taken to measure the leaf total chlorophyll and carotenoid contents using UV–VIS spectroscopy. The samples were put into 15 mL capped containers where 10 mL of 95% (*v*/*v*) ethanol was added. Afterward, to allow for the extraction of the leaf pigments, the samples were held overnight in darkness at room temperature. Measurements were done using a spectrometer (UV/VS T80+, PG Instruments Limited, UK) at wavelengths of 470, 648.6, and 664.2 nm. Total chlorophyll (a-Total-Chlo) and total carotenoids (b-TC) were estimated from the spectrometric readings using the formulae described by Lichtenthaler [19]:

(a) Total-Chlo (mg/g plant sample) = [5.24 WL664.2 − 22.24 WL648.6 × 8.1]/ weight plant sample (g)

(b) TC (mg/g plant sample) = [(4.785 WL470 + 3.657 WL664.2) − 12.76 WL648.6 × 8.1]/ weight plant sample (g).

(Note: WL470, WL648.6, and WL664.2 refer to spectrometric readings at wavelengths 470, 648.6, and 664.2 nm, respectively).

2.5. Proline Contents and Lipid Peroxidation Measurements

The proline contents were measured according to the method described by Bates et al. [20]. To homogenize the plant material, 3% aqueous sulfosalicylic acid was used. After centrifugation of the homogenate mixture at 10,000 rpm, the proline contents were determined in supernatant. To prepare the reaction mixture, 2 mL of acid ninhydrin and 2 mL of glacial acetic acid were used and then boiled at 100 °C for 1 h. Afterwards, the reaction was terminated in an ice bath. For the extraction of the reaction mixture, 4 mL of toluene was used, and then the absorbance was read at 520 nm. Membrane lipid peroxidation was characterized by the main product of lipid peroxidation, the malondialdehyde (MDA) concentration, which was determined according to the method described by Lutts et al. [21].

2.6. Leaf and Root Electrolyte Leakage Measurements

Electrolyte leakage (EL) in leaves and roots was measured according to the method described by Lutts et al. [22]. The youngest fully expanded leaves were used for the EL measurements in between 1100 and 1500 h every 48 h with three replications per treatment. Leaf disks (1 cm^2) were excised from young fully expanded leaves using a cork borer. To clean leaf surface contamination, samples were washed three times with distilled water. Afterwards, the samples were placed in individual stoppered vials containing 10 mL of distilled water.

EL determination in plant roots was done by taking fresh root tips (2 cm in length) from each treatment at the final harvest. The root samples containing 10 mL of distilled water were placed on a shaker (100 rpm) for 24 h at room temperature (25 °C) for incubation. After incubation, the first electrical conductivity (EC1) reading in the solution was performed. After a while, the same samples were placed in an autoclave at 120 °C for 20 min. After termination of the autoclave process, the samples were left at room temperature for cooling, and then the second electrical conductivity (EC2) reading was performed in the solution. The EL was expressed as EL = (EC1/EC2) × 100.

2.7. Mineral Analysis Measurements

To determine mineral element composition, 0.5 g dried leaf tissues were used. Potassium (K^+), calcium (Ca^{++}), and sodium (Na^+) contents were measured by dry ashing at 400 °C for 4.5 h. After that, the ash samples were dissolved in 5 mL of 20% (*v/v*) HCl, which was then filtered. The filtered solutions were then diluted with distilled water to a volume 50 mL. An amount of 10 mL was used for inductively coupled plasma atomic emission spectroscopy (ICP-AES) analysis. The ICP-AES results were converted into percentages (%) and parts per million (ppm). Chloride (Cl^-) was determined by precipitation as AgCl and titration according to the method described by Johnson and Ulrich [23].

2.8. Statistical Analysis

Statistical analysis of the data was performed using the PROC GLM procedure of the SAS Statistical Software (SAS for Windows 9.1, SAS Institute Inc., Cary, NC, USA). A two-factor analysis of variance was performed to study the effects of genotype or grafting combination and salt and their interactions on the variables analyzed. The levels of significance are represented at $p < 0.05$ (*), $p < 0.01$ (**), $p < 0.001$ (***), or n.s. as not significant (*F*-test and Pearson correlation coefficients). Differences between the treatments were analyzed using Duncan's multiple range test ($p < 0.05$).

3. Results

3.1. Results and Discussion

3.1.1. Changes in Shoot and Root Biomass Productions and Partitioning

The results indicated that shoot and root dry matter and the shoot-to-root ratio of melon plants were affected significantly ($p < 0.001$) by salt, graft combination, and salt × graft combination interaction (Table 1). Irrespective of the graft combinations, shoot and root growths were affected detrimentally by hydroponic salt stress, and thus significant reductions were found in shoot (49.9%) and root (17.6%) dry matter and shoot-to-root ratio (45.8%) of melon plants under salt stress as compared with the control conditions. It is well-known that crop growth decreased with rising salinity level. Corroborative results were demonstrated in several studies conducted with melon [4,9], watermelon [24,25], cucumber [8], tomato [26], eggplant [27], and pepper [10] under salt stress. Our results clearly indicated that grafting with the *C. maxima* × *C. moschata* hybrid rootstocks had pronounced positive effects on the improvement of growth of melon scions under control and particularly salt stress conditions.

Table 1. Shoot and root dry weight and shoot-to-root ratio of melon graft combinations under control (1.5 dS m^{-1}) and salt stress (8.0 dS m^{-1}) conditions.

Graft Combination	Shoot Dry Weight (g plant^{-1})		Root Dry Weight (g plant^{-1})		Shoot-to-Root Ratio (g g^{-1})	
(Scion/Rootstock)	Control	Salt	Control	Salt	Control	Salt
Altinbas	14.67 e^z	6.03 f	4.24 a	2.91 cd	3.52 ef	2.05 f
Altinbas/Nun9075	28.02 ab	13.82 e	3.02 bc	3.22 b	9.41 b	4.32 de
Altinbas/Kardosa	25.83 b	15.51 de	2.71 cd	2.81 cd	9.52 b	5.52 cd
Citirex	20.92 c	4.53 f	3.63 ab	1.23 f	5.70 cd	3.83 ef
Citirex/Nun9075	29.04 a	17.51 d	2.31 de	3.20 b	12.81 a	5.52 cd
Citirex/Kardosa	28.12 ab	16.14 de	2.11 e	2.41 de	13.43 a	6.72 c
F-test						
Graft combination	***		***		***	
Salt	***		***		***	
Graft comb. × salt	***		***		***	

^{z} Values denoted by different letters are significantly different between graft combinations within columns at $p < 0.05$. Significance of main and interaction effects *F* values: $p < 0.05$ (*), $p < 0.01$ (**), and $p < 0.001$ (***).

Significant differences were found between the two melon cultivars and their graft combinations. Nongrafted Citirex showed significantly higher shoot dry matter than nongrafted Altinbas under control conditions, whereas both melon cultivars did not differ significantly in shoot dry matter under salt stress (Table 1). However, shoot dry matter reductions of Citirex (78.5% decline) tended to be more than those of Altinbas (59.2% decline) under salt stress. This might be due to root morphological differences between the two melon cultivars. Nongrafted Altinbas showed a significantly higher root dry matter than Citirex under salt stress (Table 1). Furthermore, Altinbas exhibited similar root dry matter as Nun9075 and Kardosa rootstocks under salt stress. These indicated that Altinbas has a vigorous root system compared with Citirex. The shoot dry matter of Citirex was increased by 288.8% in Citirex/Nun9075 and 257.7% in Citirex/Kardosa graft combinations, whereas the increase in the shoot dry matter of Altinbas was 130.1% in Altinbas/Nun9075 and 158.3% in Altinbas/Kardosa graft combinations under salt stress.

This was also shown by the significantly higher shoot-to-root ratios of Citirex/Nun9075 and Citirex/Kardosa graft combinations under salt stress. The graft combination Altinbas/Nun9075 and nongrafted Altinbas showed significantly lower shoot-to-root ratios in control and salt stress conditions. All the results clearly indicated that grafting with the *Cucurbita maxima* × *C. moschata* hybrid rootstocks significantly improved the salt tolerance of both melon (scions) cultivars. However, the contribution of both rootstocks to salt tolerance was much higher for Citirex (high sensitivity) than for Altinbas (less sensitivity). Grafted plants usually have strong and vigorous root systems [24], and thus, improved crop growth performance of grafted melons might be the result of more water and nutrient uptake that caused an increase in leaf area and photosynthetic activity of leaves under salt stress.

3.1.2. Changes in Leaf Area, Photosynthesis, Chlorophyll, and Carotenoid Contents

The results indicated that the leaf area, photosynthetic activity of leaves, total chlorophyll content, and carotenoid content of melon plants were affected significantly by salt and graft combination

(Table 2). An interaction between salt and graft combination was found only in the total leaf area and carotenoid content. Irrespective of the graft combinations, similar shoot and root biomass reductions under salt stress led to a significant decline in leaf area formation (49.1%), photosynthesis (9.1%), chlorophyll content (14.8%), and carotenoid content (20.4%) of melon plants. This also explains why the shoot and root dry biomass productions (Table 1) of both melon cultivars and their graft combinations were detrimentally affected by hydroponic salt stress, since crop biomass production and yield is strongly dependent on leaf area formation and leaf photosynthetic activity [28]. Our results also correspond to those from the study of Colla et al. [24], who found that salinity decreased the photosynthesis of grafted and nongrafted watermelon plants grown in a hydroponic system. Similar results were also demonstrated with grafted and nongrafted pepper plants under saline conditions [10].

Table 2. Leaf area, photosynthesis, chlorophyll content (a+b), and carotenoid content of melon graft combinations under control (1.5 dS m^{-1}) and salt stress (8.0 dS m^{-1}) conditions.

Graft Combination	Leaf Area (cm^2 plant^{-1})		Photosynthesis (μmol m^{-2} s^{-1})		Chlorophyll (a + b) (mg gr^{-1})		Carotenoid (mg gr^{-1})	
(Scion/Rootstock)	Control	Salt	Control	Salt	Control	Salt	Control	Salt
Altinbas	3843 d^z	1767 g	43.4 c	39.2 d	1.72 bc	1.47 d	0.25 ef	0.22 g
Altinbas/Nun9075	4418 c	2459 f	46.1 ab	43.9 bc	1.87 ab	1.59 cd	0.32 ab	0.24 efg
Altinbas/Kardosa	4342 c	2634 f	46.1 ab	42.5 c	2.03 a	1.54 cd	0.35 a	0.23 efg
Citirex	4450 c	1117 h	44.8 bc	38.0 d	1.68 cd	1.49 d	0.26 de	0.22 fg
Citirex/Nun9075	5583 a	3312 e	48.0 a	47.1 ab	1.85 ab	1.62 cd	0.30 bc	0.26 ef
Citirex/Kardosa	4954 b	2749 f	46.1 ab	42.6 c	1.74 bc	1.62 cd	0.29 cd	0.24 efg
F-test								
Graft combination	***		***		*		***	
Salt	***		***		***		***	
Graft comb. × salt	***		n.s.		n.s.		***	

^{z} Values denoted by different letters are significantly different between graft combinations within columns at $p < 0.05$. Significance of main and interaction effects *F* values: $p < 0.05$ (*), $p < 0.01$ (**), and $p < 0.001$ (***), with n.s. meaning not significant.

That study showed that the photosynthetic activity of pepper leaves decreased as a result of the reduction in chlorophyll and carotenoid contents as salinity level increased in nutrient solution. Similar to our study, a substantial decline in the chlorophyll content of leaves was reported for several horticultural species, such as melon [5] and tomato [26], under salt stress conditions. Furthermore, significant variations existed between grafted and nongrafted melon plants regarding measured parameters at control and salt stress conditions (Table 2). The grafted melons produced 16.3% and 93.43% higher leaf area than the nongrafted melons under control and salt stress conditions, respectively. This clearly indicated that the rootstock contributions to leaf area development of scions (melon) were substantially higher under salt stress than under control conditions. As a result, the reduction in the total leaf area of nongrafted melons was 65.2%, whereas the reduction in grafted melons was only 42.2%. The grafted melons showed 11.7% higher photosynthetic activity than the nongrafted ones under salt stress. This might be due to higher chlorophyll (7.4%) and carotenoid contents (9.4%) of the grafted melons as compared with the nongrafted ones under salt stress. Our results corroborated those of a study that showed that the leaf area of nongrafted watermelon (cv. Tex) was significantly improved when it was grafted onto two commercial rootstocks, Macis [*Lagenaria siceraria* (Mol.) Standl.] and Ercole (*Cucurbita maxima* Duchesne × *Cucurbita moschata* Duchesne), under salt stress conditions [20].

In our study, significant genotypic variation existed regarding leaf area formation between the two nongrafted melon cultivars under salt stress. Under control conditions, Altinbas and Citirex had similar leaf areas, whereas significantly higher total leaf area was exhibited by Altinbas than Citirex under salt stress. This clearly indicated a significant genotype × salt interaction. The Altinbas total leaf area was reduced by 54.1% under salt stress, whereas the reduction in total leaf area of Citirex was 74.9%. As shown by shoot and root dry matter productions (Table 1), Altinbas can be characterized as a salt-tolerant cultivar due to maintaining high leaf area under salt stress as compared with the salt-sensitive cultivar Citirex (Table 2). On the other hand, the salt-sensitive cultivar Citirex exhibited significantly higher leaf area formation, photosynthetic activity, leaf chlorophyll content, and carotenoid content when it was grafted on Nun9075 and Kardosa rootstocks under salt stress. Although similar rootstock contributions to leaf area formation, photosynthesis, total leaf chlorophyll content, and carotenoid content were recorded with Altinbas/Nun9075 and Altinbas/Kardosa graft combinations under salt stress, the increases were lower than those in graft combinations with Citirex. These results clearly indicated that the two melon cultivars had contrasting salt tolerances (Citirex: sensitive, Altinbas: tolerant) and therefore responded significantly differently when they were grafted with both tolerant rootstocks.

3.1.3. Changes in Proline, Lipid Peroxidation, and Root and Leaf Ion Leakages

The proline content (Figure 1A), lipid peroxidation (MDA) (Figure 1B), and ion leakages in roots (Figure 1C) and leaves (Figure 1D) of melon plants were affected significantly ($p < 0.001$) by salt, graft combination, and salt × graft combination interaction. Regardless of the graft combination, salt stress led to a significant increase in proline (59.1%) and MDA (31.3%) contents and leaf (56.8%) and root (36.7%) ion leakages of salt-treated melons as compared with controls (Figure 1A–D). These are common responses of plants that usually exhibit tolerance strategies as shown in studies with melon [29], cucumber [30], pepper [10,31], and tomato [32]. However, there were significant differences between grafted and nongrafted melons regarding biochemical responses under both control and salt stress conditions (Figure 1A–D). Irrespective of the cultivars, grafted melons produced 24.4%, 2.9%, 0.53%, and 9.1% lower proline, MDA, leaf ion leakage, and root ion leakage, respectively, than nongrafted melon plants under control conditions. Similar contributions of rootstocks to the biochemical responses of melon plants were also observed under salt stress. However, the plants responded much more under salt stress, such that grafted melons produced 27.6%, 29.6%, 13.5%, and 13.2% lower proline, MDA, leaf ion leakage, and root ion leakage, respectively, than nongrafted melon plants. Our results clearly indicated that grafting with the *Cucurbita maxima* × *C. moschata* rootstocks had pronounced contributions to the biochemical responses of the scions (melon) under both control and salt stress conditions. Similar results were observed when the experiment was conducted using different Iranian melon landraces [29].

Nongrafted Altinbas showed significantly higher proline, MDA, and leaf ion leakage than Citirex under control conditions, whereas the root ion leakage of Altinbas was significantly lower than that of Citirex. Without salt stress, significantly lower root ion leakage could be the result of the vigorous root system of Altinbas (Table 1), which leads to its characterization as salt tolerant. However, under salt stress, opposite results were found between the two melon cultivars. Citirex showed significantly higher proline, slightly higher MDA, and significantly higher root ion leakage than Altinbas. This might be due to the sensitivity of the response of Citirex to salt stress. This was confirmed by the results, which revealed that Citirex had increased proline, MDA, leaf ion leakage, and root ion leakage by 231.6%, 80.3%, 91.6%, and 32.7%, respectively, whereas the increase in proline, MDA, leaf ion leakage, and root ion leakage of Altinbas was 97.3%, 43.6%, 58.2%, and 50.1%, respectively, under salt stress as compared with control conditions. Similarly, greater responses were exhibited by Citirex in shoot and root growth (Table 1), leaf area formation, photosynthesis, and photosynthetic pigment contents (Table 2) under salt stress.

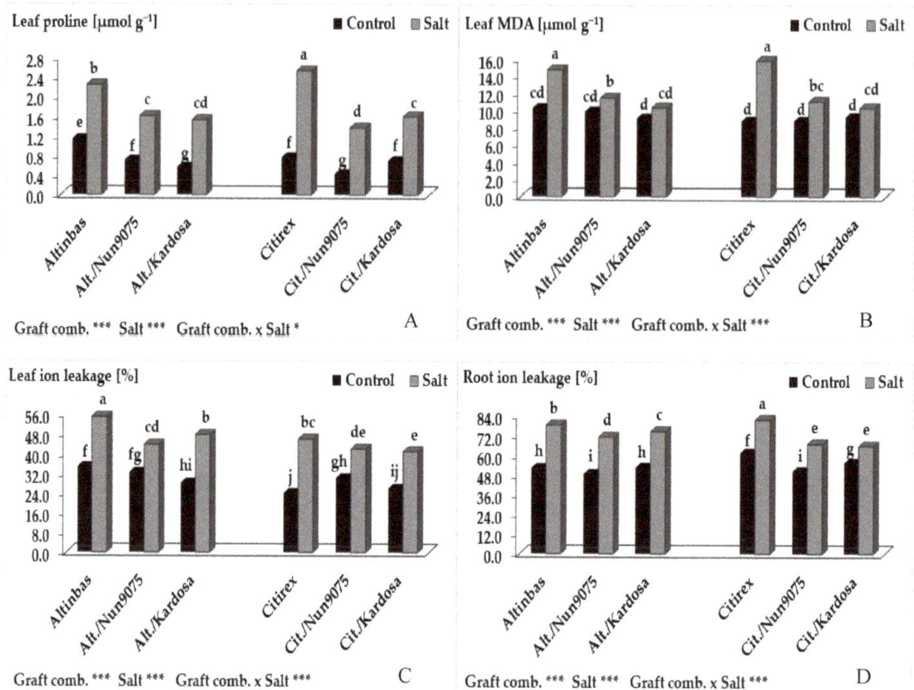

Figure 1. Leaf proline (**A**), malondialdehyde (MDA) (**B**), leaf ion leakage (**C**), and root ion leakage (**D**) of melon graft combinations under control (1.5 dS m^{-1}) and salt stress (8.0 dS m^{-1}) conditions. Values denoted by different letters are significantly different between graft combinations within columns. Significance of main and interaction effects *F* values: $p < 0.05$ (*), $p < 0.01$ (**), and $p < 0.001$ (***).

All of the results clearly indicated that Citirex is a salt-sensitive cultivar, whereas Altinbas is a salt-tolerant cultivar. Our results corroborate those from the study of Yasar et al. [27], who concluded that MDA content in leaf tissues of salt-tolerant eggplant genotypes was twofold lower than that of salt-sensitive eggplant genotypes under salt stress. Similar results were also demonstrated by the study of Lutts et al. [21], who elucidated that MDA content was lowest in salt-tolerant rice genotypes, whereas a salt-sensitive rice genotype exhibited the highest MDA content under salt stress.

Interestingly, irrespective of the cultivar, the proline, MDA, leaf ion leakage, and root ion leakage were significantly reduced when they were grafted with Nun9075 and Kardosa rootstocks under salt stress (Figure 1A–D). Although significant reductions existed when Altinbas was grafted onto both rootstocks, significantly lower proline, MDA, leaf ion leakage, and root ion leakage were exhibited only in Citirex/Nun9075 and Citirex/Kardosa graft combinations under salt stress. One of the indicators of tolerance to salt stress is low absolute or proportional ion leakages, which was demonstrated in studies conducted with rice [21], cucumber [30], pepper [31], tomato [32], and melon [29]. Our results again confirmed that grafting with tolerant *Cucurbita maxima* × *C. moschata* rootstocks had pronounced positive effects on the biochemical responses that contribute to the tolerance mechanisms of sensitive scions (melons) under salt stress.

3.1.4. Changes in Leaf Na$^+$, Cl$^-$, K$^+$, and Ca^{++} Uptakes

The leaf Na$^+$, Cl$^-$, K$^+$, and Ca^{++} uptakes of melon plants was significantly ($p < 0.001$) affected by salt, graft combination, and salt × graft combination interaction (Table 3). Irrespective of the graft combination, leaf Na$^+$ and Cl$^-$ concentrations of melon plants increased by 1137.5% and 1392.3%, respectively, under salt stress as compared with control conditions. It is well-known that Na^+ and Cl^- uptakes of leaves increase with increasing salinity level. A study by Colla et al. [8] demonstrated that the Cl^- concentration of cucumber leaves increased by 300% with salt application regardless of genotype. Similar increases in Na^+ and Cl^- concentrations in leaves have been reported in melon [33], watermelon [25], and pumpkin [6] grown under salt stress. However, we observed significant differences between grafted and nongrafted melons regarding leaf Na^+ and Cl^- uptakes under both control and salt stress conditions (Table 3).

Table 3. Leaf Na^+, Cl^-, K^+, and Ca^{++} contents of melon graft combinations under control (1.5 dS m^{-1}) and salt stress (8.0 dS m^{-1}) conditions.

Graft Combination	Leaf Na^+ (%)		Leaf Cl^- (mg gr^{-1})		Leaf K^+ (%)		Leaf Ca^{++} (%)	
(Scion/Rootstock)	Control	Salt	Control	Salt	Control	Salt	Control	Salt
Altinbas	0.36 f^z	3.29 a	18.1 f	207 c	3.30 ab	1.25 e	0.40 gh	0.94 e
Altinbas/Nun9075	0.11 g	1.51 d	17.2 f	252 a	3.37 a	3.08 bc	0.50 fg	2.33 b
Altinbas/Kardosa	0.10 g	1.71 c	14.4 f	265 a	3.20 bc	3.02 c	0.36 h	2.17 c
Citirex	0.15 g	2.77 b	15.3 f	150 e	3.36 a	2.27 d	0.40 c	1.29 d
Citirex/Nun9075	0.13 g	1.05 e	12.1 f	225 b	3.37 a	3.14 bc	0.53 f	2.79 a
Citirex/Kardosa	0.08 g	1.50 d	11.3 f	178 d	3.15 bc	3.07 bc	0.42 gh	2.68 a
F-test								
Graft combination	***		***		***		***	
Salt	***		***		***		***	
Graft comb. × salt	***		***		***		***	

^{z} Values denoted by different letters are significantly different between graft combinations within columns at $p < 0.05$. Significance of main and interaction effects F values: $p < 0.05$ (*), $p < 0.01$ (**), and $p < 0.001$ (***).

Irrespective of the cultivars, grafted melons exhibited 58.1% and 19.8% lower Na^+ and Cl^- uptakes, respectively, than nongrafted melon plants under control conditions. Similar contributions of rootstocks to Na^+ exclusion were observed, whereas an opposite response was observed in Cl^- uptake in grafted plants under salt stress. Consequently, the grafted melons exhibited 52.1% lower Na^+ uptake than the nongrafted ones under salt stress. However, under the same conditions, the grafted plants showed the opposite, a higher Cl^- uptake (28.7%) than that of the nongrafted melon plants. Excluding the toxic ion in roots and retaining salt in the root and not transporting it to shoots are known biochemical responses of salt-tolerant genotypes [34,35]. In agreement with this characterization, our results clearly confirmed that with the two tolerant *Cucurbita maxima* × *C. moschata* rootstocks, Na^+ uptake might be excluded by the roots, and thus the leaf Na^+ content of the scions (melon) was significantly reduced under salt stress. On the other hand, higher leaf Cl^- uptake of the grafted plants than that of the nongrafted plants disagrees with the salt tolerance characterization studies of Acosta-Motos et al. [34] and Zhu and Bie [35]. However, the study of Colla et al. [33] reported that grafted melon plants had higher leaf Cl^- contents than nongrafted ones under salt stress, which was corroborated by our results. In our study, this result might be due to maintenance of a higher leaf area and photosynthetic activity (Table 2) of the grafted melon plants as compared with the nongrafted ones under salt stress. Chloride can play an essential role in photosynthetic activity by controlling stomatal conductance [36] and osmoregulation [37]. Therefore, the increase in leaf area of the grafted melons with high chloride uptake may be a result of enhancement in cell division rates and cell extension [38].

Irrespective of the graft combination, the leaf K^+ concentration of the melon plants was reduced by 20.1%, whereas the leaf Ca^{++} concentration increased by 353.3% under salt stress as compared with that under control conditions. The reduction in leaf K^+ uptake under salt stress could be the result of high Na^+ uptake, which usually causes a disruption in ion activities [39] and a specific competition with K^+ for binding sites [40]. Moreover, highly significant differences were found between grafted and nongrafted melon plants regarding leaf Na^+, K^+, and Ca^{++} uptakes under salt stress. As compared with nongrafted melon cultivars, leaf Na^+ uptake was significantly reduced (52.1%), whereas leaf K^+ (75.1%) and Ca^{++} (123.2%) uptakes significantly increased in all graft combinations under salt stress. This indicates that the increase in leaf K^+ might be the result of indirect contributions of substantial Ca^{++} uptake of the grafted plants under salt stress. High Ca^{++} content can maintain membrane stability in roots and leaves by limiting the adverse effects of Na^+ ions on the membrane [41] and leads to decreased Na^+ uptake and increased K^+ uptake [42]. Yetisir and Uygur [43] reported that *Cucurbita* and *Lagenaria* rootstocks expressed mechanisms to avoid physiological damage caused by excessive accumulation of Na^+ ion in leaves and hence showed higher performance than watermelon under salinity stress. In agreement with this study, our results clearly indicated that grafting with two tolerant *Cucurbita maxima* × *C. moschata* rootstocks (Nun9075 and Kardosa) led to an increase in leaf K^+ and Ca^{++} ions and hence caused a decline in the leaf Na^+ ion of the two melon cultivars under salt stress. This might be a useful strategy for preserving membrane stability and maintaining K^+ balance for increasing the tolerance of plants to salt stress [44].

3.1.5. Correlation between Shoot and Root Growths and the other Parameters under Salt Stress

Irrespective of the graft combination, the correlation coefficients between shoot and root dry biomass productions, leaf area formation, and the other parameters of melon plants under salt stress conditions are shown in Table 4. Shoot dry weight and leaf area of salt stress plants were significantly negatively correlated with leaf proline, leaf MDA, leaf Na^+, and leaf and root ion leakages. Similar negative correlations between root dry matters were recorded only with leaf MDA and root ion leakage.

Table 4. Irrespective of the graft combination, the correlation coefficients between shoot and root dry biomass productions, leaf area formation, and other parameters of melon plants under salt stress $(8.0\ dS\ m^{-1})$ condition.

	Correlation Coefficients under Salt Stress		
Parameters	**Shoot Dry Weight**	**Root Dry Weight**	**Leaf Area**
Shoot dry weight	1.00	0.506 *	0.954 ***
Root dry weight	0.506 *z	1.00	0.655 **
Shoot-to-root ratio	0.858 ***	0.005 n.s.	0.717 ***
Leaf area	0.954 ***	0.655 **	1.00
Photosynthesis	0.860 ***	0.644 **	0.852 ***
Chlorophyll	0.653 **	0.215 n.s.	0.583 *
Carotenoid	0.681 **	0.199 n.s.	0.696 **
Leaf proline	−0.896 ***	−0.446 n.s.	−0.816 ***
Leaf MDA	−0.851 ***	−0.501 *	−0.818 ***
Leaf Na	−0.903 ***	−0.262 n.s.	−0.783 ***
Leaf Cl^-	0.541 *	0.712 ***	0.547 *
Leaf K^+	0.831 ***	0.096 n.s.	0.699 **
Leaf Ca^{++}	0.935 ***	0.298 n.s.	0.866 ***
Leaf electrolyte	−0.628 **	0.134 n.s.	−0.504 *
Root electrolyte	−0.900 ***	−0.491 *	−0.895 ***

[z] Levels of significance are represented by $p < 0.05$ (*), $p < 0.01$ (**), and $p < 0.001$ (***) with n.s. meaning not significant (Pearson correlation coefficient, $n = 18$).

On the other hand, all physiological (leaf area, photosynthesis, and chlorophyll and carotenoid contents) and nutritional (leaf K, Ca, and Cl) parameters were significantly positively correlated with shoot dry weight and leaf area under salt stress. All of these results clearly indicated that salt tolerance was closely associated with high shoot biomass production with an extensive photosynthetically active leaf area formation, but conversely with substantially lower leaf proline, leaf MDA, leaf Na^+, and leaf and root ion leakages. This might be due to common tolerance responses of grafted plants that were usually exhibited as salt tolerance strategies in studies carried out with rice [21], melon [29], cucumber [30], pepper [10,31], and tomato [32].

4. Conclusions

One of the most prevalent abiotic stress factors, salinity usually has harmful effects on crop productive capacity by decreasing yield and quality, particularly in arid and semiarid regions of the world. To solve this problem, grafting with salt-tolerant rootstocks can be an effective management strategy for improving the salt tolerance of crop plants. In this short-term hydroponic experiment, two melon cultivars were grafted onto two different commercial *Cucurbita maxima* × *C. moschata* hybrid rootstocks to assess plant growth performance under control (1.5 dS m^{-1}) and salt stress (8.0 dS m^{-1}) conditions. Results indicated that the shoot and root growths of grafted and nongrafted melon plants were detrimentally affected by salt stress. Significant reductions were recorded in some agronomic and physiological plant responses under salt stress. On the other hand, susceptible plants responded to salt stress by increasing leaf proline and malondialdehyde (MDA), ion leakage, and leaf Na$^+$ and Cla$^-$ contents. As a result, significant negative correlations existed between shoot dry biomass production and leaf proline (r: −0.89 ***), leaf MDA (r: −0.85 ***), leaf Na$^+$ (r: −0.90 ***), leaf ion leakage (r: 0.63 *), and root ion leakage (r: −0.90 ***) under salt stress. The two melon cultivars differed significantly in salt tolerance. Nongrafted Citirex tended to be more sensitive than Altinbas to salt stress. The *Cucurbita* rootstock genotypes (Nun9075 and Kardosa) significantly improved the growth and biomass production of the grafted melon scions by inducing physiological (high leaf area and photosynthesis), biochemical (low leaf proline and MDA), and nutritional (low leaf Na and ion leakages and high K$^+$ and Ca^{++}) responses under salt stress. The highest plant growth performance was exhibited by Citirex/Nun9075 and Citirex/Kardosa graft combinations. All of these suggest that these *Cucurbita* cultivars have a high rootstock potential for melon, and their significant contributions to salt tolerance were closely associated with inducing beneficial plant physiological and biochemical responses of melon scions. Consequently, these traits could be useful for the selection and breeding of salt-tolerant rootstocks for sustainable agriculture in the future.

Author Contributions: Conceptualization, A.U. and H.Y.; methodology, A.U., A.A., and H.Y.; software, A.U. and H.Y.; validation, A.A., F.U., and T.F.M.; formal analysis, A.U. and H.Y.; investigation, A.U., A.A., F.U., T.F.M., and H.Y.; resources, A.U. and H.Y.; data curation, A.U. and H.Y; writing—original draft preparation, A.U., A.A., and F.U.; writing—review and editing, A.U., F.U., and H.Y.; visualization, A.A., F.U., and T.F.M.; supervision, H.Y.; project administration, A.U. All authors have read and agreed to the published version of the manuscript.

Funding: This research received no external funding.

Acknowledgments: We would like to thank all the staff members of the Plant Physiology Laboratory of Erciyes University, Turkey, for their technical support and for supplying all facilities during the experiments.

Conflicts of Interest: The authors declare no conflict of interest. The funders had no role in the design of the study; in the collection, analyses, or interpretation of data; in the writing of the manuscript; or in the decision to publish the results.

References

1. Edelstein, M.; Plaut, Z.; Ben-Hur, M. Sodium and chloride exclusion and retention by non-grafted and grafted melon and Cucurbita plants. *J. Exp. Bot.* **2011**, *6*, 177–184. [CrossRef] [PubMed]
2. Sönmez, B. *Saline and Sodic Soils*; T.O.K.B. Rural Affairs Research Institute: Şanlı Urfa, Turkey, 1990; No.62; pp. 1–60.
3. Wang, W.; Vinocur, B.; Altman, A. Plant responses to drought, salinity and extreme temperatures: Towards genetic engineering for stress tolerance. *Planta* **2003**, *218*, 1–14. [CrossRef] [PubMed]
4. Del Amor, F.M.; Martinez, V.; Cerda, A. Salinity duration and concentration affect fruit yield and quality and growth and mineral composition of melon plants grown in perlite. *HortScience* **1999**, *34*, 1234–1237. [CrossRef]
5. Mavrogianopoulos, G.N.; Spanakis, J.; Tsikalas, P. Effect of carbon dioxide enrichment and salinity on photosynthesis and yield in melon. *Sci. Hort.* **1999**, *79*, 51–63. [CrossRef]
6. Balkaya, A.; Yıldız, S.; Horuz, A.; Dogru, S.M. Effects of salt stress on vegetative growth parameters and ion accumulations in cucurbit rootstock genotypes. *Ekin J. Crop. Breed. Genet.* **2016**, *2*, 11–24.

7.   Kurtar, E.S.; Balkaya, A.; Kandemir, D. Screening for salinity tolerance in developed winter squash (*Cucurbita maxima*) and pumpkin (*Cucurbita moschata*) lines. *Yuz. Yil Univ. J. Agric. Sci.* **2016**, *26*, 183–195. (In Turkish)

8.   Colla, G.; Rouphael, Y.; Jawad, R.; Kumara, P.; Rea, E.; Cardarellic, M. The effectiveness of grafting to improve NaCl and CaCl$_2$ tolerance in cucumber. *Sci. Hortic.* **2013**, *164*, 380–391. [CrossRef]

9.   Rouphael, Y.; Cardarelli, M.; Rea, E.; Colla, G. Improving melon and cucumber photosynthetic activity, mineral composition, and growth performance under salinity stress by grafting onto Cucurbita hybrid rootstocks. *Photosynthetica* **2012**, *50*, 180–188. [CrossRef]

10. Al Rubaye, O.M.; Yetisir, H.; Ulas, F.; Ulas, A. Growth of pepper inbred lines as affected by rootstocks with vigorous root system under salt stress conditions. *Acta Hort.* **2020**, *1273*, 479–485. [CrossRef]

11. FAO. FAOSTAT Agricultural Database. 2018. Available online: http://www.fao.org/faostat/en/#data/QC (accessed on 23 June 2020).

12. Romero, L.; Belakbir, A.; Ragala, L.; Ruiz, M.J. Response of plant yield and leaf pigments to saline conditions: Effectiveness of different rootstocks in melon plant (*Cucumis melo* L.). *Soil Sci. Plant. Nutr.* **1997**, *43*, 855–862. [CrossRef]

13. Huang, C.H.; Zong, L.; Buonanno, M.; Xue, X.; Wang, T.; Tedeschi, A. Impact of saline water irrigation on yield and quality of melon (*Cucumis melo* cv. Huanghemi) in northwest China. *Eur. J. Agron.* **2012**, *43*, 68–76. [CrossRef]

14. Pogonyi, Á.; PéK, Z.; Helyes, L.; Lugasi, A. Effect of grafting on the tomato's yield, quality and main fruit components in spring forcing. *Acta Aliment.* **2005**, *34*, 453–462. [CrossRef]

15. Rivero, R.M.; Ruiz, J.M.; Romero, L. Can grafting in tomato plants strengthen resistance to thermal stress? *J. Sci. Food Agric.* **2003**, *83*, 1315–1319. [CrossRef]

16. Schwarz, D.; Rouphael, Y.; Colla, G.; Venema, J.H. Grafting as a tool to improve tolerance of vegetables to abiotic stresses: Thermal stress, water stress and organic pollutants. *Sci. Hortic.* **2010**, *127*, 162–171. [CrossRef]

17. Kumar, P. Grafting Affects Growth, Yield, Nutrient Uptake, and Partitioning Under Cadmium Stress in Tomato. *HortScience* **2015**, *50*, 1654–1661. [CrossRef]

18. Ulas, F.; Fricke, A.; Stutzel, H. Leaf physiological and root morphological parameters of grafted tomato plants under drought stress conditions. *Fresenius Envir. Bull.* **2019**, *28*, 3423–3434.

19. Lichtenthaler, H.K. Chlorophylls and carotenoids: Pigments of photosynthetic biomembranes. *Methods Enzymol.* **1987**, *148*, 350–382.

20. Bates, L.S.; Waldren, R.P.; Tevre, I.U. Rapid determination of free proline for water stress studies. *Plant Soil* **1973**, *39*, 205–207. [CrossRef]

21. Lutts, S.; Kinet, J.M.; Bouharmont, J. NaCl-Induced senescence in leaves of rice (*Oryza sativa* L.) cultivars differing in salinity resistance. *Ann. Bot.* **1996**, *78*, 389–398. [CrossRef]

22. Lutts, S.; Kinet, J.M.; Bouharmont, J. Changes in plant response to NaCl during development of rice (*Oryza sativa* L.) varieties differing in salinity resistance. *J. Exp. Bot.* **1995**, *46*, 1843–1852. [CrossRef]

23. Johnson, C.M.; Ulrich, A. Analytical Methods for Use in Plant Analysis. *Bull. Calif. Agric. Exp. Stn.* **1959**, *766*, 26–78.

24. Colla, G.; Rouphael, Y.; Cardarelli, M.; Rea, E. Effect of salinity on yield, fruit quality, leaf gas exchange, and mineral composition of grafted watermelon plants. *Hort Sci.* **2006**, *41*, 622–627. [CrossRef]

25. Yetisir, H.; Uygur, V. Responses of grafted watermelon onto different gourd species to salinity stress. *J. Plant. Nutr.* **2010**, *33*, 315–327. [CrossRef]

26. Gong, B.; Wen, D.; VandenLangenberg, K.; Wei, M.; Yang, F.; Shi, Q.; Wang, X. Comparative effects of NaCl and NaHCO$_3$ stress on photosynthetic parameters, nutrient metabolism, and the antioxidant system in tomato leaves. *Sci. Hortic.* **2013**, *157*, 1–12. [CrossRef]

27. Yasar, F.; Ellialtioglu, S.; Kusvuran, S. Ion and lipid peroxide content in sensitive and tolerant eggplant callus cultured under salt stress. *Eur. J. Hortic. Sci.* **2006**, *71*, 169.

28. Hirasawa, T.; Hsiao, T.C. Some characteristics of reduced leaf photosynthesis at midday in maize growing in the field. *Field Crop. Res.* **1999**, *62*, 53–62. [CrossRef]

29. Sarabi, B.; Bolandnazar, S.; Ghaderi, N.; Ghashghaie, J. Genotypic differences in physiological and biochemical responses to salinity stress in melon (*Cucumis melo* L.) plants: Prospects for selection of salt tolerant landraces. *Plant Physiol. Biochem.* **2017**, *119*, 294–311. [CrossRef]

30. Mumtaz-Khan, M.; Ruqaya, S.; Al-Mas'oudi, M.; Al-Said, F.; Khan, I. Salinity effects on growth, electrolyte leakage, chlorophyll content and lipid peroxidation in cucumber (*Cucumis sativus* L.). *Int. Conf. Food Agric. Sci.* **2013**, *55*, 28–32. [CrossRef]

31. Kaya, C.; Kirnak, H.; Higgs, D. The effects of supplementary potassium and phosphorus on physiological development and mineral nutrition of cucumber and pepper cultivars grown at high salinity (NaCl). *J. Plant Nutr.* **2001**, *24*, 1457–1471. [CrossRef]

32. Kaya, C.; Kirnak, H.; Higgs, D. Enhancement of growth and normal growth parameters by foliar application of potassium and phosphorus on tomato cultivars grown at high (NaCl) salinity. *J. Plant Nutr.* **2001**, *24*, 357–367. [CrossRef]

33. Colla, G.; Rouphael, Y.; Cardarelli, M.; Massa, D.; Salerno, A.; Rea, E. Yield, fruit quality and mineral composition of grafted melon plants grown under saline conditions. *J. Hortiv. Sci. Biotechnol.* **2006**, *81*, 146–152. [CrossRef]

34. Acosta-Motos, J.R.; Ortuño, M.F.; Bernal-Vicente, A.; Diaz-Vivancos, P.; Sanchez-Blanco, J.; Jose Hernandez, A. Plant Responses to Salt Stress: Adaptive Mechanisms. *Agronomy* **2017**, *7*, 1–38. [CrossRef]

35. Zhu, J.; Bie, Z. Physiological response mechanism of different salt-resistant rootstocks under NaCl stress in greenhouse. *Trans. Chin. Soc. Agric. Eng.* **2008**, *24*, 227–231.

36. Braconnier, S.; d'Auzac, F. Chloride and stomatal conductance in coconut. *Plant Physiol. Biochem.* **1990**, *28*, 105–112.

37. Heslop-Harrison, J.S.; Reger, B.J. Chloride and potassium ions and turgidity in the grass stigma. *J. Plant Physiol.* **1986**, *124*, 55–60. [CrossRef]

38. Terry, N. Photosynthesis, growth, and the role of chloride. *Plant Physiol.* **1977**, *60*, 69–75. [CrossRef]

39. Grattan, S.R.; Grieve, C.M. Salinity-mineral nutrient relations in horticultural crops. *Sci. Hortic.* **1999**, *78*, 127–157. [CrossRef]

40. Marschner, H. *Marschner's Mineral Nutrition of Higher Plants*, 3rd ed.; Academic Press: London, UK, 2012; Volume 89, p. 651.

41. Greenway, H.; Munns, R. Mechanisms of salt tolerance in nonhalophytes. *Annu. Rev. Plant Physiol.* **1980**, *31*, 149–190. [CrossRef]

42. Cramer, G.R.; Läuchli, A.; Polito, V.S. Displacement of Ca^{++} by Na^+ from the plasmalemma of root cells. A primary response to salt stress? *Plant Physiol.* **1985**, *79*, 207–211. [CrossRef]

43. Yetisir, H.; Uygur, V. Plant Growth and Mineral Element Content of Different Gourd Species and Watermelon under Salinity Stress. *Turk J. Agric. For.* **2009**, *33*, 65–77. [CrossRef]

44. Huang, Y.; Tang, R.; Cao, Q.L.; Bi, Z.L. Improving the fruit yield and quality of cucumber by grafting onto the salt tolerant rootstock under NaCl stress. *Sci. Hortic.* **2009**, *122*, 26–31. [CrossRef]

Publisher's Note: MDPI stays neutral with regard to jurisdictional claims in published maps and institutional affiliations.

© 2020 by the authors. Licensee MDPI, Basel, Switzerland. This article is an open access article distributed under the terms and conditions of the Creative Commons Attribution (CC BY) license (http://creativecommons.org/licenses/by/4.0/).

Article

The Effects of Gibberellic Acid and Emasculation Treatments on Seed and Fruit Production in the Prickly Pear (*Opuntia ficus-indica* (L.) Mill.) cv. "Gialla"

Lorenzo Marini [1,*], Chiara Grassi [1], Pietro Fino [1], Alessandro Calamai [1], Alberto Masoni [1], Lorenzo Brilli [2] and Enrico Palchetti [1]

[1] The Department of Agriculture, Food, Environment and Forestry (DAGRI), University of Florence, Piazzale delle Cascine 18, 50144 Firenze, Italy; chiara.grassi@unifi.it (C.G.); pietro.fino@stud.unifi.it (P.F.); alessandro.calamai@unifi.it (A.C.); alberto.masoni@unifi.it (A.M.); enrico.palchetti@unifi.it (E.P.)

[2] CNR-IBE, via Giovanni Caproni 8, 50144 Firenze, Italy; lorenzo.brilli@ibe.cnr.it

* Correspondence: lo.marini@unifi.it; Tel.: +39-055-2755800

Received: 26 June 2020; Accepted: 12 August 2020; Published: 17 August 2020

Abstract: Prickly pear (*Opuntia ficus-indica* (L.) Mill. 1768) is cultivated in several dry and semi-dry areas of the world to produce fresh fruit, bioenergy, cosmetics, medicine, and forage. One of the main production constraints is the presence of many seeds within the fruit, which can negatively influence both the fresh-fruit market price and industrial transformation processes. In this study, different gibberellic acid (GA_3) concentrations were tested for their ability to produce well-formed and seedless fruits. Different application methods (injection and spraying) and concentrations of GA_3 (0, 100, 200, 250, and 500 ppm) combined with floral-bud emasculation were applied to a commercial plantation in southern Italy to evaluate their effects on the weight, length, and diameter of the fruits, total seed number, hard-coated viable seed number, and seed weight per fruit. The results indicated that the application of 500 ppm GA_3 sprayed on emasculated floral buds was the most effective method for reducing seed numbers of prickly pear fruits (−46.0%). The injection method resulted in a very low number of seeds (−50.7%) but produced unmarketable fruit. Observed trends suggest the need to investigate the impact of higher GA_3 concentrations and the applicability of a maximum threshold. Further studies are needed to increase our understanding of the physiological effects of the gibberellic acid pathway through productive tissue in terms of organoleptic and fruit quality.

Keywords: cactus pear; GA_3; injection application; spraying application; lignification

1. Introduction

Prickly pear (*Opuntia ficus-indica* (L.) Mill.) is the most cultivated plant species in the Cactaceae family due to edible fruit production [1]. It is a bushy-shaped, xerophytic, and crassulacean acid metabolism (CAM) plant originating from dry areas of Mexico [2]. The annual global production of prickly pear is approximately 500,000 tons, and Italy supplies 12% of the total market, preceded by Mexico and followed by Israel [3,4]. According to the Food and Agriculture Organization (FAO), in the last few years, prickly pear cultivation has gained a considerable amount of interest in relation to coping with food security issues in mostly dry and semi-dry regions, such as in South America, Africa, and the Mediterranean basin, thanks to its high resistance to drought and the important nutritional compounds present in the fruits [2,5]. Despite numerous species of the *Opuntia* genus being mainly cultivated to produce fresh fruit, this cultivation can play a key role in other contexts, such as environmental defense, forage and bioenergy production, the medicine and cosmetics sectors, and human health [1,6–8]. For instance, in some tropical agroforestry systems, *Opuntia elatior* (Mill.)

and *O. ficus-indica* are cultivated in association with other crops [9,10], as a productive living fence guarding against desertification [11,12]. In Africa and South America, the association of *Opuntia robusta* (J.C. Wendl) and *O. ficus-indica* var. *inermis* provide both living fences and livestock fodder [7,13]. Species such as *Opuntia maxima* (Mill.), *Opuntia heliabranoava* (Scheinvar), and *O. ficus-indica* are currently studied for biogas and fertilizer production, especially when associated with domestic plants in rural areas that are situated off of the energy grid [8,14–16]. *O. ficus-indica* has also been recently studied for medical and nutritional purposes since its juice has been found to show nutraceutical activities [17] and beneficial properties against specific types of cancer cells (bladder, ovarian, and leukemia) [18–20].

The *Opuntia ficus-indica* fruit is a false berry with an average weight of approximately 100–120 g, of which 2–10% is seeds and 60–70% is pulp [21]. *Opuntia ficus-indica* fruits present polyembrionatic seeds and 40–45% of them are aborted [22,23], while the remaining 60–55% are viable hard-coated seeds. This represents one of the main production challenges, which can influence the market price, because fruits with few or abortive seeds are more appreciated by consumers [24]. Abundant hard-coated seeds also complicate industrial processes, negatively impacting the transformation of fruit into such products as juice, nectars, jam, and food coloring [25–27], and potentially affecting consumer health by causing constipation [28–30]. However, whilst in other species such as citrus, the pulp development is not strictly linked with seed presence, in *Opuntia ficus-indica*, fruit pulp development depends on the funiculus of the seeds [31,32]. The funiculi are needed to produce a commercially acceptable pulp volume; however, a higher number of abortive seeds, characterized by their smaller size, would be more acceptable to consumers. The creation of hybrid types characterized by a good balance between these two latter aspects (i.e., high volume pulp—less number of seeds) would be feasible primarily through genetic breeding programs and agronomic techniques. The application of the first approach [31,33,34] is not economically viable and has resulted in poor fruit performance. By contrast, agronomic techniques, such as spring flushing or growth regulator treatments to inhibit seed growth, especially with auxin and/or gibberellins, may easily and more quickly provide well-formed and seedless fruits [2]. In the last decades, a few studies investigated the use of a phytoregulator on prickly pear [35–38]. For instance, Gil et al. [36] showed that the treatment of emasculated floral buds using gibberellic acid (GA$_3$) at 200 ppm increased both the development of ovular tissue and the funiculus, but also the hard-coated abortive seeds. Barbera et al. [35] indicated that at least 200 ppm of GA$_3$ injected into *Opuntia*'s stem (cladode) underneath the fruit was able to decrease the percentage of regular seeds. Mejía et al. [38], comparing the use of GA$_3$ by injection and spray application at different maturation stages, indicated the best performances using a 100 ppm GA$_3$ injection in pre- and postblooming. Kaaniche-Elloumi [23] also reported that the number and timing of GA$_3$ applications can affect fruit and seed development.

In this study, we tested the application of two methods (injection and spraying) of gibberellic acid (GA$_3$) on cactus prickly pear both at pre- and postblooming in order to obtain well-formed seedless fruits in emasculated flowers. Increased GA$_3$ concentrations and floral-bud emasculation techniques were also applied to evaluate fruit weight, length, and diameter; and seed weight, the total number of seeds, and the number of hard-coated viable seeds per fruit.

2. Materials and Methods

2.1. Study Area

The experiments were conducted in the spring–summer of 2016 in a prickly pear orchard located in the Apulia region, southern Italy (41°35′58″ N, 15°45′25″ E). The soil is sub-alkaline and shallow, with a calcareous bedrock substrate [39]. The regional climate is typically Mediterranean, with dry summers and mild winters. Average yearly rainfall is approximately 400 mm, with the lowest precipitation occurring in July and August. Air temperature maximums occur in August and July (~30 °C) and the minimum in January and February (~3 °C). During the experiments, the recorded maximum daily temperature was 39.1 °C, while the lowest rainfall (21 mm) was observed in July [40].

The field experiment was a 4-hectare, 10-year-old orchard with a density of 2.000 plants/ha, characterized by globe-shaped growth and a north–south row-oriented axis. The orchard was under organic management, with no irrigation and a permanent grass cover between rows.

2.2. Experimental Design

The experiment was conducted on *O. ficus-indica* cv. "Gialla", an Italian cultivar [41]. Combination treatments consisted of two randomized blocks to investigate both the effect of flower emasculation and the application of different concentrations of GA$_3$. Floral buds (N° = 360) were treated and examined, considering five floral buds for each plant. Fifty percent of the floral buds were emasculated (EM), while the rest were left intact (IN). Emasculation was performed 24 h before the first gibberellic acid application, in the morning (6:00–8:30 a.m.), by cutting stamens with a scalpel and then isolating the flower buds with a non-woven fabric cover to prevent natural pollination [38]. Both EM and IN floral buds were then subdivided into two groups and exposed to the two different application methods of gibberellic acid (Berelex® 40SG—Sumitomo Chemical, Saint Didier au Mont d'Or, Lion, France (GA$_3$)).

The GA$_3$ was injected (INJ) or sprayed (SPY) on floral buds using 1 mL of GA$_3$ solution in the following concentrations: for INJ, 0, 100, and 200 ppm; for SPY, 0, 250, and 500 ppm. Control floral buds were injected and/or sprayed with distilled water. The different dose regimes of INJ and SPY were selected because injection treatment is more efficient than spraying [38]. Doses will hereafter be indicated as control level (0 ppm for both INJ and SPY), low level (100 ppm for INJ and 250 ppm for SPY), and high level (200 ppm for INJ and 500 ppm for SPY). Following the methodology proposed by Mejía and Cantwell [38] and De La Barrera and Nobel [42], the GA$_3$ was applied twice on each bud at two different times corresponding to different phenological stages: 1–2 days before blooming (i.e., floral-bud diameter of 1.3–1.5 cm), and 20 days after blooming. Other management options (i.e., irrigation and fertilization) were not applied during the experiment.

2.3. Fruit and Seed Analyses

All fruits were harvested at the end of August when control fruits reached commercial maturity. At harvest time, fruit weight, diameter, and length were measured, and fruits were immediately stored in plastic bags at −20 °C. While frozen, each fruit was peeled and centrifuged (Girmi il Naturista mod. CE25 500W, Omegna, Italy) to separate pulp and seeds. Seeds were collected, washed with tap water, dried at 30 °C for 24 h, and then weighed, counted, and separated by seed type (hard-coated viable seeds or soft-coated aborted seeds) [38]. The fruit parameters considered were length, weight, and diameter, while the seed parameters considered were number, weight, and presence of viable seeds.

2.4. Statistical Analyses

Analysis of variance (ANOVA) was carried out by applying a mixed model on a complete factorial design to evaluate both the effects of each factor and all interactions, considering the GA$_3$ levels between INJ and SPY as equivalent. In the model, blocks were considered as random factors, while emasculation (EM/IN), application methods (INJ/SPY), and GA$_3$ levels were considered as fixed factors. Data that did not fulfill ANOVA assumptions were square-root transformed before running the model. All the analyses were performed using SPSS v.25 software (IBM Corp., New York, NY, USA). Tukey's post-hoc test was also calculated with Bonferroni correction ($p \leq 0.05$).

3. Results

3.1. Fruits Characterization

In general, the lowest average values for diameter, length, and weight were those of EM plants (Table 1). Considering the combined effect of only INJ/SPY and EM/IN (not GA$_3$ level) on fruit diameter, the highest mean value was found in the IN+SPY group. No significant difference in diameter was found between IN+SPY and IN+INJ combined treatment (−3%), while lower diameters were detected using the combined treatments EM+SPY (−16%) and EM+INJ (−18.5%). The GA_3 levels somewhat affected the results. Higher diameters were found in IN+SPY under all GA_3 levels and the IN+INJ control, while the lower values were found in the EM group control using both methods (i.e., INJ and SPY). However, the highest statistical significance was observed between control levels of emasculated fruits (INJ: 3.35 ± 0.33 cm, SPY: 3.29 ± 0.33 cm) and control levels of intact fruits (INJ: 4.95 ± 0.24 cm, SPY: 4.74 ± 0.32 cm).

Table 1. Effect of gibberellic acid (GA_3) application treatments on fruit variables.

Method	GA_3 Level	Fruit Diameter (cm)		Fruit Length (cm)		Fruit Weight (g)	
		EM [z]	IN	EM	IN	EM	IN
INJ	control [y]	3.35 ± 0.33 f [x]	4.95 ± 0.24 a	5.10 ± 0.58 de	6.65 ± 0.47 a	23.84 ± 6.51 e	82.24 ± 9.71 a
	low	4.07 ± 0.38 de	4.27 ± 0.49 de	5.78 ± 0.62 bc	6.21 ± 0.74 ab	54.93 ± 14.28 cd	59.42 ± 17.25 cd
	high	3.95 ± 0.37 de	4.38 ± 0.34 cd	5.63 ± 0.55 cd	6.27 ± 0.51 ab	49.41 ± 11.95 d	65.34 ± 12.76 bc
SPY	control	3.29 ± 0.33 f	4.74 ± 0.32 ab	5.00 ± 0.53 e	6.43 ± 0.52 a	24.11 ± 7.26 e	74.46 ± 12.97 ab
	low	4.10 ± 0.33 de	4.60 ± 0.28 bc	5.79 ± 0.67 bc	6.66 ± 0.42 a	53.14 ± 14.9 d	72.67 ± 10.87 ab
	high	4.34 ± 0.31 cd	4.61 ± 0.33 bc	6.23 ± 0.54 ab	6.64 ± 0.68 a	60.35 ± 8.89 cd	73.39 ± 16.79 ab
INJ	mean	3.79 ± 0.06 B	4.53 ± 0.05 A	5.50 ± 0.08 B	6.38 ± 0.07 A	42.70 ± 2.10 B	68.97 ± 1.75 A
SPY	mean	3.91 ± 0.06 B	4.65 ± 0.04 A	5.67 ± 0.08 B	6.58 ± 0.06 A	45.83 ± 2.01 B	73.43 ± 1.57 A
All	mean	3.85 ± 0.06	4.59 ± 0.04	5.59 ± 0.08	6.48 ± 0.07	44.27 ± 2.06	71.20 ± 1.67

[z] EM: emasculated fruits; IN: intact fruits. Application methods were injection (INJ) and spraying (SPY). [y] The levels of GA_3 were control, low levels (100 ppm and 250 ppm) and high levels (200 ppm and 500 ppm). [x] Different letters indicate significant differences using Tukey's post-hoc test with Bonferroni correction ($p \leq 0.05$). Lower case letters (a–f) are for comparison of individual treatment means and upper case (A,B) are for main effect means.

Considering the combined effect of only INJ/SPY and EM/IN (without GA_3 levels) on fruit length, the highest mean was found within the IN+SPY combined treatment, while the lowest was observed using EM+INJ. The application of GA_3 resulted in statistically significant differences in fruit lengths between intact and emasculated fruits. Specifically, greater fruit lengths were found in the intact fruits under all GA_3 levels and application methods. In contrast, low statistical significance was observed in emasculated fruits among all GA_3 treatments and methods, and the only exception was for the highest level of the SPY method in line with the results of intact fruits.

Considering the combined effect of INJ/SPY and EM/IN (without GA_3 level) on fruit weight, the highest mean was found using the IN+SPY combined treatment, while the lowest was observed using EM+INJ. No significant difference in weight was found between EM+INJ and IN+INJ combined treatment (−6%), while with respect to the remaining combinations (EM+SPY and EM+INJ), considerable weight differences were observed (−38% and −42%, respectively). The application of GA_3 resulted in statistical significance between intact and emasculated fruits. The higher values were found using IN+SPY at all GA_3 levels, with similar weights for all treatments, and in the IN+INJ control group. In contrast, the lowest values were found specifically for the control of the emasculated fruits using both INJ and SPY. The highest statistical significance was observed, indeed, between the control group of the IN+INJ and control group of EM+INJ combination treatment.

3.2. Fruit Defects

At harvest time, 39% of total fruit displayed defects. These defects were defined as (Table 2, Figure 1) (a) lignification on pulp tissue; (b) lignification of ovular tissue; (c) recalcitrant fruits (i.e., fruits that have not reached maturity).

Table 2. Relative frequencies of harvested fruits and the related defects per combination treatment. Frequency is based on the observation of thirty floral buds per combination treatment.

Method	GA_3 Level	Lignification on Pulp Tissue (%)		Lignification on Ovary Tissue (%)		Recalcitrant Fruits (%)		Healthy Fruits (%)	
	- - -	EM	IN	EM	IN	EM	IN	EM	IN
INJ	control	-	-	20.00	-	80.00	-	-	100.00
	low	53.30	63.90	13.30	6.70	3.00	13.30	30.40	16.10
	high	60.00	43.30	3.30	-	20.00	-	16.70	56.70
SPY	control	-	-	30.30	-	53.30	-	16.40	100.00
	low	-	-	-	-	3.00	-	97.00	100.00
	high	-	-	-	-	-	-	100.00	100.00
INJ	mean	37.77	35.73	12.20	2.23	34.33	4.43	15.70	57.61
SPY	mean	-	-	10.10	-	18.77	-	71.13	100.00
All	mean	18.88	17.87	11.15	1.12	26.55	2.22	43.42	78.79

Figure 1. Main defects of harvested fruits: (**a**) lignification on pulp tissue; (**b**) lignification of ovary tissue; (**c**) healthy fruits; (**d**) recalcitrant fruits and (**e**) its longitudinal section; (**f**) longitudinal section of floral bud. Scale bar = 2 cm.

Lignification on pulp tissue was only found in EM+INJ and IN+INJ combined treatments, while under EM+SPY and IN+SPY, this defect was not observed. Specifically, this defect was observed when GA_3 treatments were applied. In particular, the highest and lowest defect percentages were found in the intact fruits using the injection method for low and the high GA_3 levels, respectively.

Lignification of ovular tissue was observed in three of the four combined treatments, specifically EM+INJ, EM+SPY (10.10%), and IN+INJ. Only within the IN+SPY group was this defect not found. This defect was observed to a greater extent in the EM fruits, with the highest percentage observed within the control group of both the INJ (20.00%) and SPY methods. In contrast, in intact fruit, lignification of ovular tissue was observed only for the INJ method for the low GA_3 level (6.70%).

Finally, recalcitrant fruits were found under EM+INJ, EM+SPY, and IN+INJ, while no defect was found using IN+SPY. Fruit recalcitrance was observed in the EM group, with the highest percentage observed in the control groups of both the injected and sprayed methods. The lignification of ovular tissue in IN fruits was observed only in the low GA_3 group using the INJ method (13.30%).

The only group free of defects was IN+SPY under all GA_3 levels. In contrast, the highest presence of defects was observed using the EM+INJ combined treatment, where only 15.70% of harvested fruits were healthy.

3.3. Seed Characterization

In all groups, the EM flowers showed the lowest average values for all seed variables: the number of seeds was 83.83 ± 8, the number of hard seeds was 12.17 ± 2, and the weight of seeds was 0.16 ± 0.02 g (Table 3).

Table 3. Effect of GA_3 application treatments on seed variables.

Method	GA₃ Level	Seeds Per Fruit (N°) EM [z]	Seeds Per Fruit (N°) IN	Hard Seeds Per Fruit (N°) EM	Hard Seeds Per Fruit (N°) IN	Weight of Seeds Per Fruit (g) EM	Weight of Seeds Per Fruit (g) IN
INJ	control [y]	16.00 ± 43 f [x]	232.00 ± 58 a	1.00 ± 1 d	220.00 ± 55 a	0.02 ± 0.03 e	2.81 ± 0.56 a
	low	118,00 ± 81 cd	140.00 ± 86 bcd	25.00 ± 15 d	71.00 ± 44 c	0.28 ± 0.20 de	0.61 ± 0.44 d
	high	103.00 ± 73 de	185.00 ± 65 ab	16.00 ± 8 d	167.00 ± 59 b	0.24 ± 0.21 de	1.01 ± 0.55 c
SPY	control	45.00 ± 50 ef	195.00 ± 55 ab	0.00 ± 0 d	185.00 ± 53 ab	0.04 ± 0.04 e	2.47 ± 0.78 a
	low	106.00 ± 57 d	174.00 ± 47 abc	21.00 ± 10 d	166.00 ± 44 b	0.23 ± 0.32 de	2.05 ± 0.58 b
	high	115.00 ± 42 d	181.00 ± 51 ab	10.00 ± 4 d	177.00 ± 50 b	0.15 ± 0.07 e	1.99 ± 0.67 b
INJ	mean	79.00 ± 9 B	185.67 ± 9 A	14.00 ± 2 C	152.67 ± 9 B	0.18 ± 0.02 C	1.48 ± 0.11 B
SPY	mean	88.67 ± 7 B	183.33 ± 7 A	10.33 ± 1 C	176 ± 6 A	0.14 ± 0.02 C	2.17 ± 0.08 A
All	mean	83.83 ± 8	184.5 ± 8	12.17 ± 2	164.33 ± 8	0.16 ± 0.02	1.82 ± 0.1

[z] EM: emasculated fruits; IN: intact fruits. Application methods were injection (INJ) and spraying (SPY). [y] The levels of GA_3 were control, low levels (100 ppm and 250 ppm) and high levels (200 ppm and 500 ppm). [x] Different letters indicate significant differences using Tukey's post-hoc test with Bonferroni correction ($p \leq 0.05$). Lower case letters (a–f) are for comparison of individual treatment means and upper case (A,B) are for main effect means.

Without considering the GA_3 levels, the combination treatment IN+INJ showed the highest average number of seeds. Only a slight difference was found between IN+INJ and the IN+SPY combined treatment (−1.3%), while, on average, a considerably lower number of seeds was found with EM+SPY (−52%) and EM+INJ (−57.5%). The GA_3 levels resulted in statistical significance between intact and emasculated fruits. Specifically, more seeds were found in the IN fruits under most GA_3 levels and application methods, with the only exception being the low GA_3 level using the INJ method. In contrast, low values were observed using EM fruits under all GA_3 treatments and methods, particularly in the control groups.

Regardless of GA_3 levels, the highest number of hard seeds per fruit was in the IN+SPY combined treatment. Few differences were observed between IN+SPY and IN+INJ (−13.3%), while large differences were found under EM+SPY (−94.1%) and EM+INJ (−92%) combined treatments. The GA_3 levels resulted in statistical significance between intact and emasculated fruits. In particular, the highest level of significance was found in the control groups of both injection and sprayed methods.

Finally, the average highest seed weight per fruit, without considering GA_3 levels, was within IN+SPY. The lower average weight of seeds per fruit was found using IN+INJ (−32%) with respect to IN+SPY, while large differences in seed weight were observed using EM+SPY (−93.5%) and EM+INJ (−91.7%). Similarly, the GA_3 levels for the seed weight per fruit resulted in statistical significance between intact and emasculated fruits. In particular, the highest level of significance was found under the control of both the INJ and SPY methods.

3.4. ANOVA Results

ANOVA assumptions showed that only the seed variables needed to be transformed. Afterwards, the statistical analysis showed the significant effects of fixed factors and their interactions on both fruit and seed variables (Table 4). Emasculation treatment significantly influenced each variable for both fruits and seeds. The combined treatment SPY/INJ influenced fruit size, seed weight, and viability, but not the number of seeds per fruit. GA3 levels strongly influenced all variables and only moderately impacted the seed number per fruit. Interaction between all factors (EM/IN × SPY/INJ × GA) revealed that there was a strong effect among factors for all the variables.

Table 4. ANOVA significance results of the single and combined effects on fruit and seed dimensions.

Source	Abbr.	Fruit Diameter	Fruit Length	Fruit Weight	Seeds	Hard Seeds	Seed Weight
				Significance			
Emasculation	EM/IN	** z	**	**	**	**	**
Application methods	SPY/INJ	**	*	*	ns	*	**
GA3 levels	GA	**	**	**	*	**	**
EM/IN × SPY/INJ × GA	-	**	**	**	**	**	**

z "ns" means not significant ($p \geq 0.05$); "*" low significance ($0.01 < p < 0.05$); "**" high significance ($p \leq 0.01$). These results were obtained from data presented in Tables 1 and 3.

4. Discussion

The use of different application methods and GA3 concentrations in prickly pears to obtain well-formed and seedless fruits in *O. ficus-indica* (L.) Mill. "Gialla" provided several curious results.

The objective of obtaining well-formed fruits with few seeds was only partially achieved in this study. More well-formed fruits were obtained from the IN rather than the EM treatment, but the IN treatment produced a higher seed content. This was also observed in the aforementioned study by Mejía and Cantwell [38], which found the emasculated fruits were generally smaller and had lower numbers of hard seeds (viable seeds) than the intact ones. This difference could be explained by the lack of stamens (emasculation), which contributed to the lack of or low development of the flower tissues. The external tissue of the anthers is the main gibberellin biosynthesis site, and thus the main regulating factor for the development of the remaining floral parts. This was deduced from Inglese et al. [37], who observed that *Opuntia ficus-indica* flowers with removed anthers show lower levels of endogenous gibberellin than pollinated ones. This behavior has been observed in other crops such as rice [43] and *Arabidopsis* [44].

The emasculated fruits showed a general decrease in all the analyzed variables (i.e., diameter, length, and weight) compared to the intact ones, regardless of the GA3 treatment applied. In the prickly pear, pulp development originates from the funiculus, which connects the seed to the ovular tissue, and thus seeds are needed for fruit development [32,38,45]. On this basis, if the development of the ovular tissue and funiculus was inhibited, the *Opuntia*'s fruit may have difficulty in developing properly [23]. Despite the plausibility of the hypothesis, the gibberellin transport mechanism in the developing organs of the flower—from the male part (stamen) to the female part (ovary tissue and funiculus)—is not currently fully understood [46,47].

Generally, it has been observed that treatment with GA3 may improve pulp development and reduce seed numbers in emasculated prickly pear fruits as a result of the replacement of endogenous gibberellins. This was also suggested by the fact that in *Opuntia ficus-indica*, the highest levels of GA3 in the flowers were found during blooming [37], and these, in turn, are responsible for the development of fruits and natural pollination [42]. This has also been observed in other plant species, i.e., the *Citrus* genus [48], where although gibberellic acid is not the only factor emulating the effects of natural pollination, the contribution of both pollination and exogenous GA3 application can improve fruit development.

However, the response of fruits and seeds to growth regulator treatments in this study depended on the specific application method. More specifically, while the spraying of GA_3 (both low and high levels) generally enhanced the performance of all the analyzed variables of the treated fruits, the injection method showed the opposite pattern, especially for the intact fruits. One of the effects caused by gibberellic acid is control over the elongation of cellular tissues in plants [49–51]. Generally, the exogenous sprayed GA_3 is able to spread through the elongation of cells in plant tissues, thus facilitating the absorption and avoiding the direct negative effect of gibberellic acid within the parenchyma tissue [42,52].

Whilst several studies have demonstrated that GA_3 application can increase the presence of defects in *Arabidopsis* [44,53], *Coriadrum sativum* L. [54], *Oryza sativa* L. [46], *Zea mays* L. [52], and *Daucus carota* L. [55], to our knowledge, only a few studies have investigated the effects of GA_3 on *Opuntia* [35,56]. In this study, the main defects were lignification of the ovular and pulp tissue, and the presence of recalcitrant fruits. Lignification on pulp tissue was observed in both EM and IN fruits, particularly in the INJ groups. Specifically, most pulp tissue lignification was found corresponding to the needle entry hole for the GA_3 injection. This condition may have been caused by two different factors. First, the needle was not able to reach the ovary, thus spreading the GA_3 solution into the pulp. It is also possible that the injected compounds returned to the entry hole. This was partially deduced by Nobel et al. [56], who observed that in *Opuntia ficus-indica* tissues, injected gibberellic acid likely came into contact with expanding parenchyma tissue, thus leading to an excessive accumulation of dry matter in the tissue around the needle entry hole. This has been confirmed in several studies [42,52], which proposed that the gibberellic acid in prickly pear fruits can cause a sink effect promoting dry matter accumulation in parenchyma tissue. This pattern was also indirectly supported by Jedidi Neji et al. [57], in which injection of gibberellic acid into the ovary of the *Opuntia* flower through the stigma and not the pulp did not result in lignification effects on the pulp tissue in fruits.

The lignification on ovular tissue, mostly recorded in EM fruits, was highest when the GA_3 was sprayed rather than injected. Ortiz Hernandez et al. [58] reported a similar behavior in an *O. amyclaea* study: when the flowers were emasculated and treated with different growth regulators, the fruits showed ovarian tissue lignification. This result was likely driven by the flower emasculation rather than the method of GA_3 application since the absence of stamens can cause a lack or little development of the remaining flower tissues. Gupta et al. [47] suggested that, generally, even a short-distance movement of GA_3 from the stamen to the other floral organs and the pedicel may be sufficient for flower development.

Recalcitrant fruits were found mostly in EM fruits. The highest incidence was observed within the control groups for EM+INJ (80%) and EM+SPY (53%). This result was likely because emasculation does not allow for the development of full fruit maturity, thus creating smaller fruits. These findings are consistent with a similar study carried out by Kaaniche-Elloumi et al. [23], in which prickly pear emasculated fruit reached maturity after GA_3 treatment. Besides, when comparing the IN control group fruits with the different GA_3 treatments, an overall decrease in the number of seeds, the number of hard-coated seeds, and the weight of seeds was observed. This may confirm that exogenous GA_3 on *Opuntia* can reduce the number of hard-coated seeds [35,36,38]. Furthermore, seed abortion could be related to the effect of GA_3 on chromosomal DNA, which may lead to the incomplete development of the endosperm [57].

5. Conclusions

Prickly pear cultivation is important in several dry and semi-dry areas of the world owing to its diverse uses (e.g., as fresh fruit, in bioenergy, cosmetics, and medicine production, and as forage). The results of GA_3 application on fruits indicated that 500 ppm of GA_3 sprayed on emasculated floral buds was the most effective technique for reducing the number of seeds within prickly pear fruits. The spraying of the GA_3 (both low and high levels) enhanced the growth performance of all the analyzed variables of the treated fruits, while the injection method, though capable of reducing the number of seeds, can increase the presence of defects, making the fruit unmarketable. The results suggested the need to further investigate the impact of higher GA_3 concentrations on fruit production, and particularly, GA_3 application methods, especially regarding industrial production. The GA_3 spraying method would indeed be easier to apply in large-scale production than the injection method, whilst manual emasculation may be better replaced by chemical emasculation, which provides similar results. Given the scarcity of studies on prickly pear cultivation and the repercussions of its industrial processes, future studies should focus on these aspects by conducting experiments that directly address the application of these treatments in industrial-scale processes. Moreover, further studies should focus on the maximum thresholds of GA_3 applicability and the physiological effects of the gibberellic acid pathway through productive tissue, thus elucidating the economic viability of this cultivation technique and the changes in fruit quality and organoleptic properties.

Author Contributions: Conceptualization, C.G. and E.P.; Formal analysis, L.M. and A.M.; Investigation, C.G. and P.F.; Methodology, C.G., P.F. and E.P.; Resources, E.P.; Writing—original draft, L.M.; Writing—review and editing, A.C., L.B. and E.P. All authors have read and agreed to the published version of the manuscript.

Funding: This research received no external funding.

Conflicts of Interest: The authors declare no conflict of interest.

References

1. Nazareno, M.A. Nutritional properties and medicinal derivative of fruits and cladodes. In *Crop Ecology, Cultivation and Uses of Cactus Pear*; Inglese, P., Mondragon, C., Nefzaoui, A., Saenz, C., Eds.; FAO: Rome, Italy, 2017; pp. 151–158.
2. Ochoa, M.J.; Barbera, G. History and economic and agro-ecological importance. In *Crop Ecology, Cultivation and Uses of Cactus Pear*; Inglese, P., Mondragon, C., Nefzaoui, A., Saenz, C., Eds.; FAO: Rome, Italy, 2017; pp. 2–11.
3. Sáenz, C.; Berger, H.; Rodríguez-Félix, A.; Arias, E.; Galletti, L.; Corrales García, J.; Sepúlveda, E.; Varnero, M.T.; García de Cortázar, V.; Cuevas, R.; et al. *Agro-Industrial Utilization of Cactus Pear*; FAO, Ed.; FAO: Rome, Italy, 2013; p. 150.
4. Timpanaro, G.; Urso, A.; Spampinato, D.; Foti, V.T. Cactus pear market in Italy: Competitiveness and perspectives. *Acta Hortic.* **2015**, *1067*, 407–415. [CrossRef]
5. Sarbojeet, J. Nutraceutical and functional properties of Cactus pear (*Opuntias* spp.) and its utilization for food applications. *J. Eng. Res. Stud.* **2012**, *3*, 60–66.
6. Jacobo, C.M.; de Gallegos, S.J.M. Nopalitos or vegetable cactus production and utilization. In *Crop Ecology, Cultivation and Uses of Cactus Pear*; Inglese, P., Mondragon, C., Nefzaoui, A., Saenz, C., Eds.; FAO: Rome, Italy, 2017; pp. 93–103.
7. Jose, C.B.D., Jr.; Hichem, S.B.; Nefzaoui, A. Forage production and supply for animal nutrition. In *Crop Ecology, Cultivation and Uses of Cactus Pear*; Inglese, P., Mondragon, C., Nefzaoui, A., Saenz, C., Eds.; FAO: Rome, Italy, 2017; pp. 73–91.
8. Varnero, M.T.; Homer, I. Biogas production. In *Crop Ecology, Cultivation and Uses of Cactus Pear*; Inglese, P., Mondragon, C., Nefzaoui, A., Saenz, C., Eds.; FAO: Rome, Italy, 2017; pp. 187–193.
9. Zimmermann, H. Global invasions of cacti (*Opuntia* sp.): Control, management and conflicts of interest. In *Crop Ecology, Cultivation and Uses of Cactus Pear*; Inglese, P., Mondragon, C., Nefzaoui, A., Saenz, C., Eds.; FAO: Rome, Italy, 2017; pp. 172–185.
10. Pardini, A.; Massolino, F.; Grassi, C. *Opuntia ficus-indica* (L.) Mill. Growing in soil and containers for urban agriculture in developing areas. *Adv. Hortic. Sci.* **2017**, *31*, 289–294. [CrossRef]
11. Nefzaoui, A.; Ben Salem, H. *Opuntia spp.—A Strategic Fodder and Efficient Tool for Combat Desertification in the Wana Region*; Mondragon, C., Perez, S., Eds.; FAO: Rome, Italy, 2002; ISBN 92-5-104705-7.
12. Mulas, M.; Mulas, G. *The Strategic Use of Atriplex and Opuntia to Combat Desertification*; University of Sassari: Sassari, Italy, 2004.
13. Vieira, E.L.; Batista, Â.M.V.; Guim, A.; Carvalho, F.F.; Nascimento, A.C.; Araújo, R.F.S.; Mustafa, A.F. Effects of hay inclusion on intake, in vivo nutrient utilization and ruminal fermentation of goats fed spineless cactus (*Opuntia ficus-indica* Mill) based diets. *Anim. Feed Sci. Technol.* **2008**, *141*, 199–208. [CrossRef]

14. Nobel, P.S. *Cacti: Biology and Uses*; University of California Press: Los Angeles, CA, USA; London, UK, 2002; Volume 40, pp. 199–210.

15. Quintanar-Orozco, E.T.; Vázquez-Rodríguez, G.A.; Beltrán-Hernández, R.I.; Lucho-Constantino, C.A.; Coronel-Olivares, C.; Montiel, S.G.; Islas-Valdez, S. Enhancement of the biogas and biofertilizer production from *Opuntia heliabravoana* Scheinvar. *Environ. Sci. Pollut. Res.* **2018**, *25*, 28403–28412. [CrossRef] [PubMed]

16. Ramos-Suárez, J.L.; Martínez, A.; Carreras, N. Optimization of the digestion process of *Scenedesmus* sp. and *Opuntia maxima* for biogas production. *Energy Convers. Manag.* **2014**, *88*, 1263–1270. [CrossRef]

17. Gupta, R.C. A Wonder Plant; Cactus Pear: Emerging Nutraceutical and Functional Food. In *Chemistry of Phytopotentials: Health, Energy and Environmental Perspectives*; Khemani, L.D., Srivastava, M.M., Srivastava, S., Eds.; Springer: Heidelberg, Germany; Dordrecht, The Netherlands, 2012; pp. 183–187.

18. Zou, D.M.; Brewer, M.; Garcia, F.; Feugang, J.M.; Wang, J.; Zang, R.; Liu, H.; Zou, C. Cactus pear: A natural product in cancer chemoprevention. *Nutr. J.* **2005**, *4*, 1–12. [CrossRef]

19. Feugang, J.M.; Ye, F.; Zhang, D.Y.; Yu, Y.; Zhong, M.; Zhang, S.; Zou, C. Cactus pear extracts induce reactive oxygen species production and apoptosis in ovarian cancer cells. *Nutr. Cancer* **2010**, *62*, 692–699. [CrossRef]

20. Sreekanth, D.; Arunasree, M.K.; Roy, K.R.; Chandramohan Reddy, T.; Gorla, V.R.; Reddanna, P. Betanin a betacyanin pigment purified from fruits of *Opuntia ficus-indica* induces apoptosis in human chronic myeloid leukemia Cell line-K562. *Phytomedicine* **2007**, *14*, 739–746. [CrossRef]

21. Lamghari, R.; Kossori, E.L.; Villaume, C.; El Boustani, E.; Sauvaire, Y.; Méjean, L. Composition of Pulp, Skin and Seeds of Prickly Pears Fruit (*Opuntia ficus indica* sp.). *Plant Foods Hum. Nutr.* **1998**, *52*, 263–270.

22. Prat, L.; Nicolás, F.; Sudzuki, F. Morphology and anatomy of Platyopuntiae. In *Crop Ecology, Cultivation and Uses of Cactus Pear*; Inglese, P., Mondragon, C., Nefzaoui, A., Saenz, C., Eds.; FAO: Rome, Italy, 2017; pp. 21–28.

23. Kaaniche-Elloumi, N.; Jedidi, E.; Mahmoud, K.B.; Chakroun, A.; Jemmali, A. Gibberellic acid application and its incidence on in vitro somatic embryogenesis and fruit parthenocarpy in an apomictic cactus pear (*Opuntia ficus-indica* (L.) Mill.) clone. *Acta Hortic.* **2015**, *1067*, 225–229. [CrossRef]

24. Vardi, A.; Levin, I.; Carmi, N. Induction of Seedlessness in Citrus: From Classical Techniques to Emerging Biotechnological Approaches. *J. Am. Soc. Hortic. Sci.* **2008**, *133*, 117–126. [CrossRef]

25. Mora, M. Marketing and communication constraints and strategies. In *Crop Ecology, Cultivation and Uses of Cactus Pear*; Inglese, P., Mondragon, C., Nefzaoui, A., Saenz, C., Eds.; FAO: Rome, Italy, 2017; pp. 195–202.

26. Pandolfini, T. Seedless fruit production by hormonal regulation of fruit set. *Nutrients* **2009**, *1*, 168–177. [CrossRef]

27. Sáenz, C. Processing and utilization of fruit cladodes and seeds. In *Crop Ecology, Cultivation and Uses of Cactus Pear*; Inglese, P., Mondragon, C., Nefzaoui, A., Saenz, C., Eds.; FAO: Rome, Italy, 2017; pp. 135–150.

28. Bartha, G.W. Fecal Impaction Containing Cactus Seeds. *JAMA J. Am. Med. Assoc.* **1976**, *236*, 2390–2391. [CrossRef]

29. Eitan, A.; Bickel, A.; Katz, I.M. Fecal Impaction in Adults: Report of 30 Cases of Seed Bezoars in the Rectum. *Dis. Colon Rectum* **2006**, *49*, 1768–1771. [CrossRef] [PubMed]

30. Eitan, A.; Katz, I.M.; Sweed, Y.; Bickel, A. Fecal impaction in children: Report of 53 cases of rectal seed bezoars. *J. Pediatric Surg.* **2007**, *42*, 1114–1117. [CrossRef]

31. Felker, P.; Inglese, P. Short-Term and Long-Term Research Needs for *Opuntia ficus-indica* (L.) Mill. Utilization in Arid Areas. *J. Prof. Assoc. Cactus Dev.* **2003**, *5*, 131–152.

32. Pimienta-Barrios, E.; Engleman, M.E. Desarrollo de la pulpa y proporcion en volumen de los componetes de loculo maduro en tuna (*Opuntia ficus-indica* [L] Miller). *Agrociencia* **1985**, *62*, 51–56.

33. Inglese, P.; Liguori, G.; De La Barrera, E. Ecophysiology and reproductive biology of cultivated cacti. In *Crop Ecology, Cultivation and Uses of Cactus Pear*; Inglese, P., Mondragon, C., Nefzaoui, A., Saenz, C., Eds.; FAO: Rome, Italy, 2017; pp. 29–40.

34. Weiss, J.; Nerd, A.; Mizrahi, Y. Vegetative Parthenocarpy in the Cactus Pear *Opuntia ficus-indica* (L.) Mill. *Ann. Bot.* **1993**, *72*, 521–526. [CrossRef]

35. Barbera, G.; Carimi, F.; Inglese, P. Effects of GA3 and shading on return bloom of prickly pear (*Opuntia ficus-indica* (L.) Miller). *J. S. Afr. Soc. Hortic. Sci.* **1993**, *3*, 9–10.

36. Gil, G.S.; Espinoza, A.R. Fruit development in the princly pear (*Opuntia ficus indica*, Mill.) with preanthesis application of gibberellin and auxin. *Cienc. E Investig. Agrar.* **1980**, *7*, 141–147. [CrossRef]

37. Inglese, P.; Chessa, I.; Nieddu, G.; La Mantia, T. Evolution of endogenous gibberellins at different stages of flowering in relation to return bloom of cactus pear (*Opuntia ficus-indica* L. Miller). *Sci. Hortic.* **1998**, *73*, 45–51. [CrossRef]

38. Mejía, A.; Cantwell, M. Prickly Pear Fruit Development and Quality in Relation to Gibberellic Acid Applications to Intact and Emasculated Flower Buds. *J. Prof. Assoc. Cactus Dev.* **2003**, *6*, 72–85.

39. ISIS Italian Soil Maps. In *ISIS Feature Dataset: Soil Regions, Soil Systems and Climatic Rasters (Google Earth)*; CREA: Rome, Italy, 2010.

40. ARPA-Puglia Dati Climatici Regione Puglia. Available online: http://www.arpa.puglia.it/web/guest/serviziometeo (accessed on 30 September 2019).

41. Inglese, P.; Barbera, G.; La Mantia, T. Research strategies for the improvement of cactus pear (*Opuntia ficus-indica*) fruit quality and production. *J. Arid Environ.* **1995**, *29*, 455–468. [CrossRef]

42. De La Barrera, E.; Nobel, P.S. Carbon and water relations for developing fruits of *Opuntia ficus-indica* (L.) Miller, including effects of drought and gibberellic acid. *J. Exp. Bot.* **2004**. [CrossRef]

43. Kaneko, M.; Itoh, H.; Inukai, Y.; Sakamoto, T.; Ueguchi-Tanaka, M.; Ashikari, M.; Matsuoka, M. Where do gibberellin biosynthesis and gibberellin signaling occur in rice plants? *Plant J.* **2003**, *35*, 104–115. [CrossRef]

44. Song, S.; Qi, T.; Huang, H.; Xie, D. Regulation of stamen development by coordinated actions of jasmonate, auxin, and gibberellin in arabidopsis. *Mol. Plant* **2013**, *6*, 1065–1073. [CrossRef]

45. Mahmoody, M.; Noori, M. Effect of gibberellic acid on growth and development plants and its relationship with abiotic stress. *Int. J. Farming Allied Sci.* **2014**, *3*, 717–721.

46. Hirano, K.; Aya, K.; Hobo, T.; Sakakibara, H.; Kojima, M.; Shim, R.A.; Hasegawa, Y.; Ueguchi-Tanaka, M.; Matsuoka, M. Comprehensive transcriptome analysis of phytohormone biosynthesis and signaling genes in microspore/pollen and tapetum of rice. *Plant Cell Physiol.* **2008**, *49*, 1429–1450. [CrossRef]

47. Gupta, R.; Chakrabarty, S.K. Gibberellic acid in plant still a mystery unresolved. *Plant Signal. Behav.* **2013**, *8*, e25504. [CrossRef]

48. Ben-Cheikh, W.; Perez-Botella, J.; Tadeo, F.R.; Talon, M.; Primo-Millo, E. Pollination increases gibberellin levels in developing ovaries of seeded varieties of citrus. *Plant Physiol.* **1997**, *114*, 557–564. [CrossRef] [PubMed]

49. Raven, P.H.; Evert, R.F.; Eichhorn, S.E. *Biologia Delle Piante*, 5th ed.; Zanichelli: Bologna, Italy, 2002; p. 720.

50. Taiz, L.; Zeiger, E. *Fisiologia Delle Piante*, 1st ed.; Piccin Nuova Libraria, S.p.A.: Padova, Italy, 2013; p. 825.

51. McComb, A.J. The Stability and Movement of Gibberellic Acid in Pea Seedlings. *Ann. Bot.* **1964**, *28*, 669–687. [CrossRef]

52. Almeida, G.M.; Rodrigues, J.G.L. Development of plants by interference auxins, cytokinins, gibberellins and ethylene. *Braz. J. Appl. Technol. Agric. Sci.* **2016**, *9*, 111–117.

53. Hu, J.; Mitchum, M.G.; Barnaby, N.; Ayele, B.T.; Ogawa, M.; Nam, E.; Lai, W.C.; Hanada, A.; Alonso, J.M.; Ecker, J.R.; et al. Potential sites of bioactive gibberellin production during reproductive growth in Arabidopsis. *Plant Cell* **2008**, *20*, 320–336. [CrossRef]

54. Kalidasu, G.; Sarada, C.; Reddy, P.V.; Reddy, T.Y. Use of male gametocide: An alternative to cumbersome emasculation in coriander (*Coriandrum sativum* L.). *J. Hortic. For.* **2009**, *1*, 126–132.

55. Wang, G.L.; Que, F.; Xu, Z.S.; Wang, F.; Xiong, A.S. Exogenous gibberellin enhances secondary xylem development and lignification in carrot taproot. *Protoplasma* **2017**, *254*, 839–848. [CrossRef]

56. Nobel, P.S. Shading, Osmoticum, and Hormone Effects on Organ Development for Detached Cladodes of *Opuntia ficus-indica*. *Int. J.* **1996**, *157*, 722–728. [CrossRef]

57. Jedidi Neji, E.; Ben Mahmoud, K.; Bouhlal Ben Hadj Alouane, R.; Jemmali, A. Effect of exogenous application of gibberellic acid (GA 3) on seed and fruit development in a Tunisian spineless clone of the prickly pear cactus *Opuntia ficus-indica* (L.). Mill. *J. New Sci. Agric. Biotechnol.* **2017**, *43*, 2344–2351.

58. Ortiz Hernandez, Y.D.; Barrientos Perez, F.; Colinas Leon, M.; Teresa, B.; Martinez Garza, A. Effect of gibberelic acid and auxins in *Opuntia* fruit (*Opuntia amyclaea* T.). *Int. Inf. Syst. Agric. Sci. Technol.* **1991**, *2*, 17–32.

© 2020 by the authors. Licensee MDPI, Basel, Switzerland. This article is an open access article distributed under the terms and conditions of the Creative Commons Attribution (CC BY) license (http://creativecommons.org/licenses/by/4.0/).

 *horticulturae*

Article

Water Use and Yield Responses of Chile Pepper Cultivars Irrigated with Brackish Groundwater and Reverse Osmosis Concentrate

Gurjinder S. Baath [1,*], Manoj K. Shukla [2], Paul W. Bosland [2], Stephanie J. Walker [3], Rupinder K. Saini [4] and Randall Shaw [5]

1   Department of Plant and Soil Sciences, Oklahoma State University, 371 Agricultural Hall, Stillwater, OK 74078, USA
2   Department of Plant and Environmental Sciences, New Mexico State University, MSC 3Q, P.O. Box 30003, Las Cruces, NM 88003, USA; shuklamk@nmsu.edu (M.K.S.); pbosland@nmsu.edu (P.W.B.)
3   Department of Extension Plant Sciences, New Mexico State University, MSC 3AE, P.O. Box 30003, Las Cruces, NM 88003, USA; swalker@nmsu.edu
4   Department of Plant and Soil Science, Texas Tech University, Lubbock, TX 79409, USA; r.saini@ttu.edu
5   Brackish Groundwater National Desalination Research Facility, 500 La Velle Road, Alamogordo, NM 88310, USA; rshaw@usbr.gov
*   Correspondence: gbaath@okstate.edu

Received: 3 March 2020; Accepted: 20 April 2020; Published: 6 May 2020

Abstract: Freshwater availability is declining in most of semi-arid and arid regions across the world, including the southwestern United States. The use of marginal quality groundwater has been increasing for sustaining agriculture in these arid regions. Reverse Osmosis (RO) can treat brackish groundwater, but the possibility of using an RO concentrate for irrigation needs further exploration. This greenhouse study evaluates the water use and yield responses of five selected chile pepper (*Capsicum annuum* L.) cultivars irrigated with natural brackish groundwater and RO concentrate. The four saline water treatments used for irrigation were tap water with an electrical conductivity (EC) of 0.6 dS m^{-1} (control), groundwater with EC 3 and 5 dS m^{-1}, and an RO concentrate with EC 8 dS m^{-1}. The evapotranspiration (ET) of all chile pepper cultivars decreased and the leaching fraction (LF) increased, particularly in the 5 dS m^{-1} and 8 dS m^{-1} irrigation treatments. Based on the water use efficiency (WUE) of the selected chile pepper cultivars, brackish water with an EC $\leq$ 3 dS/m could be used for irrigation in scarce freshwater areas while maintaining the appropriate LFs. A piecewise linear function resulted in a threshold soil electrical conductivity (EC$_e$) ranging between 1.0–1.3 dS m^{-1} for the tested chile pepper cultivars. Both piecewise linear and sigmoid non-linear functions suggested that the yield reductions in chile peppers irrigated with Ca^{2+} rich brackish groundwater were less than those reported in studies using an NaCl-dominant saline solution. Further research is needed to understand the role of supplementary calcium in improving the salt tolerance of chile peppers.

Keywords: *Capsicum annuum*; salinity; evapotranspiration; leaching fraction; calcium

---

1. Introduction

Freshwater is an integral resource for all ecological and social activities, including food and energy production, industrial growth, and human health. As freshwater resources are unevenly and irregularly distributed [1], many arid and semi-arid parts of the world are facing acute water shortages. Similar water shortages affect the southwestern United States due to low rainfall and high evapotranspiration [2]. As agriculture is the largest consumer of freshwater [3], the use of marginal quality water resources, including brackish groundwater, has been increasing [4,5]. About 75% of the groundwater aquifers in the southwestern United States have brackish water, with an electrical conductivity (EC) of > 3 dS/m [6,7]. Additionally, the desalination of brackish groundwater through Reverse Osmosis (RO) produces potable, low saline water and high saline–sodic wastewater known as RO concentrate [7]. The application of desalinated water for irrigation can promote soil hydrological functions [8]. However, the disposal of RO concentrate from an inland desalination system can be problematic, and its sustainable management is a major environmental challenge that restricts the widespread application of RO for groundwater desalination. RO concentrate could serve as a potential source of irrigation for the production of salt-tolerant crops, along with brackish water available from natural saline aquifers [9,10], which will consequently encourage desalination through RO in freshwater scarce-areas.

Continued irrigation with brackish groundwater can lead to salt accumulation in soil which can lower yields, although plants differ extensively in their response to soil salinity. Most crop plants are glycophytes, which can be affected by even a moderate level of soil salinity [11]. Instead of accumulating salts, most glycophytes produce some chemicals (sugars and organic acids) to raise the concentration of constituents in the root cell. This process requires more energy, and thus their crop growth and yield are more susceptible to damage compared to halophytes [12]. Moreover, salt tolerance within the glycophytes group varies widely [13]. Sugarbeet (*Beta vulgaris* L.) and wheat (*Triticum aestivum* L.) are considered salt-tolerant; potato (*Solanum tuberosum* L.), sunflower (*Helianthus annuus* L.), maize (*Zea mays* L.), soybean (*Glycine max* L. Merrill.), and tomato (*Solanum lycopersicum* L.) are moderately salt-sensitive; and chickpea (*Cicer arietinum* L.) and lentil (*Lens culinaris* Medic.) are salt-sensitive [14].

Chile pepper (*Capsicum annuum* L.), also a glycophyte, is an important cash crop of the southwestern United States, cultivated over an area of 20,000 acres annually [15]. It is classified as moderately salt-sensitive, with a saturated soil paste extract EC (EC_e) threshold value of 1.5 dS m^{-1} [16]. Studies have also reported threshold values between an EC_e of 0–2 dS/m for peppers [17,18]. To the best of our knowledge, most studies on peppers have used NaCl as the sole or the dominant salinizing agent [17–21]. However, Na$^+$, Ca^{2+}, and Mg^{2+} are the dominant cations, and Cl$^-$, SO$_4^{2-}$, and HCO$_3^-$ are the dominant anions in most groundwater across the world [22]. It has been suggested that the adoption of salinizing solutions with a single salt may result in ambiguous and erroneous interpretations about plant responses to salinity [23]. Only a few accounts are available involving the use of natural brackish groundwater for growing chile peppers [24]. Therefore, more research on the use of natural brackish groundwater and RO concentrate for irrigating chile pepper cultivars is needed.

The use of brackish groundwater often brings risks and obligations to an agricultural system. The application of insufficient water quantities causes a lowering of the osmotic potential of soil water, ultimately causing stress to the plants [25], whereas over-application is economically ineffectual and could exacerbate salinity problems, including groundwater contamination [26]. There is very limited information available on the evapotranspiration (ET) responses of the chile pepper to varying irrigation salinity. An understanding of water uptake by the chile pepper under contrasting saline water treatments would allow the exploration of irrigation scheduling protocols for regions utilizing brackish groundwater. Therefore, the objectives of this study were to (1) quantify the influences of brackish groundwater and RO concentrate irrigation on the leaching fraction and water use of five chile pepper cultivars; and (2) determine their yield responses to the resulting soil salinity.

2. Materials and Methods

2.1. Experimental Set Up

The study was conducted in a greenhouse located at the Fabian Garcia Science Center of New Mexico State University (NMSU), Las Cruces, New Mexico (32.2805° N latitude and 106.770° W longitude at an elevation of 1186 m above sea level), consistent with the potential of greenhouse chile pepper production in New Mexico [27] and the New Mexico Department of Agriculture regulation of no land application of water with an EC > 4 dS m^{-1}. The chile pepper cultivars selected for this study were AZ1904 (Curry Chile and Seed, Pearce, AZ, USA), Paprika LB25 (Biad Chile, Leasburg, NM,

USA), Paprika 3441 (Olam, Las Cruces, NM, USA), and two NMSU varieties: NuMex Joe E. Parker and NuMex Sandia Select. The natural brackish groundwater and RO concentrate provided by the Brackish Groundwater National Desalination Research Facility (BGRNDRF), Alamogordo, were used in the irrigation treatments (Table 1). Sandy loam soil (78.7% sand, 11% silt, and 10.3% clay) with an initial EC_e of 0.87 dS/m was air-dried, crushed, and sieved through a 4 mm sieve. A soil mix was prepared by mixing soil, sand, and organic peat in the ratio 8:1:1 on a volume basis. The soil mix was sterilized in an oven at 80 °C for at least 30 min. The cylindrical pots used in the experiment were 0.14 m in diameter and 0.25 m in depth. The bottom of each pot was perforated and covered with cheesecloth and then gravels to allow free drainage. The soil packing was done in 5 cm depth increments to obtain a bulk density of 1.36 g/cm^3. The average day and night temperatures recorded during the study period (148 days) were 31.8 ± 0.2 °C and 24.4 ± 0.1°C.

Table 1. Mean (standard error) for chemical properties of the four saline water treatments over the growing period.

	EC dS m^{-1}	Ion Concentration (meq/L)					
		Mg	**Ca**	**Na**	**K**	**Cl**	**SO$_4$**
Tap water	0.6	0.75 (0.01)	2.28 (0.01)	2.73 (0.37)	0.15 (0.01)	1.64 (0.07)	1.58 (0.02)
Well 1	3	8.65 (0.01)	11.90 (0.38)	8.94 (0.18)	0.16 (0.00)	11.91 (0.05)	18.70 (0.79)
Well 2	5	15.24 (0.28)	17.60 (2.08)	19.04 (1.92)	0.21 (0.02)	16.86 (1.72)	38.78 (3.56)
RO conc.	8	25.81 (0.16)	29.43 (2.69)	33.51 (2.96)	0.37 (0.06)	31.23 (5.04)	67.15 (7.43)

Tap water is the control; EC: electrical conductivity; RO conc.: reverse osmosis concentrate.

2.2. Saline Irrigation Treatments

The four irrigation water treatments selected were tap water with the EC 0.6 dS m^{-1}, brackish groundwater with the EC 3 and 5 dS m^{-1}, and an RO concentrate with the EC 8 dS m^{-1}. Before planting, the soil was washed three times with tap water to remove any pre-existing salts, and then the soil salinity was raised to the saline treatment level by irrigating twice with each of the saline water treatments. Four seeds of each chile pepper were sown in pots at a soil depth of 1–2 cm. After emergence, the seedlings were thinned, and only one vigorous seedling was retained in each pot. The irrigation water treatments were continuously applied at an interval of 3–4 days during the experiment period, based on the change in weights of some reference pots. The plants were fertigated using a water-soluble synthetic fertilizer (Miracle-Gro®; 15-30-15) at 2 g L^{-1} every six weeks.

2.3. Data Collection

The same amount of irrigation (I) was applied manually to each pot, and the deep percolation (D) was measured by collecting all the water coming out of the bottom of each pot. The ET was calculated using the following water balance equation:

$$ET = P + I - D - R - \Delta S \tag{1}$$

where ET is the actual crop evapotranspiration (cm), P is the precipitation (cm), I is the irrigation amount (cm), D is the deep percolation (cm), R is the runoff (cm), and ΔS is the change in soil water storage (cm). As the experiments were carried out in a greenhouse, the precipitation and runoff were zero. The change in soil water storage (ΔS) was determined from the difference in weights of the pots at planting and final harvest. The leaching fraction (LF) was calculated for every irrigation as the ratio of D and I. The pods were hand-harvested at the horticultural green mature stage, and the fresh pod weights were measured. The water use efficiency (WUE) was calculated as the ratio of the total yield to total crop ET.

At the end of the experiments, the top 10 cm layer of soil was collected from each pot and saturated soil paste extracts were prepared using composite samples and analyzed for their EC_e, magnesium

(Mg), calcium (Ca), and sodium (Na) ion concentrations [28]. The sodium adsorption ratio (*SAR*) was determined using the following equation:

$$SAR = \frac{[Na^+]}{\sqrt{\frac{([Ca^{2+}]\,[Mg^{2+}])}{2}}} \tag{2}$$

2.4. Salinity Yield Response Equations

The relative yield (Y_r) was obtained as the ratio of actual total yield and maximum total yield for each cultivar. The relationship between the EC_e at the end of the growing season and the relative yield was predicted using the piecewise linear function [16]:

$$Y_r = 1 - b\,(EC_e - a) \tag{3}$$

where a = the salinity threshold (dS m^{-1}); b = the yield reduction, or slope (per dS m^{-1}); and EC_e = the EC of saturated soil extracts from the root zone (dS m^{-1}).

Similarly, the relationships between the Y_r and EC_e of each cultivar were best-fitted with the sigmoid non-linear function [29]:

$$Y_r = \frac{1}{\left(1 + \frac{c}{c_{50}}\right)^p} \tag{4}$$

where Y_r = relative yield; c = the EC of saturated soil extracts from the root zone (dS m^{-1}), c_{50} = root zone EC_e at which the yield had declined by 50% (dS m^{-1}) and p is the exponential constant.

2.5. Statistical Analysis

The cylindrical pots for the experiments were arranged in a completely randomized factorial design with eight replicates of each cultivar and a saline water treatment combination. A two-way analysis of variance (ANOVA) was used to identify the significant differences at alpha 5% applying general linear model procedure (PROC GLM) for ET, D, ΔS, and WUE [30]. The means were separated using the least significance difference (LSD) post hoc test at a 5% significance level ($p \leq 0.05$). The relationships of the EC_e with the concentrations of Mg, Ca, and Na ions and SAR were tested for linear, quadratic and exponential functions using Sigmaplot version 14 (Systat Software Inc., San Jose, CA, USA), and the best fit was selected based on regression statistics. The relative yield response to the EC_e was best fitted to the piecewise linear and sigmoid non-linear functions using the 'nls2' package in R [31].

3. Results and Discussion

3.1. Leaching Fraction over Growing Season

The leaching fractions (LF) for AZ 1904, NuMex Joe E. Parker, NuMex Sandia Select, Paprika LB25, and Paprika 3441 over the growing season are shown in Figure 1a–e, respectively. For almost one month after planting, LFs were similar for cultivars grown using the four saline water treatments. Over time, variations in the LF appeared and became a function of the irrigation water salinity for all five cultivars. Among the four irrigation water treatments, the LFs for the 0.6 dS m^{-1} (control) treatment were the least, while they were the most for the 8 dS m^{-1} RO irrigation treatment throughout the growing season. The differences in LFs of between 0.6 dS m^{-1} and 3 dS m^{-1} were considerably smaller compared to the other two treatments in all five cultivars.

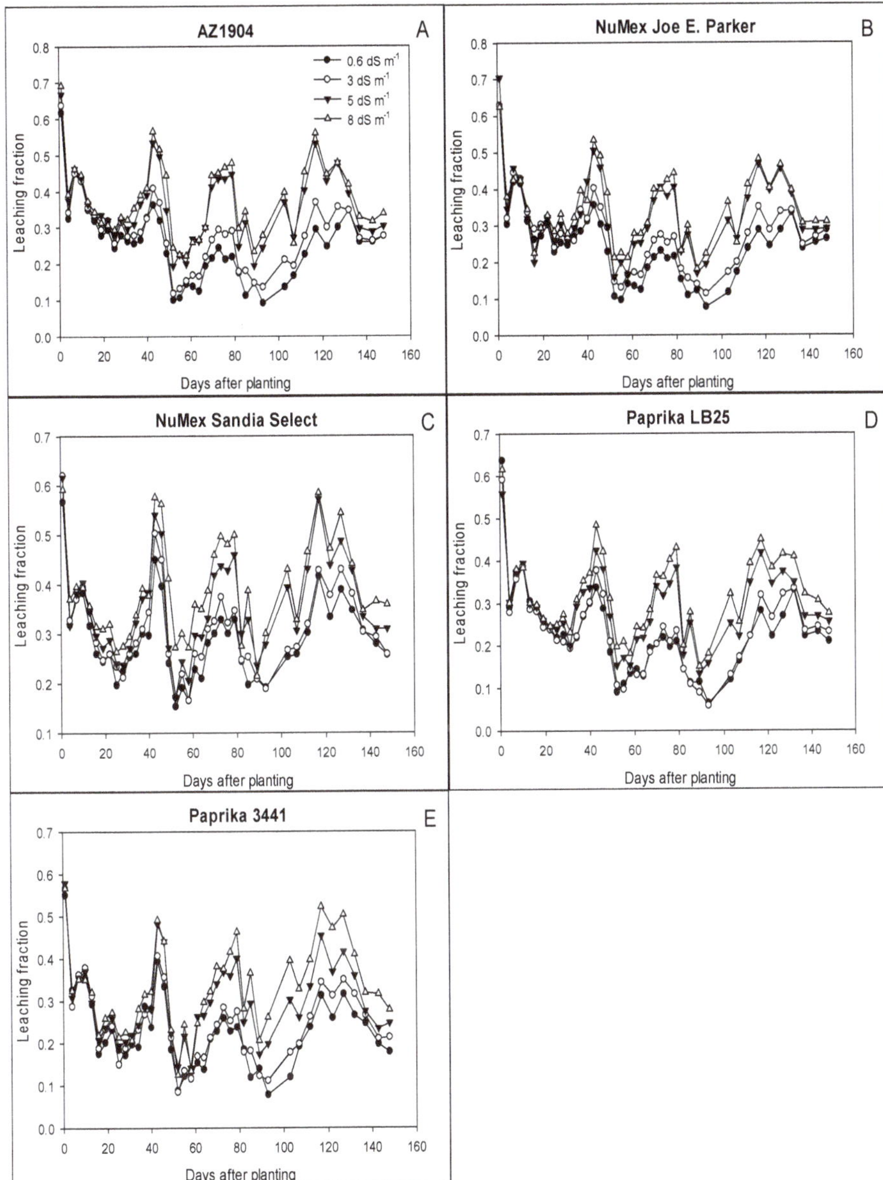

Figure 1. Effect of different saline water treatments on the leaching fractions of (**A**) AZ1904, (**B**) NuMex Joe E. Parker, (**C**) NuMex Sandia Select, (**D**) Paprika LB25, and (**E**) Paprika 3441 over the growing season.

The observed higher LFs under the saline treatments could be due to the self-adjusting nature of the plants under water and osmotic stresses. In response to saline irrigation water, the transpiration rate of chile pepper plants would have decreased due to the reduction in water potential caused by accumulated salts at the root zone [32]. Similar increases in LFs at a given irrigation rate occurred due to the reduction in transpiration rates for the bell pepper (*Capsicum annuum* L.) [33]. In the areas with a shallow water table, more deep percolation could cause secondary salinization [34]. Therefore, it is advisable to explore irrigation scheduling protocols before the application of the concentrate in a field to maintain the soil and groundwater quality [7].

3.2. Water Balance

A total irrigation of 106.3 cm was applied to each pot during the experiment period. The influence of irrigation water salinity on the total crop ET, ΔS and D are shown in Table 2. There was no significant interaction ($p > 0.05$) between the saline treatments and cultivars for D, ΔS, and ET, while the significant main effects of both the saline treatments and cultivars were observed. The total ET of five chiles showed a significant decrease ($p \leq 0.05$) with increasing irrigation water salinity. The highest cumulative ET of the five cultivars was noted at 0.6 dS m^{-1} (control), which was only 4% greater than 3 dS m^{-1}; however, it was around 12% and 17% greater compared to the 5 dS m^{-1} and 8 dS m^{-1} treatments, respectively. The total deep percolation was inversely related to the total crop ET and significantly increased from 24% of the total irrigation amount in the 0.6 dS m^{-1} (control) to 35% in the 8 dS m^{-1} (RO concentrate).

Table 2. Effect of irrigation water salinity on the total deep percolation, change in soil water storage, and evapotranspiration of five chile pepper cultivars.

Treatment	Deep Percolation (cm)	Change in Storage (cm)	Evapotranspiration (cm)
Salinity (S; dS m^{-1})			
0.6	25.06 d	1.56 a	76.99 a
3	28.06 c	1.48 ab	74.07 b
5	33.26 b	1.42 bc	68.93 c
8	36.73 a	1.34 c	65.55 d
LSD (0.05)	1.36	0.09	1.37
Cutivars (C)			
AZ 1904	32.55 b	1.42 a	69.65 c
NuMex Joe E. Parker	30.80 c	1.45 a	71.37 b
NuMex Sandia Select	34.77 a	1.50 a	67.34 d
LB 25	27.56 d	1.45 a	74.60 a
3441	28.22 d	1.43 a	73.97 a
LSD (0.05)	1.52	0.11	1.53
C X S	NS	NS	NS

† Values within each column followed by same letter(s) are not significantly different according to the least significance difference test ($p \leq 0.05$). NS = non-significant at $p \leq 0.05$. Irrigation amount applied was 103.6 cm for all of the treatments.

The reduction in ET with increasing water salinity could be attributed to retarded plant growth and a decrease in bioavailable water under saline soil conditions. The water uptake of plants, through apoplastic and symplastic pathways at roots, is largely regulated by the osmotic and matric potentials of the root zone [35]. Under saline soil conditions, the reduced osmotic potential affects the free energy of water and decreases the root water uptake by plants, which leads to a reduction in the plant growth and ET and thus an increase in leaching [36]. In addition, a salt crust formed at the top soil layer due to saline irrigation could reduce evaporation from the soil surface [37]. Therefore, the surface crusting (visual observations) could also have played some role in reducing the total ET of the chile pepper.

In contrast to the total ET, NuMex Sandia Select had the greatest D, while Paprika LB 25 and 3441 had the minimum among the five cultivars. The differences noticed in the cumulative ET among the cultivars could be attributed to natural variations in the growth of the cultivars. The overall change in soil water storage was small in all of the pots, but it decreased significantly across irrigation treatments from 1.56 cm in the 0.6 dS m^{-1} to 1.34 cm in the 8 dS m^{-1} irrigation water treatment.

3.3. Water Use Efficiency

No significant interaction between the saline treatments and cultivars ($p > 0.05$) was observed for the WUE, while significant reductions ($p \leq 0.05$) in the WUE were noted with the increasing salinity of the irrigation water (Figure 2A). The reduction in the WUE was only 9% in the 3 dS m^{-1} compared to the control treatment, while it was 38% and 42% in the 5 and 8 dS m^{-1} water treatments, respectively. The WUE is generally treated as an important physiological indicator of crops that are grown in water-scarce conditions. As the WUE of chile peppers irrigated with 3 dS/m was not much different from those irrigated with 0.6 dS/m, a slightly brackish groundwater (<3 dS m^{-1}) might be considered for irrigating chile peppers if brackish groundwater is the only available source of irrigation, while simultaneously monitoring salts in the leachate water and soil. However, significant reductions can occur with a further increase in the salinity of the irrigation waters. A similar reduction in the WUE with an increased irrigation water salinity was reported in tomato [38,39]. The average WUE of the five chile pepper cultivars in this study was similar and was in agreement with results reported by Reina-Sanchez et al. (2005) for four tomato cultivars irrigated with saline water (Figure 2B) [40].

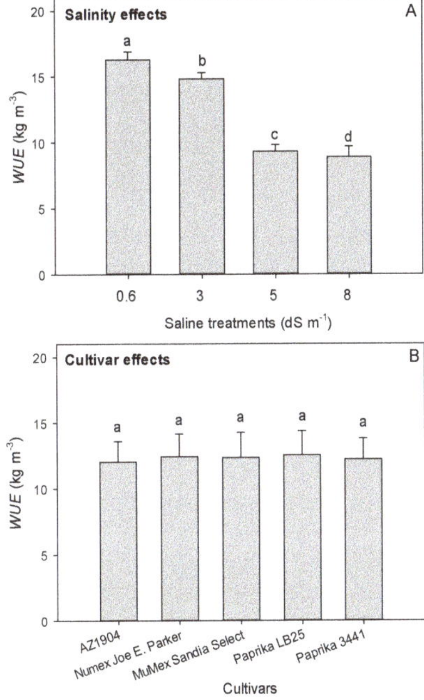

Figure 2. (**A**) Water use efficiency (WUE) of the chile pepper cultivars under four saline irrigation waters, and (**B**) the WUE of five chile pepper cultivars across saline irrigation waters. Bars with the same letters are not significantly different according to the least significance difference test at $p \leq 0.05$.

3.4. Accumulation of Mg^{2+}, Ca^{2+} and Na^{+} Cations in Soil

There were no significant differences among the cultivars ($p > 0.05$) for the magnesium, calcium and sodium concentrations and the sodium adsorption ratios of saturated soil paste extracts (data not presented). Although all three Mg^{2+}, Ca^{2+}, and Na^{+} cation concentrations increased significantly ($p \leq 0.05$) with the EC$_e$, different responses were noted, especially at EC$_e$ higher than 9 dS m^{-1} (Figure 3A). A linear relationship of Mg^{2+} concentration was obtained with the EC$_e$, while the response of Na^{+} and Ca^{2+} was positive exponential and negative exponential, respectively. It was observed that the Na/Ca ratio of the soil paste extract increased from 1.13 in the 0.6 dS m^{-1} to 1.87 in the 8 dS m^{-1}

treatment. The reason could be the displacement of Ca^{2+} by Na^+ and the subsequent Ca^{2+} leaching under high Na^+ concentrations in soil [41]. The SAR increased linearly ($p \leq 0.05$) with the increasing EC_e (Figure 3B), which could be well explained by a greater increase in Na^+ than in Ca^{2+} concentration under high salinity.

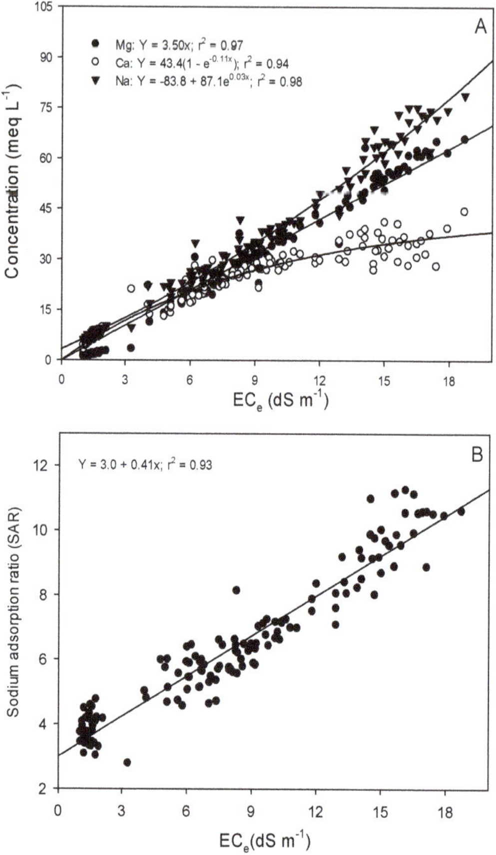

Figure 3. Relationships of (**A**) magnesium (Mg), calcium (Ca), and sodium (Na) ions buildup and (**B**) the sodium adsorption ratio (SAR) with the salinity level (EC_e) of soil irrigated with brackish water and reverse osmosis concentrate.

3.5. Yield Responses to Root-Zone Salinity

The relative yield responses of five chile pepper cultivars to the EC_e were similar and considerably well explained by both piecewise linear and sigmoid non-linear functions; though the sigmoid function resulted in a slightly better fit for each of the five cultivars, as evident by their higher coefficient of determination (r^2; 0.87–0.91) and lower residual sum of squares (RSS) values (Table 3). Likewise, the overall yield responses of chile pepper cultivars were slightly better explained by the sigmoid function compared to the piecewise function. The threshold value (*a*) estimated using the piecewise linear function ranged between 1.04–1.33 dS m^{-1}, with Numex Joe E. Parker and Paprika LB25 having the lowest and greatest *a* value, respectively, among the cultivars. Whereas, both Paprika 3441 and NuMex Sandia Select resulted in threshold values (1.09 & 1.12 dS m^{-1}) close to the observed value of 1.10 dS m^{-1} for all the cultivars. The *a* values obtained in this study were lower than the earlier threshold values of 1.5–1.8 dS m^{-1} suggested for the peppers [16,17,42]. Additionally, the determined slope (*b*) values of 0.038–0.046 were also lower as compared to the earlier reported values of 0.14 [43] and 0.12 [42].

Table 3. Regression statistics for two response functions applied to yield responses of five chile pepper cultivars against soil salinity.

	Piecewise Linear Function				
	a (dS m^{-1})	*b* (dS m^{-1})$^{-1}$	r^2	RSS	N
AZ1904	1.19	0.044	0.88	0.21	32
NuMex Joe E. Parker	1.04	0.045	0.90	0.24	32
Numex Sandia Select	1.12	0.045	0.89	0.25	32
Paprika LB25	1.33	0.046	0.85	0.25	32
Paprika 3441	1.09	0.038	0.89	0.17	32
All cultivars	1.10	0.043	0.87	1.18	160
	Sigmoid non-linear function				
	c_{50} (dS m^{-1})	*p*	r^2	RSS	N
AZ1904	12.22	2.110	0.89	0.18	32
NuMex Joe E. Parker	11.61	1.633	0.91	0.19	32
Numex Sandia Select	10.75	1.262	0.89	0.32	32
Paprika LB25	12.01	1.761	0.87	0.16	32
Paprika 3441	13.55	1.537	0.90	0.12	32
All cultivars	12.11	1.618	0.88	0.94	160

a: salinity (EC$_e$) threshold; *b*: slope; c_{50}: EC$_e$ at which yield is reduced by 50%; *p*: regression constant for sigmoid function.

The 50% yield reduction (c_{50}) estimations from sigmoid non-linear functions were ranged between 10.75–13.55 dS m^{-1}, which was in agreement with the range (12.15–14.21 dS m^{-1}) predicted using piecewise linear equations for the chile pepper cultivars. The lowest c_{50} (10.75 dS m^{-1}) was noted for NuMex Sandia Select, while Paprika 3441 had the greatest c_{50} of 13.55 dS m^{-1}. The other three varieties showed similar yield reductions with an increase in soil salinity, and their c_{50} values were ranged between 12.01–12.22 dS m^{-1}. The observed c_{50} values of all the chile peppers were much higher than the 6 dS m^{-1} proposed for the peppers [42,44]. Furthermore, the constant *p* values ranging between 1.26–2.11 were comparatively lower than the value of 3.0 suggested for most of the crops, including peppers [45].

Lower yield reductions in chile peppers against the soil salinity compared to previous reports could be attributed to the calcium dominated brackish groundwater used in this study. The considerable amount of calcium in the natural saline irrigation treatments has been reported to ameliorate the salinity's impact on plants [46]. Calcium plays regulatory roles in the metabolism, water transport, and root hydraulic conductivity of plants under salt stress [47,48]. Moreover, high calcium levels can shield the cell membrane from detrimental salinity effects [49].

4. Conclusions

This study evaluated the effects of natural brackish groundwater and RO concentrate irrigation on the water use, leaching fraction, and yield responses of chile pepper cultivars. Saline irrigation caused a reduction in the water uptake of the chile peppers and increased LFs, particularly in the 5 dS m^{-1} and the 8 dS m^{-1}. The WUE was not substantially different between 0.6 and 3 dS m^{-1} but decreased significantly in the other two higher salinity treatments. Therefore, irrigating chile peppers with up to 3 dS m^{-1} brackish water could be possible by maintaining appropriate leaching fractions to sustain chile pepper production in freshwater-scare areas, where brackish groundwater is the only available source of irrigation. The yield response curves showed that the yield reductions in the chile peppers irrigated with natural brackish water were lesser compared to those of NaCl-dominant solution studies.

Low yield reductions could be related to significant Ca^{2+} concentrations in the brackish groundwater and RO concentrate. However, there is further need to investigate the effects of different Na^+/Ca^{2+} concentrations on plant physiology, water transport, ion content and transport, growth, nutrition, and yields for improving the salt tolerance of chile peppers.

Author Contributions: Conceptualization, G.S.B. and M.K.S.; methodology, G.S.B., M.K.S., P.W.B., and S.J.W.; formal analysis, G.S.B. and R.K.S.; writing—original draft preparation, G.S.B.; writing—review and editing, M.K.S., R.K.S., P.W.B., S.J.W. and R.S.; visualization, G.S.B. and R.K.S.; supervision, M.K.S.; project administration, M.K.S. All authors have read and agreed to the published version of the manuscript.

Funding: This work was partially supported by the USDA National Institute of Food and Agriculture, Hatch project 1006850, and Nakayama Chair endowment.

Acknowledgments: Authors would like to acknowledge New Mexico State University Agricultural Experiment Station for the financial support through Graduate Research Award and Brackish Groundwater National Desalination Research Facility in Alamogordo, NM for providing the groundwater and RO concentrate. Authors thank Barbara Hunter and Jacob Pino for their assistance in analyzing soil and water samples.

Conflicts of Interest: The authors declare no conflict of interest.

References

1. O'Neill, M.P.; Dobrowolski, J.P. Water and agriculture in a changing climate. *HortScience* **2011**, *46*, 155–157. [CrossRef]

2. Singh, M.; Saini, R.K.; Singh, S.; Sharma, S.P. Potential of integrating biochar and deficit irrigation strategies for sustaining vegetable production in water-limited regions: A review. *HortScience* **2019**, *54*, 1872–1878. [CrossRef]

3. Wallace, J. Increasing agricultural water use efficiency to meet future food production. *Agric. Ecosyst. Environ.* **2000**, *82*, 105–119. [CrossRef]

4. Pasternak, D.; De Malach, Y. Irrigation with brackish water under desert conditions X. Irrigation management of tomatoes (Lycopersicon esculentum Mills) on desert sand dunes. *Agric. Water Manag.* **1995**, *28*, 121–132. [CrossRef]

5. Ayars, J.E.; Schoneman, R.A. Irrigating field crops in the presence of saline groundwater. *Irrig. Drain.* **2006**, *55*, 265–279. [CrossRef]

6. Hibbs, B.J.; Boghici, R.N.; Hayes, M.E.; Ashworth, J.B.; Hanson, A.N.; Samani, Z.A.; Kennedy, J.F.; Creel, R.J. *Transboundary Aquifers of the El Paso/Ciudad Juarez/Las Cruces Region*; Contract Report; Texas Water Development Board, Austin & New Mexico Water Resources Research Institute: Las Cruces, NM, USA, 1997; p. 148.

7. Flores, A.M.; Shukla, M.K.; Daniel, D.; Ulery, A.L.; Schutte, B.J.; Picchioni, G.A.; Fernald, S. Evapotranspiration changes with irrigation using saline groundwater and RO concentrate. *J. Arid Environ.* **2016**, *131*, 35–45. [CrossRef]

8. Assouline, S.; Russo, D.; Silber, A.; Or, D. Balancing water scarcity and quality for sustainable irrigated agriculture. *Water Resour. Res.* **2015**, *51*, 3419–3436. [CrossRef]

9. Kankarla, V.; Shukla, M.; VanLeeuwen, D.; Schutte, B.; Picchioni, G. Growth, evapotranspiration, and ion uptake characteristics of alfalfa and triticale irrigated with brackish groundwater and desalination concentrate. *Agronomy* **2019**, *9*, 789. [CrossRef]

10. Ozturk, O.F.; Shukla, M.K.; Stringam, B.; Picchioni, G.A.; Gard, C. Irrigation with brackish water changes evapotranspiration, growth and ion uptake of halophytes. *Agric. Water Manag.* **2018**, *195*, 142–153. [CrossRef]

11. Glenn, E.P.; Brown, J.J.; Blumwald, E. Salt tolerance and crop potential of halophytes. *Crit. Rev. Plant Sci.* **1999**, *18*, 227–255. [CrossRef]

12. Läuchli, A.; Epstein, E. Plant responses to saline and sodic conditions. *Agric. Salin. Assess. Manag.* **1990**, *71*, 113–137.

13. Läuchli, A.; Grattan, S. Plant growth and development under salinity stress. In *Advances in Molecular Breeding toward Drought and Salt Tolerant Crops*; Springer: Dordrecht, Netherlands, 2007; pp. 1–32.

14. Katerji, N.; Van Hoorn, J.; Hamdy, A.; Mastrorilli, M. Salinity effect on crop development and yield, analysis of salt tolerance according to several classification methods. *Agric. Water Manag.* **2003**, *62*, 37–66. [CrossRef]

15. Nass, U. *Quick Stats*; USDA-NASS: Washington, DC, USA, 2016.

16. Maas, E.V.; Hoffman, G.J. Crop salt tolerance—Current assessment. *J. Irrig. Drain. Div.* **1977**, *103*, 115–134.

17. Chartzoulakis, K.; Klapaki, G. Response of two greenhouse pepper hybrids to NaCl salinity during different growth stages. *Sci. Hortic.* **2000**, *86*, 247–260. [CrossRef]
18. Navarro, J.; Garrido, C.; Carvajal, M.; Martinez, V. Yield and fruit quality of pepper plants under sulphate and chloride salinity. *J. Hortic. Sci. Biotechnol.* **2002**, *77*, 52–57. [CrossRef]
19. Cornillon, P.; Palloix, A. Influence of sodium chloride on the growth and mineral nutrition of pepper cultivars. *J. Plant Nutr.* **1997**, *20*, 1085–1094. [CrossRef]
20. Niu, G.; Rodriguez, D.S.; Crosby, K.; Leskovar, D.; Jifon, J. Rapid screening for relative salt tolerance among chile pepper genotypes. *HortScience* **2010**, *45*, 1192–1195. [CrossRef]
21. Aktas, H.; Abak, K.; Cakmak, I. Genotypic variation in the response of pepper to salinity. *Sci. Hortic.* **2006**, *110*, 260–266. [CrossRef]
22. Grattan, S.; Grieve, C. Salinity-mineral nutrient relations in horticultural crops. *Sci. Hortic.* **1998**, *78*, 127–157. [CrossRef]
23. Carter, C.T.; Grieve, C.M. Mineral nutrition, growth, and germination of Antirrhinum majus L.(snapdragon) when produced under increasingly saline conditions. *HortScience* **2008**, *43*, 710–718. [CrossRef]
24. Baath, G.S.; Shukla, M.K.; Bosland, P.W.; Steiner, R.L.; Walker, S.J. Irrigation water salinity influences at various growth stages of Capsicum annuum. *Agric. Water Manag.* **2017**, *179*, 246–253. [CrossRef]
25. Shani, U.; Dudley, L. Field studies of crop response to water and salt stress. *Soil Sci. Soc. Am. J.* **2001**, *65*, 1522–1528. [CrossRef]
26. Tripler, E.; Ben-Gal, A.; Shani, U. Consequence of salinity and excess boron on growth, evapotranspiration and ion uptake in date palm (Phoenix dactylifera L., cv. Medjool). *Plant Soil* **2007**, *297*, 147–155. [CrossRef]
27. Sharma, H.; Shukla, M.K.; Bosland, P.W.; Steiner, R.L. Physiological responses of greenhouse-grown drip-irrigated chile pepper under partial root zone drying. *HortScience* **2015**, *50*, 1224–1229. [CrossRef]
28. Gavlak, R.G.; Horneck, D.A.; Miller, R.O. *Plant, Soil, and Water Reference Methods for the Western Region*; Western Rural Development Center: Fairbanks, AK, USA, 1994.
29. Van Genuchten, M.T.; Hoffman, G.F. Analysis of crop salt tolerance data. In *Soil Salinity under Irrigation: Processes and Management*; Ecology studies; Shainberg, I., Shalhevet, J., Eds.; Springer-Verlag: New York, NY, USA, 1984; pp. 258–271.
30. Institute, S. *Base SAS 9.4 Procedures Guide*; SAS Institute: Cary, NC, USA, 2015.
31. Grothendieck, G. Non-linear regression with brute force. R package nls2. Available online: http://CRAN.R-project.org/package=nls2 (accessed on 30 December 2019).
32. Acosta Motos, J.; Ortuño, M.; Bernal-Vicente, A., Diaz-Vivancos, P., Sanchez-Blanco, M.; Hernandez, J. Plant responses to salt stress: Adaptive mechanisms. *Agronomy* **2017**, *7*, 18. [CrossRef]
33. Ben-Gal, A.; Ityel, E.; Dudley, L.; Cohen, S.; Yermiyahu, U.; Presnov, E.; Zigmond, L.; Shani, U. Effect of irrigation water salinity on transpiration and on leaching requirements: A case study for bell peppers. *Agric. Water Manag.* **2008**, *95*, 587–597. [CrossRef]
34. Beltrán, J.M.N. Irrigation with saline water: Benefits and environmental impact. *Agric. Water Manag.* **1999**, *40*, 183–194. [CrossRef]
35. Bhantana, P.; Lazarovitch, N. Evapotranspiration, crop coefficient and growth of two young pomegranate (Punica granatum L.) varieties under salt stress. *Agric. Water Manag.* **2010**, *97*, 715–722. [CrossRef]
36. Homaee, M.; Schmidhalter, U. Water integration by plants root under non-uniform soil salinity. *Irrig. Sci.* **2008**, *27*, 83–95. [CrossRef]
37. Zhang, J.; Xu, X.; Lei, J.; Li, S.; Hill, R.; Zhao, Y. The effects of soil salt crusts on soil evaporation and chemical changes in different ages of Taklimakan Desert Shelterbelts. *J. Soil Sci. Plant Nut.* **2013**, *13*, 1019–1028. [CrossRef]
38. Al-Karaki, G.N. Growth of mycorrhizal tomato and mineral acquisition under salt stress. *Mycorrhiza* **2000**, *10*, 51–54. [CrossRef]
39. Yurtseven, E.; Kesmez, G.; Ünlükara, A. The effects of water salinity and potassium levels on yield, fruit quality and water consumption of a native central anatolian tomato species (Lycopersiconesculantum). *Agric. Water Manag.* **2005**, *78*, 128–135. [CrossRef]
40. Reina-Sánchez, A.; Romero-Aranda, R.; Cuartero, J. Plant water uptake and water use efficiency of greenhouse tomato cultivars irrigated with saline water. *Agric. Water Manag.* **2005**, *78*, 54–66. [CrossRef]
41. Hanson, B.; Grattan, S.R.; Fulton, A. Agricultural salinity and drainage. In *University of California Irrigation Program*; University of California: Davis, CA, USA, 1999.

42. Rhoades, J.D.; Kandiah, A.; Marshali, A.M. *The Use of Saline Waters for Crop Production*; FAO: Rome, Italy, 1992.
43. Maas, E.V.; Poss, J.A.; Hoffman, G.J. Salt tolerance of plants. *Appl. Agric. Res.* **1986**, *1*, 12–26.
44. De Pascale, S.; Ruggiero, C.; Barbieri, G.; Maggio, A. Physiological responses of pepper to salinity and drought. *J. Am. Soc. Hortic. Sci.* **2003**, *128*, 48–54. [CrossRef]
45. Van Genuchten, M.T.; Gupta, S. A reassessment of the crop tolerance response function. *J. Indian Soc. Soil Sci.* **1993**, *41*, 730737.
46. Ehret, D.; Redmann, R.; Harvey, B.; Cipywnyk, A. Salinity-induced calcium deficiencies in wheat and barley. *Plant Soil* **1990**, *128*, 143–151. [CrossRef]
47. Cramer, G.R.; Spurr, A.R. Salt reponses of lettuce to salinity. II. Effect of calcium on growth and mineral status. *J. Plant Nutr.* **1986**, *9*, 131–142. [CrossRef]
48. Cabañero, F.J.; Martínez-Ballesta, M.C.; Teruel, J.A.; Carvajal, M. New evidence about the relationship between water channel activity and calcium in salinity-stressed pepper plants. *Plant Cell Physiol.* **2006**, *47*, 224–233. [CrossRef]
49. Busch, D. Calcium regulation in plant cell and his role in signalling. *Annu. Rev. Plant Physiol.* **1995**, *46*, 95–102. [CrossRef]

© 2020 by the authors. Licensee MDPI, Basel, Switzerland. This article is an open access article distributed under the terms and conditions of the Creative Commons Attribution (CC BY) license (http://creativecommons.org/licenses/by/4.0/).

 *horticulturae*

Article

Alterations in the Chemical Composition of Spinach (*Spinacia oleracea* L.) as Provoked by Season and Moderately Limited Water Supply in Open Field Cultivation

Christine Schlering [1,2,*,†], Jana Zinkernagel [2], Helmut Dietrich [1], Matthias Frisch [3] and Ralf Schweiggert [1]

[1] Department of Beverage Research, Geisenheim University, 65366 Geisenheim, Germany; Helmut.Dietrich@hs-gm.de (H.D.); Ralf.Schweiggert@hs-gm.de (R.S.)
[2] Department of Vegetable Crops, Geisenheim University, 65366 Geisenheim, Germany; Jana.Zinkernagel@hs-gm.de
[3] Department of Biometry and Population Genetics, Justus Liebig University, 35392 Giessen, Germany; matthias.frisch@uni-giessen.de
* Correspondence: Christine.Schlering@hs-gm.de
† Present address: Department of Soil Science and Plant Nutrition, Department of Microbiology and Biochemistry, Geisenheim University, 65366 Geisenheim, Germany.

Received: 6 March 2020; Accepted: 7 April 2020; Published: 10 April 2020

Abstract: The current use and distribution of agricultural water resources is highly prone to effects of global climate change due to shifting precipitation patterns. The production of vegetable crops in open field cultivation often requires demanding water applications, being impaired in regions where climate change will increasingly evoke water scarcity. To date, increasingly occurring precipitation-free periods are already leading to moderate water deficits during plant growth, e.g., in southern Europe. Among all vegetable crops, leafy vegetables such as spinach (*Spinacia oleracea* L.) are particularly vulnerable to limited water supply, because leaf expansion is highly dependent on water availability. Besides biomass production, water limitation might also affect the valuable nutritional composition of the produce. Therefore, we investigated the impact of moderately reduced water supply on the chemical composition of spinach, cultivated in the open field in three consecutive years. Two different water supply treatments, full and reduced irrigation, were used in a randomized block design consisting of three sets of six plots each. In the reduced water supply treatment, the total amount of supplied water, including both irrigation and natural precipitation, amounted to 90%, 94% and 96% in 2015, 2016 and 2017, respectively, of the full, optimal water supply treatment. Spinach grown under limited water supply showed significantly higher fresh biomass-based contents of polyols (e.g., inositol, glycerol), ascorbic acid, potassium, nitrogen, phosphorous, zinc and manganese, as well as total flavonoids and carotenoids. Increased dry biomass-based levels were found for total inositol, zinc and manganese, as well as decreased levels for malic acid, fumaric acid, phosphate and chloride. Furthermore, we report a high seasonal variation of several minor phytochemicals, such as single flavonoids. Spinacetin derivatives, spinatoside-glucoside as well as a rather unusual hexuronylated methylenedioxy flavonoid showed highest amounts when grown under relatively low irradiation in autumn. Levels of patuletin derivatives tended to increase under high irradiation conditions during spring. In summary, the chemical composition of spinach was shown to be highly sensitive to moderately reduced water supply and seasonal variation, but the overall nutritional quality of fresh marketable spinach was only marginally affected when considering health-related constituents such as minerals, trace elements, flavonoids and carotenoids.

Keywords: vegetables; water deficit; climate change; polyols; minerals; flavonoids; carotenoids

1. Introduction

The ongoing climate change as well as the associated side effects, such as rising temperatures and substantial shifts in precipitation patterns, may lead to less favorable conditions for the cultivation of the respective, currently grown crop plants. Future projections of precipitation indicate even more severe conditions for almost the entire European continent [1]. Changing climatic conditions do not only boost temperature and rainfall fluctuations, but also consequently influence soil evaporation and plant transpiration [2]. Adverse distributions of rainfall during the cultivation period are already likely to create short-term water deficiencies in horticultural crops, which lead to sub-optimal conditions for plant growth. These temporary stress-related conditions may induce complex interactions in plant metabolism, especially under heterogenic requirements in open-field cultivation, where a combination of different stress factors simultaneously occurs. The results in terms of physiological and biochemical responses can be very different as a function of growth stage, severity and duration of stress [3].

In addition, plant responses to stress are dependent on the tissue or organ affected by the stress [4]. The reaction of plants to drought consists of numerous coordinated processes to alleviate both cellular hyperosmolarity and ion disequilibria [5]. Plants respond to drought stress with physiological and biochemical changes, aiming at the retention of water against a high external osmotic pressure and the maintenance of photosynthetic activity, while stomatal opening is reduced to counteract water loss [5]. While severe drought stress often results in clearly unmarketable crops, mild and moderate limitations in water supply induce more subtle changes and have earlier been considered to even enhance the formation of health-promoting antioxidant constituents in leafy vegetables [6]. However, this effect is highly plant-specific. For instance, experiments with moderately induced drought stress towards the end of the cultivation period followed by re-watering did not lead to significant increases in the concentrations of antioxidant compounds such as carotenoids and tocopherols in spinach (*Spinacia oleracea* L.), but did increase the levels of the aforementioned compounds in rocket [7]. Therefore, smart irrigation strategies are already used in horticultural systems to reduce the consumed irrigation water and even improve harvest quality [8]. However, severe water deficits evoked by fewer precipitation events or scarcity of irrigation water can be hard to manage and thereby cause short-term stress events during plant development.

Leafy vegetables such as spinach are usually highly sensitive to water deficits, because transpiration is affected due to stomatal closure. Stomatal closure as well as leaf growth inhibition are among the earliest responses to drought, protecting the plants from extensive water loss [3]. Since the biological function of plant leaves, i.e., photosynthesis, requires their exposure to sunlight and air, they might be expected to most sensitively react to a number of stress factors. The sensitivity of the photosynthetic apparatus is the basis of chlorophyll fluorescence measurements for recognizing plant stress prior to other physiological and even macroscopically visible stress responses [9]. Among the latter, the total leaf area is determinative for biomass accumulation and crop yield and also represents a factor highly correlated with detrimental effects on crop growth under stressful conditions [10]. Mild environmental stress was shown to lead to a significant decrease in the yield of freshly harvested spinach due to diminution of the relative water content, which recovered to control values within three days after re-watering [7]. However, plant growth remained affected. Thus, attempts to enhance levels of health-promoting phytochemicals like antioxidants by manipulating environmental factors may be burdened by a drop in yield [7]. Water deficit inhibits plant growth by reducing water uptake into the expanding cells, and enzymatically alters the robustness and plastic properties of the cell wall [4]. A crop's need for water varies considerably among species, but water requirements for horticultural crops are generally high [11]. Especially in the case of leafy vegetables, constant amounts of available water are crucial [11].

Spinach is a leafy vegetable belonging to the long-day plants, which flower, or bolt, if the light period is as long as or longer than a 'critical day length' [11]. In northern regions, spinach is usually grown under short-day conditions in spring or fall in order to avoid bolting [11] and unfavorable textural alterations like fibrousness and stringiness. In brief, bolting renders spinach unmarketable. Similar to other leafy vegetables like lettuce or cabbage, spinach can tolerate lower light levels than fruiting vegetable crops [11]. Spinach is viewed as a vegetable with high nutritional quality [12], providing many health-promoting antioxidant constituents like carotenoids, flavonoids and other phenolic compounds [13], as well as considerable amounts of minerals, trace elements and vitamins, like vitamin C [14]. As compared to other flavonoid-rich vegetables such as Swiss chard (2.700 mg/kg fresh matter (FM) [15]) or red lettuce (1.400 mg/kg FM [16]), spinach showed similarly high contents of flavonoids (1.000 mg/kg FM [17]). Spinach is also known to accumulate high levels of rather undesired substances such as nitrate (547–3.350 mg/kg FM) and oxalate (2.309–10.108 mg/kg FM) [18]. With respect to its high nutritional quality, we studied how moderately reduced water supply might influence the chemical composition and, thus, the product quality of field-grown spinach. By evaluating this during three different cultivation years, this study also demonstrated the seasonal variability of the above-mentioned nutritionally relevant substances in spinach.

2. Materials and Methods

2.1. Chemicals

All reagents and solvents used were at least of analytical or HPLC quality, unless specified differently. Folin–Ciocalteu's phenol reagent and sodium carbonate were purchased from Merck (Darmstadt, Germany), L-(+)-ascorbic acid from Carl Roth (Karlsruhe, Germany), (+)-catechin-hydrate, *myo*-inositol (>99.5%, HPLC), anhydrous glycerol (≥99.5%) and *meso*-erythritol (≥99%) from Sigma Aldrich (Steinheim, Germany). Quercetin-4'-O-glucoside (spiraeoside), (all-*E*)-lutein and (all-*E*)-β-carotene were obtained from Extrasynthese (Genay Cedex, France). L-aspartic acid (Ph. Eur., USP) was received from AppliChem (Darmstadt, Germany).

2.2. Plant Material and Field Experimental Design

Spinach cv. 'Silverwhale' seeds were purchased from Rijk Zwaan Welver (Welver, Deutschland) and were grown under open field conditions on a sandy loam at Geisenheim University, Germany (49°59′ N, 7°58′ E). The experimental field design has been reported earlier in detail by Schlering et al. [19]. In brief, one cultivation set per year was conducted in a plot installation between the years 2015 and 2017, resulting in three years, i.e., cultivation replicates, being grown as shown in Table 1. Each set consisted of six circular plots (marked by A, C, F, H, M and P) with an inner diameter of 11.9 m. Each plot was subdivided into four quarters (subplots) for the simultaneous cultivation of different vegetable crops with an annual crop rotation. Each subplot quarter was again divided into two further segments for the implementation of the different irrigation treatments following a randomized block design (cf. Schlering et al. [19]). The size of each segment was at least 6.5 m², respectively. A third segment located outwardly was not harvested to avoid boundary effects but cultivation and irrigation took place as described in the next section. The irrigation levels of the subplots described below remained unchanged during the full experimental period, irrespective of the annual crop rotation.

Table 1. Cultivation and climate data of the spinach years during the different experimental periods. DAS: days after sowing. Irrigation (mm): Total irrigation amount including irrigation during initial stage. Total water amount (%) RWS: water amount (%) of the reduced variants including watering after sowing.

Growing Season	2015	2016	2017
Sowing date (Year-month-day)	2015-09-07	2016-04-19	2017-04-20
Beginning of water supply differentiation (DAS)	23	23	26
Harvest date (DAS)	51	44	39
Temperature sum (°C)	626.0	593.2	549.8
Daily mean air temperature (°C)	12.0	13.2	13.8
Mean relative air humidity (%)	78.8	68.7	67.8
Global irradiation sum (MJ/m^2)	453.5	799.9	756.9
Daily mean global irradiation (MJ/m^2)	8.7	17.8	18.9
Wind speed sum (m/s) at height of 2 m	46.0	69.3	50.8
Evapotranspiration sum (mm) [z]	68.7	126.9	126.6
Precipitation sum (mm)	65	101	68
Number of differentiated irrigation events	4	5	2
Total irrigation (mm) [y]	39	95	45
Total water amount (mm) incl. precipitation	104	184	113
Total water amount of RWS (% of CTR)	**90%**	**94%**	**96%**

[z] Reference evapotranspiration (ETo) using grass and the FAO56 Penman–Monteith method. [y] Includes irrigation during the treatment period (6.2 L/m^2 x irrigation events) and the irrigation during initial growth stages.

2.3. Cultivation and Water Supply

Spinach seeds were sown by a manual seed-drilling machine with row spacing of ca. 0.25 m and a sowing distance of ca. 0.013 m, corresponding to a sowing density of ca. 308 seeds per m^2. Uniform fertilization with calcium ammonium nitrate was carried out according to commercial standard specifications for the cultivation of standard-quality spinach for the fresh food market (135 kg/N ha^{-1}) based on mineralized N ($NO_3^- $-N) in 0–30 cm soil depth.

Crop protection was applied equally to all plots, whereby application depended on the growing set, described as follows: Chemical and biological insecticides were used depending on pest occurrence. NeemAzal®-T/S (Trifolio-M GmbH, Lahnau, Germany) was used against leafminer flies (Agromyzidae) and Fastac ME (BASF-SE, Limburgerhof, Germany) was applied against aphids (Sternorrhyncha). Goldor®Bait (BASF, Ludwigshafen, Germany) had been brought into the soil before sowing once per cultivation set against wireworms (Elateridae). Weed control was done manually.

Water supply by drip irrigation was generally activated when the soil suction tension fell below −20 kPa at a 10 cm depth, as controlled by a tensiometer with an electronic pressure sensor (Tensio-Technik, Bambach, Geisenheim and Deutschland). During the initial growth stages, i.e., until the appearance of the seedlings, the soil moistures of both the well-watered control (CTR) and the reduced water supply (RWS) treatments were kept evenly moist by irrigation. Depending on the year, this period of identical water supply lasted for 19 days (2015), 20 days (2016) and 17 days (2017). Then, the well-irrigated CTR segments were provided 6.2 L/m^2 per irrigation, whereas the RWS treatment was reduced to ca. 50% of that of the CTR treatment. Because of the natural precipitation, both the volume of irrigation water and also the ultimate total water volume varied strongly between years. As a result, the total water volume of the RWS was 90% (2015), 94% (2016) and 96% (2017) of that of the CTR treatments during the respective years (Table 1).

2.4. Harvest

A total of 40 plants per segment was harvested randomly when most of the leaves were unfolded, representing the commercially targeted maturation stage for fresh-marketed spinach. To avoid border effects, outer rows were omitted from harvest. Specifically, harvest dates were 51, 44 and 39 days after sowing years in 2015, 2016 and 2017, respectively (Table 1). After harvest, plants were cleaned twice manually with fresh tap water to remove adhered soil particles and roots were removed with a knife, before leaves were spun using a commercial salad spinner to remove remaining water. Subsequently, the fresh spinach leaves were frozen at −80 °C until further analyses.

2.5. Climatic Data

Local climate data were supplied by the local weather station, which was located at 100 m distance from the experimental field site in Geisenheim. Climatic parameters for each cultivation period are summarized in Table 1, while detailed weather conditions can be obtained from Figures S1–S4 in the Supplementary Materials. Based on global irradiation and total water amount (Table 1), the three different growing periods were characterized and assigned as follows: 2015 (less irradiation, dry), 2016 (more irradiation, moist) and 2017 (more irradiation, dry). More precisely, the cultivation periods 2016 and 2017 were characterized by high global irradiation and evapotranspiration (Supplementary Figures S2 and S4), but relative air humidity was considerably lower during the vegetation period in 2016 (Supplementary Figure S3) and the temporal development of the temperature profiles were opposite between the cultivation periods. While the temperature constantly decreased in 2015, it increased substantially during the growth periods in 2016 and 2017. Altogether, temperature sum and mean air humidity did not differ strongly between years, whereas their profiles during plant growth varied strongly (Supplementary Figures S1–S4). In addition, the precipitation profiles were quite different between years (data not shown). While natural precipitation was relatively high during the first half of the growing period in 2015 and 2017, when additional irrigation was mainly initiated in the last two weeks of cultivation, the opposite was true for the cultivation set in 2016, when natural precipitation was very high in the second half of the cultivation period.

2.6. Sample Preparation

For all analyses, except for the determination of ascorbic acid, 150–200 g frozen spinach leaves of each sample were lyophilized (BETA 2-8 LDplus, Martin Christ Gefriertrocknungsanlagen GmbH, Osterode am Harz, Germany) prior to grinding with a laboratory mill (IKA M 20; IKA-Werke, Staufen, Germany). Dry biomass content was determined gravimetrically using fresh and freeze-dried material. D-Glucose, D-fructose, titratable acidity, L-malic acid and fumaric acid, as well as inorganic anions like nitrate, phosphate, sulfate and chloride, were determined from aqueous extracts prepared from the lyophilized powder. For this purpose, an aliquot of 8–10 g of lyophilized plant material was thoroughly homogenized with 500 mL ultrapure water at room temperature and rigorously mixing for ca. 10 s using a stainless-steel food blender (setting 2, Waring Blender, Waring Commercial, Torrington, CT 06790, USA). After transferring the extract including solids into an 800 mL beaker using another 150 mL of added ultrapure water, extraction continued for 10 min under continuous magnetic stirring, followed by a single ultrasound-assisted extraction step in an ultrasound water bath for another 5 min. After centrifuging for 5 min at 4596× g to separate liquid and solid phases, the supernatants were collected, filtered and stored at −25 °C until analyses. Extraction procedures for all other target analytes are given below.

2.7. Chemical Analyses

Unless otherwise noted, IFU-methods (International Fruit Juice Union, Paris, France) were used for determination of routine parameters in aqueous extracts, such as sugars, total acidity, organic acids and inorganic anions.

2.7.1. Sugars and Polyols

D-Glucose and D-fructose were determined spectrophotometrically using enzymatic kits (R-Biopharm, Darmstadt, Germany) and a Konelab 20 Xti analyzer (ThermoFisher, Dreieich, Germany).

Determination of polyols excluding the aforementioned sugars was carried out in duplicates as described by Schlering et al. [19]. Quantitation was carried out with linear external calibrations of *myo*-inositol, glycerol and erythritol.

2.7.2. Organic Acids

Titratable acidity, calculated as citric acid, was measured potentiometrically after titration to pH 8.1 with 0.3 M NaOH (Titroline alpha, Schott, Mainz, Germany). L malic acid was determined spectrophotometrically using enzymatic kits (R-Biopharm, Darmstadt, Germany). Fumaric acid was determined by HPLC with UV detection and ascorbic acid by iodometric titration, as previously described [19].

2.7.3. Inorganic Anions

Nitrate, sulfate and phosphate were determined by ion chromatography and chloride by potentiometric titration with an AgCl-electrode, as described previously [19].

2.7.4. Total Carbon and Nitrogen

Elemental analyses of spinach samples for carbon and nitrogen were carried out in duplicate by the Dumas combustion method (Vario MAX CNS, Elementar Analysensysteme GmbH, Langenselbold, Germany), combusting 300 mg of lyophilized and powdered plant material at 950 °C. L-aspartic acid was used as a reference substance.

2.7.5. Minerals and Trace Elements

Minerals and trace elements were analyzed by inductively coupled plasma optical emission spectrometry (ICP-OES) (SPECTRO ARCOS, SPECTRO Analytical Instruments, Kleve, Germany) after Kjeldahl digestion with the Gerhardt Turbotherm rapid digestion unit (C. Gerhardt GmbH & Co. KG, Königswinter, Germany), as reported earlier [19].

2.7.6. Total Phenols

An aliquot of 500 mg of lyophilized powdered plant material was extracted twice with 12 mL of 70% aqueous methanol under ultrasonication for 30 min. After centrifuging for 10 min at $12,857 \times g$, the combined supernatants were made up to 25 mL with extraction solvent and stored at -28 °C until spectrophotometric analyses with the Folin–Ciocalteu reagent. A linear (+)-catechin calibration was used as described previously [20].

2.7.7. Flavonoids

Flavonoids were extracted from lyophilized powdered material as described for total phenols (see above) in duplicates. Extracts were filtered through a 0.2 µm membrane using a polytetrafluoroethylene (PTFE) syringe filter (Macherey-Nagel, Düren, Germany), before extracts were analyzed with reversed-phase HPLC equipped with a photodiode array detector (RP-HPLC/PDA) with negative electrospray ionization mass spectrometry (ESI-MS) on an Accela/LXQ system (ThermoFisher, Dreieich, Germany). Chromatographic separation was achieved on a 150×2 mm, 3 mm C18 (2) Luna column (Phenomenex, Aschaffenburg, Germany) protected with a guard cartridge of the same material. Injection volume was 4 µL and elution conditions were the following: flow rate 250 mL/min at 40 °C, solvent A was water/formic acid (95:5, v/v), solvent B was methanol, 1 min isocratic conditions with 10% B, linear gradient from 10% to 55% B in 25 min, followed by washing with 100% B and re-equilibrating the column. MS scan range was set at m/z 250–1050 (negative mode). The MS settings were: ESI source voltage +4.5 kV, capillary voltage 32 V, capillary temperature 275 °C, collision energy for MS^n experiments 30%. Flavonoids were identified by comparison of their retention times, UV/Vis and mass spectral data with those of authentic standards or published data [21,22]. Flavonoids were quantified at 360 nm using an external spiraeoside standard (Extrasynthese, Lyon, France) in a linear range of 5–100 mg/L.

2.7.8. Carotenoids

An aliquot of 75 mg of lyophilized powdered spinach was combined with ca. 0.4 g of $NaHCO_3$ in a glass tube prior to extraction with 4 mL of a mixture of n-hexane/acetone (2:3, v/v) using vortex stirring for 10 s. Then, the extracts were stored on ice in the dark for 10 min using an intermediate stirring step after 5 min. After centrifuging at 4596× g for 10 min at 4 °C, the supernatant was collected in 25 mL amber glass volumetric flasks previously flushed with N_2. The remaining solids were extracted further four times and the obtained extracts were combined and then transferred quantitatively to amber round-bottomed flasks prior to evaporation to dryness under reduced pressure at 25 °C (Rotavopar R-210, Büchi, Switzerland). The dried extract was re-dissolved in HPLC solvent A (see below), transferred to a 10 mL volumetric flask, made up to 10 mL, and filtered through a 0.2 μm membrane filter (Chromafil O-20/25 PTFE syringe filter; Macherey-Nagel, Düren, Germany) into an HPLC vial. Carotenoids contained in the extract were separated with a HPLC-PDA system consisting of a Dionex P 680 HPLC pump, a Dionex STH 585 column oven and Dionex PDA-100 Photodiode Array Detector (Dionex/Thermo Fisher Scientific, Germany), which was mounted with a YMC Carotenoid (C30)-column (4.6 × 250 mm, 5 μm particle size, YMC, Kyoto, Japan) protected with a guard column of the same material. Column temperature was 20 °C. Solvent A was a quaternary mixture of methanol (MeOH), methyl *tert*-butyl ether (MTBE), water and a 1 M ammonium acetate solution (AAc) (MeOH/MTBE/H_2O/AAc, 88:5:2:5, v/v/v/v) and solvent B a ternary mixture of MeOH/MTBE/AAc (20:78:2, v/v/v). A linear gradient from 0% to 85% B in 45 min followed by a linear gradient to 100% B in 5 min was applied, followed by an isocratic equilibration step with 100% B for 10 min. Flow rate was 1 mL/min. Total run time was 60 min. Injection volume was 20 μL. Compounds were identified by comparing their retention times and UV/Vis absorption spectra to those of authentic reference compounds ((all-*E*)-lutein, (all-*E*)-β-carotene) and literature (violaxanthin, neoxanthin, (*Z*)-isomers of β-carotene, Britton [23]). Carotenoids were quantified at 450 nm. An external linear calibration of authentic (all-*E*)-lutein was used for quantitation of violaxanthin, neoxanthin and (all-*E*)-lutein, while authentic (all-*E*)-β-carotene calibrations were used for quantitating (9*Z*)-β-carotene and (all-*E*)-β-carotene.

2.8. Statistical Analyses

Analytical results were calculated based on both dry and fresh biomass (i) to enable evaluations regardless of different water contents in the plant material by dry biomass-related data and (ii) to assess the nutritional values of the harvested edible plant material by fresh biomass-related data. Data were analyzed by fitting a linear mixed-effect model using the lmer-function within the lme4-package [24] of the statistical software R [25] in RStudio [26]. The evaluation of the single years was based on a model which aimed at correcting for random plot-effects (Equation (1)), while the total dataset was evaluated with respect to the interaction of plot and year (Equation (2)):

$$y \sim H_2O + (1|\text{plot}) \tag{1}$$

$$y \sim H_2O + (1|\text{plot:year}) \tag{2}$$

Comparisons of means derived from different treatments were considered significantly different if $p < 0.10$. Pairwise comparison of least-squares means was carried out with lsmeans-package [27] to estimate fixed-effects of the treatment as well as random effects for the plot and season. Significances of random effects were calculated by the lmer test-package [28]. The results for the treatment (CTR: control, RWS: reduced) were evaluated on the basis of adjusted data generated from the model mentioned above. Principal Component Analysis (PCA) was carried out by using the R-packages FactoMineR [29] and factoextra [30]. Within PCA analysis, individual samples were visualized in a score plot, while corresponding chemical components were represented in a loading plot inside a correlation circle presenting the relationship between the variables. To avoid an overweight of flavonoids and carotenoids by representation of individual components, only total amounts were considered for PCA.

3. Results and Discussion

3.1. Evaluation of Years and Season

According to the PCA of data that had been corrected for plot effects (Equation (1)), the three years derived from three consecutive growing years (2015, 2016 and 2017) were clearly differentiated by their chemical composition, highlighting the strong influence of the year on the product quality. Similar results were already shown for radish grown within the same experimental field site [19]. The first two principal components (PCs) accounted for 63.7% of the total variance considering dry biomass-related data (Figure 1) and for 65.6% considering fresh biomass-related data (data not shown). Similar year-to-year variations have been reported earlier for other vegetable crops [31,32]. While the two sets from 2015 and 2016 were clearly distinguishable over the first principal component (Dim1, Figure 1), the set from 2017 was additionally separated over the second principal component (Dim 2). The year 2015 was characterized by comparably low sunlight exposure (global irradiation: 453.5 MJ/m^2) as compared to 2016 (799.9 MJ/m^2) and 2017 (756.9 MJ/m^2). Consequently, evapotranspiration was higher in 2016 (126.9 mm) and 2017 (126.6 mm) in contrast to 2015 (68.7 mm). The years 2016 and 2017 were different when considering the total water amount the plants received. In 2016, total water supply was higher (184 mm) than 2017 (113 mm), with the latter being similar to that in 2015 (104 mm, Table 1).

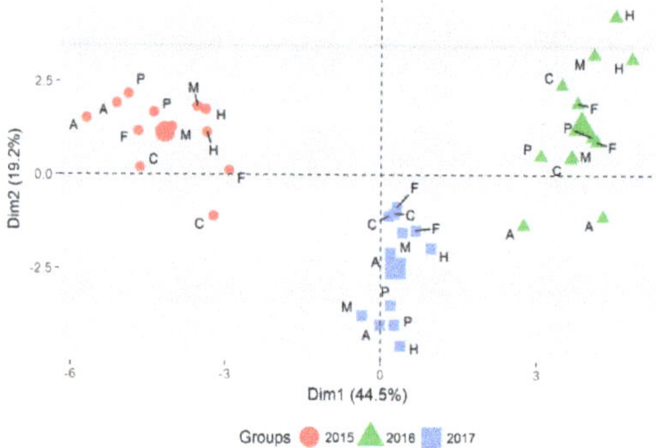

Figure 1. Principal Component Analysis (PCA) of the original, plot-adjusted dataset including all spinach years from 2015 (red circles), 2016 (green triangles) and 2017 (blue squares). Score plot represents the individual samples of each cultivation set (separated by color) and plot (marked by the letters A, C, F, H, M and P) based on dry biomass-related data.

3.2. Effects of Reduced Water Supply on the Chemical Composition of Spinach Biomass

The PCA of the entire dataset adjusted by the interaction of plot and year (Equation (2)) revealed that well-irrigated spinach samples were mainly clustered in sectors with negative PC2, while those grown under mildly reduced water supply were mainly located in sectors with positive PC2 (Figure 2A). Most of the individuals were separated by the second principal component PC2 (Dim2), which, however, accounted for only 14.1% of the total variance. The first two principal components (Dim1 + Dim2) accounted for 38.7% of the total variance in the dataset. As shown in the corresponding loading plot (Figure 2B), the most important variables with high contribution to Dim2 were dry-matter-based contents in trace elements like manganese (Mn), zinc (Zn), iron (Fe), total phenols, flavonoids, polyols, as well as nitrate, nitrogen (N), potassium (K) and carotenoids. Dim1 was mainly influenced by contents of sugars (glucose, fructose), organic acids (malic acid, fumaric acid) and anions (chloride, sulfate). Apart from these multivariate analyses, a univariate statistical evaluation shown in Table 2 was conducted to underpin the multivariate estimate. While CTR samples were characterized by higher contents of organic acids like malic and fumaric acid, as well as certain anions like chloride, phosphate and sulfate, the content of inositol and trace elements such as Zn and Mn as well as flavonoids was lower than in samples derived from RWS treatments (Table 2). However, the univariate results also pointed out that the levels of glucose, fructose, quantitatively abundant elements (K, N, P, Ca and Mg) and carotenoids remained widely unaffected by mild water reduction. Only in 2015, when the highest relevant water reduction was achieved, i.e., 10% less than the well-watered control, did some minerals and carotenoids show increased dry biomass-based contents in the RWS samples (Table 2). In our earlier study on red radish root tubers, the identical environmental conditions had not led to an increase in the corresponding levels [19]. In this study, leaves of spinach reacted to the identical water deficits by both the increase of polyols and the accumulation of certain trace elements such as Mn and Zn.

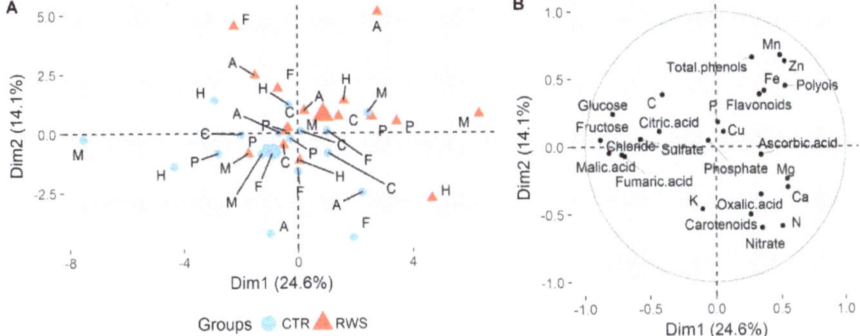

Figure 2. Multivariate evaluation by PCA (PC1 + PC2) of the total spinach dataset including all years (2015, 2016 and 2017) based on dry biomass-data. Score plot (**A**) represents all individual plot samples (marked by the letters A, C, F, H, M and P) classified according to the control (CTR, blue circles) and reduced water supply (RWS, red triangles). The corresponding loading plot (**B**) shows the related variables determined by the chemical analyses.

The distinction of both water supply treatments by PCA with fresh biomass-related data was significantly clearer (Figure 3A) than that based on dry biomass-related data (Figure 2A). With very few exceptions, fresh biomass-based data allowed for distinguishing both groups clearly over the first principal component (Dim1), explaining 28.8% of the total variance. Upon addition of the second principal component (Dim2), a share of 46.7% of the total variance was explained. In contrast to red radish root tubers grown under control and reduced water supply conditions [19], the fresh biomass-based discrimination of spinach from well-watered versus reduced water treatments was not mainly based on primary metabolites like polyols, total carbon and ascorbic acid, but also on secondary metabolites such as total phenols, flavonoids and carotenoids, as well as selected minerals and trace elements (Figure 3). In brief, fresh biomass-related contents of most of the targeted chemical components, especially those of polyols, minerals, trace elements and secondary metabolites such as flavonoids and carotenoids, were increased in spinach samples grown under reduced water supply. The univariate evaluation shown in Table 3 confirmed these results. The fresh biomass-based contents of all of the studied carotenoids, i.e., violaxanthin, lutein and ß-carotene, showed significant and rather uniform increases (Table 3, Supplementary Table S1).

Table 2. Influence of reduced water supply on the dry biomass (DM)-related levels of constituents of spinach leaves from different years (2015, 2016 and 2017). Linear mixed model, t-test, p-values: < 0.1, < 0.05 (*), < 0.01 (**) and < 0.001 (***); nd = not determined, CTR: Control treatment with full water supply, RWS: Reduced water supply treatment.

	Spinach 2015			Spinach 2016			Spinach 2017			Spinach Total		
	CTR	RWS	p-Value	CTR	RWS	p-Value	CTR	RWS	p-Value	CTR	RWS	p-Value
Leaf water potential Ψ (-MPa)	nd	nd	nd	0.14	0.18	0.1136	0.07	0.093	**0.0842**	nd	nd	nd
Dry biomass (%)	8.54	8.60	0.6186	5.40	5.48	0.6736	6.13	6.83	**0.0032 **	6.69	6.97	**0.0216 ***
Sugars and Polyols (mg/g)												
Glucose	56.71	59.55	0.3342	72.92	62.70	0.1521	50.92	57.26	0.3902	60.18	59.84	0.9207
Fructose	43.35	43.59	0.9073	37.58	30.94	**0.0721**	27.70	29.41	0.6028	36.21	34.65	0.3965
Polyols total	5.91	5.91	0.9565	3.87	4.09	**0.0302 ***	4.96	4.91	0.4331	4.91	4.97	0.2929
Inositol	1.66	1.68	0.3461	0.99	1.06	**0.0830**	1.48	1.54	0.1647	1.38	1.42	**0.0119 ***
Glycerol	2.56	2.58	0.3724	1.67	1.70	0.2236	2.31	2.28	0.3984	2.18	2.19	0.7136
Polyol unknown	1.37	1.38	0.7343	0.71	0.79	**0.0530**	0.67	0.61	0.1224	0.91	0.93	0.5387
Erythritol	0.33	0.28	0.5110	0.51	0.55	0.2218	0.50	0.47	0.2676	0.44	0.43	0.6779
Organic acids (mg/g)												
Oxalic acid	49.89	44.51	**0.0863**	62.56	62.56	1.0000	53.70	52.46	0.8734	55.38	53.18	0.4438
Total acidity	16.73	18.73	0.1270	17.20	15.71	0.5452	16.87	18.26	0.2514	16.93	17.57	0.5028
Malic acid	12.81	12.63	0.7896	22.08	18.47	**0.0103 ***	14.08	13.31	0.6206	16.33	14.80	**0.0389 ***
Ascorbic acid	9.41	9.90	**0.0493 ***	4.42	4.84	0.4642	6.85	6.70	0.7087	6.89	7.15	0.2866
Citric acid	6.19	6.14	0.8849	3.18	3.15	0.8647	2.94	3.40	0.1219	4.11	4.23	0.4872
Fumaric acid	0.61	0.57	**0.0761**	2.26	1.69	0.1530	1.30	1.11	0.5026	1.39	1.12	**0.0803**
Anions (mg/g)												
Nitrate	8.57	7.73	0.2291	28.16	28.60	0.5373	12.98	12.53	0.8169	16.57	16.29	0.6854
Phosphate	13.65	11.08	**0.0377 ***	16.69	17.13	0.1913	13.12	12.44	0.2995	14.49	13.55	**0.0594**
Sulfate	3.44	2.64	**0.0192 ***	7.73	6.56	**0.0910**	3.30	3.88	0.1040	4.82	4.36	0.1151
Chloride	5.38	4.95	0.1268	7.28	5.83	**0.0692**	8.61	7.98	0.1430	7.09	6.25	**0.0053 ***

Table 2. *Cont.*

	Spinach 2015			Spinach 2016			Spinach 2017			Spinach Total		
	CTR	RWS	*p*-Value	CTR	RWS	*p*-Value	CTR	RWS	*p*-Value	CTR	RWS	*p*-Value
Elements (mg/g)												
Carbon	381.17	380.38	0.7578	360.43	356.70	0.3442	359.68	362.83	0.3491	367.09	366.64	0.8316
Macronutrients												
Potassium	78.27	81.38	**0.0297** *	87.88	85.60	0.2540	90.68	89.89	0.7354	85.61	85.62	0.9932
Nitrogen	50.92	51.64	**0.0682**	38.98	38.07	0.7042	42.47	41.91	0.7088	44.12	43.87	0.7769
Phosphorous	5.59	5.93	**0.0596**	5.58	5.72	0.4096	5.22	5.00	0.1205	5.46	5.55	0.3650
Calcium	11.69	10.85	0.1553	14.78	14.94	0.8400	14.03	13.78	0.7268	13.50	13.19	0.4079
Magnesium	7.77	7.22	**0.0196** *	6.20	6.29	0.7254	7.57	7.45	0.7784	7.18	6.99	0.2684
Micronutrients (µg/g)												
Iron	323.89	273.68	0.2516	300.40	406.83	0.0586	348.97	336.95	0.6028	324.42	339.15	0.5711
Zinc	73.63	80.85	**0.0234** *	127.00	142.33	**0.0245** *	101.87	111.03	**0.0458** *	100.83	111.40	**0.0001** ***
Manganese	60.93	58.30	0.1549	85.46	100.41	**0.0039** **	76.31	81.83	0.0711	74.23	80.18	**0.0138** *
Copper	15.21	15.75	0.8240	12.92	13.70	**0.0112** *	13.70	12.64	0.2121	13.94	14.03	0.9141
Phenolic compounds (mg/g)												
Total phenols	10.31	10.09	0.4565	11.26	11.80	0.3630	11.42	11.38	0.9294	11.00	11.08	0.7317
Flavonoids	9.91	10.08	0.9490	9.91	10.56	0.1919	9.82	10.16	0.4886	10.08	9.88	10.27
Carotenoids (mg/g)												
Violaxanthin	1.13	1.18	0.1067	0.82	0.84	0.5267	1.78	1.71	0.5954	1.24	1.24	0.9793
Neoxanthin	0.38	0.39	0.1106	0.27	0.26	0.3837	0.65	0.64	0.8515	0.43	0.43	0.8136
Lutein	1.35	1.39	0.1431	0.94	0.90	0.6399	2.53	2.54	0.9492	1.61	1.61	0.9091
ß-Carotin	1.00	1.04	0.0558	0.91	0.95	0.3429	2.05	2.09	0.7092	1.32	1.36	0.2836
9-cis-ß-Carotin	0.11	0.12	0.0756	0.097	0.102	0.2031	0.248	0.250	0.9031	0.15	0.16	0.3772
Total carotenoids	3.98	4.11	0.0972	3.03	3.05	0.8677	7.25	7.22	0.9587	4.75	4.80	0.7542

Table 3. Influence of reduced water supply on the fresh biomass (FM)-related levels of constituents of spinach leaves from different years (2015, 2016 and 2017). Linear mixed model, t-test, p-values: < 0.1, < 0.05 (*), < 0.01 (**) and < 0.001 (***); nd = not determined, CTR: Control treatment with full water supply, RWS: Reduced water supply treatment.

	Spinach 2015			Spinach 2016			Spinach 2017			Spinach Total		
	CTR	RWS	p-Value	CTR	RWS	p-Value	CTR	RWS	p-Value	CTR	RWS	p-Value
Leaf water potential Ψ (-MPa)	nd	nd	nd	0.14	0.18	0.1136	0.07	0.093	**0.0842**	nd	nd	nd
Dry biomass (%)	8.54	8.60	0.6186	5.40	5.48	0.6824	6.13	6.83	**0.0032 ****	6.69	6.97	**0.0216 ***
Sugars and Polyols (mg/100 g)												
Glucose	536.71	563.58	0.3345	402.68	346.22	0.1522	318.31	357.98	0.3903	345.00	372.50	0.4335
Fructose	410.34	412.52	0.9085	207.51	170.87	**0.0721**	173.19	183.83	0.6033	223.89	240.56	0.3937
Polyols total	49.83	49.99	0.8777	20.66	23.29	**0.0065 ****	32.81	34.80	**0.0408 ***	34.43	36.03	**0.0048 ****
Inositol	14.16	14.29	0.5232	5.26	6.02	**0.0127 ***	9.09	10.53	**0.0170 ***	9.50	10.28	**0.0013 ****
Glycerol	21.73	21.97	0.5287	8.91	9.67	**0.0022 ****	14.16	15.56	**0.0115 ***	14.93	15.73	**0.0013 ****
Polyol unknown	11.63	11.76	0.7103	3.78	4.48	**0.0088 ****	4.09	4.22	0.5732	6.50	6.82	**0.0476 ***
Erythritol	2.32	1.96	0.4704	2.72	3.12	**0.0809**	5.46	4.49	**0.0254 ***	3.50	3.19	0.2038
Organic acids (mg/100 g)												
Oxalic acid	472.07	421.23	**0.0863**	345.47	345.47	1.0000	335.73	327.99	0.8732	384.42	364.90	0.3067
Total acidity	158.32	177.21	0.1271	94.97	86.74	0.5455	105.46	114.17	0.2515	119.58	126.04	0.3662
Malic acid	121.21	119.54	0.7926	121.93	101.96	**0.0103 ***	88.05	83.21	0.6203	110.39	101.57	**0.0528**
Ascorbic acid	80.49	85.30	**0.0568**	23.62	27.50	0.2745	41.96	46.02	0.1822	48.69	52.94	**0.0093 ****
Citric acid	58.61	58.10	0.8880	17.57	17.42	0.8771	18.39	21.29	0.1219	31.52	32.27	0.5873
Fumaric acid	5.72	5.12	0.1202	11.91	9.48	0.2684	7.81	7.62	0.9169	8.48	7.41	0.2225
Anions (mg/100 g)												
Nitrate	72.08	65.80	0.2778	150.48	162.49	**0.0238 ***	79.17	84.78	0.6029	100.58	104.35	0.4025
Phosphate	115.97	95.26	**0.0421 ***	88.92	97.02	**0.0027 ****	80.25	84.77	0.3201	95.04	92.35	0.5248
Sulfate	29.12	22.64	**0.0189 ***	41.28	37.07	0.2234	20.18	26.39	**0.0371 ***	30.20	28.70	0.4426
Chloride	46.13	43.02	0.1800	39.00	32.69	0.1556	52.82	54.48	0.6515	45.98	43.40	0.1913

Table 3. Cont.

	Spinach 2015			Spinach 2016			Spinach 2017			Spinach Total		
	CTR	RWS	p-Value	CTR	RWS	p-Value	CTR	RWS	p-Value	CTR	RWS	p-Value
Elements (mg/100 g)												
Carbon	3.26	3.28	0.8151	1.94	1.95	0.9071	2.20	2.48	**0.0105** *	2.47	2.57	**0.0687**
Macronutrients												
Potassium	662.16	694.89	**0.0185** *	468.56	485.03	0.1523	555.12	613.09	**0.0103** *	561.95	597.67	**0.0002** ***
Nitrogen	407.22	410.14	0.7606	176.17	202.64	**0.0065** **	238.58	262.42	**0.0097** **	273.99	291.73	**0.0013** **
Phosphorous	47.65	50.96	**0.0984**	29.77	32.39	**0.0135** *	31.94	34.09	**0.0297** *	36.45	39.15	**0.0004** ***
Calcium	99.76	93.89	0.1860	79.13	84.88	0.2441	86.12	94.28	0.1152	88.34	91.02	0.3357
Magnesium	66.71	62.63	**0.0117** *	33.16	35.67	0.1156	46.48	50.92	0.1760	48.78	49.74	0.4889
Micronutrients (mg/100 g)												
Iron	2.75	2.34	0.2768	1.60	2.31	**0.0314** *	2.14	2.30	0.2856	2.16	2.31	0.4017
Zinc (µg/100 g)	632.02	699.59	**0.0021** **	678.36	807.78	**0.0030** **	625.07	760.62	**0.0064** **	645.15	756.00	**0.0001** ***
Manganese (µg/100 g)	519.79	501.32	0.2620	456.93	568.91	**0.0009** ***	467.35	558.30	**0.0112** *	481.35	542.84	**0.0022** **
Copper (µg/100 g)	128.94	135.43	0.7665	68.96	77.63	**0.0004** ***	84.02	86.30	0.6883	93.97	99.79	0.4075
Phenolic compounds (mg/100 g)												
Total phenols	88.00	86.23	0.4172	60.78	64.93	0.4388	70.09	77.85	0.1354	72.95	76.34	0.1695
Flavonoids	86.52	86.93	0.8607	54.45	59.17	0.3192	62.81	73.40	**0.0495** *	67.93	73.17	**0.0308** *
Carotenoids (mg/100 g)												
Violaxanthin	9.66	10.05	0.1835	5.13	5.99	**0.0462** *	11.15	12.09	0.3774	8.65	9.38	**0.0439** *
Neoxanthin	3.18	3.28	0.1826	1.70	1.86	0.1231	4.07	4.55	0.1484	2.98	3.23	**0.0289** *
Lutein	11.52	11.87	0.2574	5.83	6.39	0.2920	15.91	18.00	0.1553	11.09	12.09	**0.0474** *
ß-Carotene	8.54	8.88	0.1166	5.63	6.73	**0.0311** *	12.85	14.81	0.1022	9.02	10.14	**0.0074** **
9-cis-ß-Carotene	0.98	1.00	0.4962	0.60	0.71	**0.0123** *	1.55	1.77	0.1128	1.04	1.16	**0.0138** *
Total carotenoids	33.88	35.07	0.1866	18.94	21.68	**0.0680**	45.53	51.21	0.1728	32.78	35.99	**0.0233** *

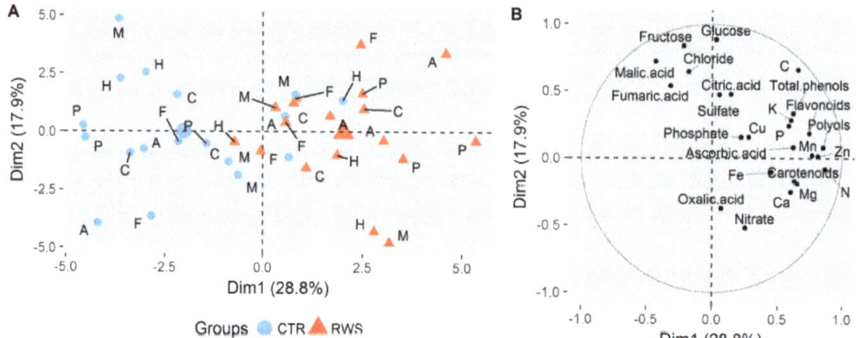

Figure 3. Multivariate evaluation by PCA (PC1 + PC2) of the total spinach dataset including all years (2015, 2016 and 2017) based on fresh biomass-data. Score plot (**A**) represents all individual plot samples (marked by the letters A, C, F, H, M and P) classified according to the control (CTR, blue circles) and reduced water supply (RWS, red triangles). The corresponding loading plot (**B**) shows the related variables, i.e., contents of constituents as determined by the chemical analyses.

It is noteworthy that the ratio of individual carotenoids in chloroplast-containing plant tissues is rather constant due to their pivotal role in photosynthesis, commonly 20%–25% β-carotene, 40%–45% lutein, 10%–15% violaxanthin and 10%–15% neoxanthin [33]. Occasionally, the levels of xanthophyll cycle carotenoids, such as antheraxanthin, were reported to be increased upon exposure of the plant to abiotic stress [34]. Findings similar to those of our study were found earlier for antioxidant compounds such as carotenoids and tocopherols in rocket (*E. sativa* Mill. var. Golden line) in the case of moderate drought stress followed by re-watering at the end of the cultivation period [7]. However, the same experiment did not result in higher levels of such antioxidants in spinach [7]. Altogether, fresh biomass-based contents of many other constituents in spinach showed a significant increase, possibly being related to higher dry biomass content of reduced watered plants (Table 2), which was also shown for cabbage (*Brassica oleracea* L.) subjected to drought stress just during head development [35].

3.3. Sugars and Polyols

The content of sugars such as glucose and fructose was hardly affected by moderate water reduction, but was characterized by high variations between the years irrespective of the water supply treatment. The dry biomass-related data emphasized that glucose was highest in 2016 (62.70–72.92 mg/g dry matter (DM)) in contrast to other years (50.92–59.55 mg/g DM), while fructose levels were highest in 2015 (43.35–43.59 mg/g DM) compared to 2016 (30.94–37.58 mg/g DM) and 2017 (27.70–29.41 mg/g DM, Table 2). As shown in Table 1, the 2015 set was characterized by comparably low irradiation (453.5 MJ/m^2) in contrast to 2016 (799.9 MJ/m^2) and 2017 (756.9 MJ/m^2), but the growing period was considerably longer in 2015 (51 days) in contrast to 2016 (44 days) and 2017 (39 days). These conditions, typically occurring during late growing seasons in autumn, resulted in the highest levels of fructose in 2015, while high light intensities, which are known to increase soluble carbohydrates in spinach [36], led to the highest contents of glucose in 2016. While both tended to decrease with mild water reduction in 2016 ($p = 0.1521$ and $p = 0.0721$ respectively, Table 2), when global irradiation and evapotranspiration were high, values were unaffected or increased in 2015 and 2017 due to mild water reduction. These trends were found for both dry and fresh biomass-related data.

In general, contents of dry biomass were significantly higher in 2015 (8.54%–8.60%) compared to other years (5.40%–6.83%), which automatically leads to enhanced values in the fresh biomass-related results (Table 3). By analogy, the overall glucose and fructose levels were higher in 2015 (537–564 mg and 410–413 mg per 100 g FM, respectively) than in other sets (318–403 mg glucose and 171–208 mg fructose per 100 g FM) (Table 3) without significant differences between the treatments.

Analogous to glucose and fructose contents, inconsistent effects of mild water reduction were found regarding the content of polyols in spinach dry biomass. Interestingly, the levels of polyols, particularly those of *myo*-inositol and glycerol, were lowest in 2016, although glucose levels were highest. These findings were in contrast to those found for radish, which showed congruently high levels of glucose and polyols [19]. Total polyols were generally lowest in 2016 (3.87 mg/g DM), being significantly increased to 4.09 mg/g DM upon reduced water supply ($p = 0.0302$) (Table 2). In contrast, levels in 2015 were higher (5.91 mg/g DM), but no changes in concentration were observed upon reduced water supply, although total water reduction was highest (−10%). These findings indicated that factors other than water supply had overruled a potential, marginal effect of the water supply. In our earlier study on red radish, a water reduction of 15%–20% led to significant increases of polyols [19]. Probably, reduced photosynthesis and the associated lowered growth under comparably "low light" conditions in 2015 (daily mean global irradiation: 8.7 MJ/m^2) have not influenced the assimilation and allocation of photosynthetic products as much as in 2016 and 2017, when plant growth was accelerated by higher levels of daily irradiation (17.8 and 18.9 MJ/m^2, respectively). In this context, Quick et al. [37] has shown that the levels of soluble carbohydrates in leaves of two annual crops (*Lupinus albus* L. and *Helianthus annuus* L.) were maintained when exposed to water stress under field conditions, even though photosynthesis was strongly inhibited. Maintenance of soluble sugars probably occurs because partitioning was altered in water-stressed plants [37]. In brief, the accumulation of polyols in spinach leaves seemed to be more sensitive to moderately reduced water supply than that in radish root tubers, which only showed increased levels after more severe reductions in water supply (≥15% lower than full water supply), as described in Schlering et al. [19].

3.4. Organic Acids

Organic acids' contents in spinach were inconsistently influenced by moderate water reduction, depending on the cultivation period (year). While dry biomass-based contents of ascorbic acid significantly increased in 2015, those of oxalic and fumaric acid decreased (Table 2). Differences in malic acid contents were insignificant comparing the treatments in 2015, but they were significantly reduced by RWS in 2016. Significant differences for malic acid were also found considering the entire dataset. Cutler and Rains [38] observed significantly increased accumulations of malic acid in leaves of cotton (*Gossypium hirsutum* L.) exposed to water stress, explained by osmotic adjustment and the importance of malate in turgor regulation of stomatal guard cells [38,39]. However, in our study on spinach, rather inconsistent changes in malic acid levels were observed, presumably because mild water reduction applied in this experiment had not been severe enough to cause sufficient stress to alter the plants' metabolism.

The significant increase of dry biomass-related ascorbic acid levels in spinach grown in 2015 ($p = 0.0493$) was in agreement with findings of Koyama et al. [6], who demonstrated an augmentation of ascorbic acid by lower water supply in hydroponically grown leafy vegetables such as lettuce and spinach. In contrast to our study, Koyama et al. [6] did not observe significant alterations in the water content. In our study, by analogy to polyols, ascorbic acid levels were highest in the relatively "low light, dry" cultivation set in 2015, both in dry and fresh biomass, possibly being related to lower global irradiation, which has previously been shown to enhance ascorbic acid contents in spinach [40]. The comparably high content of dry biomass in 2015 contributed to the increased fresh biomass-based levels of ascorbic acid (80.49–85.30 mg/100 g FM) as compared to other years (23.62–46.02 mg/100 g FM).

3.5. Inorganic Anions

The dry biomass-related contents of anions such as phosphate (PO_4^{3-}) and sulfate (SO_4^{2-}) were significantly decreased by RWS in 2015. By analogy, red radish root tubers studied within the same field experiment exhibited significant decreases of PO_4^{3-} upon similar water reduction, i.e., to 80% of full water supply [19]. Phosphate uptake into the roots might have been diminished under RWS [41], because its mobility depends on soil moisture [10]. On the other hand, the mineralization of organic matter by microbial activity, responsible for the release of ions such as $H_2PO_4^-$ and HPO_4^{2-} [11], might have been negatively affected due to water reduction. A significant decrease of dry biomass-related levels of SO_4^{2-} was measurable in 2015 ($p = 0.0192$), when total water reduction was highest compared to other years, but the contents of SO_4^{2-} ($p = 0.0910$) and chloride (Cl^-) ($p = 0.0692$) were also reduced in 2016, when water reduction was lower (Table 2). When considering the overall dataset, a dry biomass-related decline of PO_4^{3-} ($p = 0.0594$) and Cl^- ($p = 0.0053$) was observed in plants that had received slightly reduced water amounts, but the content of SO_4^{2-} also declined ($p = 0.1151$). In contrast, Abdel Rahman et al. [42] observed an accumulation of Cl^- in grasses and legumes as a result of decreasing soil moisture.

In contrast to the above-mentioned anions, the influence of RWS on the dry biomass-related nitrate (NO_3^-) levels in spinach was insignificant. However, an unexpected increase of NO_3^- in fresh biomass-based levels upon limited water supply was observed in 2016 ($p = 0.0238$, Table 3). The opposite was found in previously reported studies of fresh biomass-related nitrate levels of lettuce (*Lactuca sativa*) [6] and carrots (*Daucus carota* L.) [43] exposed to drought stress. In principle, NO_3^--flux from roots to leaves is believed to generally decrease during drought stress [44], being accompanied by a reduced expression of the nitrate reductase enzyme [45]. This did not apparently occur in spinach grown under RWS in our study, since dry biomass-related levels did not differ between water supply treatments. It is noteworthy that there was also a strong impact of the cultivation year. NO_3^- contents were substantially higher in the moist year 2016 (28.16–28.60 mg/g DM) compared to other comparably drier years 2015 and 2017 (7.73–12.98 mg/g DM), as shown in Table 2. Presumably due to frequent and high precipitation events towards the end of the cultivation period in 2016 (data not shown), a high absorption of NO_3^- might have led to higher NO_3^- levels. This hypothesis warrants further study. Nevertheless, in agreement with our study, a positive correlation was found between NO_3^- and water contents in spinach [46]. Ćustić et al. [47] also observed increased NO_3^- accumulation under warm and wet conditions in chicory (*Cichorium intybus*). It is known that NO_3^- accumulation varies with the season [47] and is greatly affected by environmental factors [48]. Santamaria et al. [49] observed more than doubled NO_3^- levels in rocket, accompanied by a decrease in dry biomass, when temperature increased from 10 to 20 °C. An inverse correlation of fresh biomass-based NO_3^- and oxalate levels with those of dry biomass content was also confirmed by our results. Therefore, the markedly increased NO_3^- levels in 2016 might have occurred due to high temperature (Supplementary Figure S1) and a simultaneously high water supply (data not shown) along with decreasing irradiation (Supplementary Figure S2) by the end of the cultivation period. It is well known that low irradiation promotes NO_3^- accumulation in spinach [11,50].

3.6. Carbon

The dry biomass-related content of total carbon (C) remained unaffected by RWS in spinach, as shown in Table 2. Fresh biomass-related levels were significantly higher in less irrigated plants (2.48 mg/g FM) in contrast to well-irrigated samples (2.20 mg/g FM) only in 2017, when, simultaneously, the dry biomass content also significantly increased, despite the comparably low water reduction (by 4%). Nevertheless, multivariate analysis supports the contribution of C to the separation of both groups in the PCA plot on a fresh biomass basis (Figure 3).

3.7. Nitrogen

Dry biomass-related total nitrogen (N) was not affected, but fresh biomass-related data was significantly influenced by RWS when considering the data across all three years (Tables 2 and 3). For instance, dry biomass-related N levels remained unchanged in 2016 and 2017, irrespective of the water supply treatment ($p = 0.7042$ and 0.7088, respectively), but they only marginally increased from 50.92 to 51.64 mg/g DM in 2015 ($p = 0.0682$). Upon reduced water supply, fresh biomass-related N levels significantly increased from 176.2 to 202.6 mg/100 g FM in 2016 ($p = 0.0065$) and from 238.58 to 262.42 mg/100 g FM in 2017 ($p = 0.0097$). A similar increase from 273.99 to 291.73 mg/100 g FM was found considering the whole dataset ($p = 0.0013$), but not when considering data of 2015 only (407.22 versus 410.14 mg/100 g FM for CTR and RWS respectively, Table 3). Multiple irrigation events in 2015 might have evoked drying–re-wetting conditions, which are known to increase soil-available N and, therefore, might have alleviated the negative effect of drought on reduced N mineralization [51].

In agreement with our fresh biomass-related data, dry biomass-related total N levels were highest in 2015 (50.92–51.64 mg/g DM) as compared to the other years (38.07–42.47 mg/g DM). These findings might be explained by the fact that, in 2015, spinach was cultivated in autumn and therefore harvested after a longer growth period (51 days) as compared to 2016 and 2017 (44 and 39 days, respectively), as caused by lower irradiation in 2015. The lowest N levels were found in 2016 when total water amount as well as irradiation sum and temperature sum were highest compared to other years (Table 1) and precipitation was more frequent just before harvest (data not shown). Different results were found earlier for red radish with highest contents of nitrogen under moist conditions [19].

3.8. Potassium

Water supply limitation resulted in significantly increased levels of potassium (K^+) in spinach dry biomass in 2015 only (Table 2), when water reduction was most pronounced. According to fresh biomass-related PCA analyses (Figure 3), where most of the minerals were contributing to the separation of the treatments, significantly higher levels of K^+ were found in spinach grown under reduced water conditions in 2015 and 2017 (Table 3). While dry biomass content of reduced watered spinach was significantly higher in 2017, which might explain the increase in K^+ based on fresh biomass, the dry biomass content did not differ in 2015 with reduced water supply. Anyway, the availability of K^+ from the soil was shown to decrease with declining soil water content due to decreasing mobility of K^+ [52]. Our results indicated that K^+ uptake, which is often reduced under drought conditions [53,54], was not significantly affected by moderate water reduction. The highest absolute K^+ levels in spinach were found in 2016 (85.60–87.88 mg/g DM) and 2017 (89.89–90.68 mg/g DM), when global irradiation and evapotranspiration rates were high and temperatures continuously increased during the cultivation period (Supplementary Figures S1, S2, S4).

3.9. Phosphorous

In accordance with other macronutrients, dry biomass-related levels of phosphorous (P) were higher in less irrigated than in well-watered spinach in 2015 ($p = 0.0596$), but not in other years (Table 2). Similar to N, fresh biomass-related P levels were significantly increased in all years as well as in the overall dataset ($p = 0.0004$, Table 3), which probably can be traced back to higher contents of dry biomass in samples obtained from the reduced watered plants. This result was also visible in the PCA on fresh biomass-based data, where P clearly contributed to the differentiation of both treatments. In contrast, red radish root tubers grown within the same experimental field site revealed a significant decrease of P based on dry biomass, while fresh biomass-related data showed no effect [19]. In general, the fresh biomass-based P levels differed strongly between the cultivation set of 2015 (47.65–50.96 mg/g DM) and those of 2016 and 2017 (29.77–34.09 mg/g DM). One possible explanation might be the prolonged cultivation time in 2015 (51 days) and the associated high content of dry biomass (8.54%–8.60%) in contrast to other years (5.40%–6.83%). The shortened cultivation times in 2016 (44 days) and 2017 (39 days) might have resulted in lower absolute nutrient uptake from the soil.

3.10. Calcium

Calcium (Ca^{2+}) content of spinach was independent of the water supply. Generally, lower Ca^{2+} levels were found in 2015 (10.85–11.69 mg/g DM) compared to the other years 2016 and 2017 (13.78–14.94 mg/g DM). This result is in contrast to those found for N, which displayed the highest levels in 2015. Since Ca^{2+} is a readily available element for plant uptake from the soil, high Ca^{2+} levels have earlier been suggested to be associated with high concentrations in the soil rather than with potential relationships with uptake efficiency or transpiration velocity of the plant [41,55]. In agreement with our findings, Sánchez-Rodríguez et al. [56] did not find any difference in the Ca^{2+} accumulation of tomato leaves exposed to moderate drought stress, even though its uptake was significantly reduced. These findings are in further agreement with those from Hu and Schmidhalter [57], stating that Ca^{2+} accumulation was much less sensitive to drought than that of K and phosphate. Basically, Ca^{2+} has structural functions and acts as an important second messenger [58]. Since important plant functions are controlled by very small, but physiologically active pools of Ca^{2+} within the cytoplasm [59], the observed insignificant effects of moderate water reduction on the Ca^{2+} levels in spinach had been expected.

3.11. Magnesium

The dry biomass-based content of magnesium (Mg^{2+}) in spinach was significantly reduced by RWS in 2015, but not in 2016 and 2017. Thus, multivariate analysis by PCA did not show a strong contribution of Mg^{2+} to the differentiation of both groups (Figure 2). The same result was found for fresh biomass-based analyses (Table 3). In agreement with our findings, effects of moderate water reduction had no effects on Mg^{2+} levels in leaves of cherry tomatoes and were dependent on the cultivar [56]. In contrast, Pulupol et al. [60] found higher fresh biomass-based Mg^{2+} levels in tomato fruits grown under deficit irrigation, while dry biomass-related amounts were influenced insignificantly. Also, radish cultivated under RWS within the same experiment exhibited no significant changes in fresh biomass-related Mg^{2+} levels [19].

3.12. Micronutrients (Fe, Zn, Mn, Cu)

Significant effects on the dry biomass-related contents of iron (Fe) and copper (Cu) were not observed in our study, irrespective of water supply and cultivation year. However, dry biomass-based zinc (Zn) and manganese (Mn) contents were significantly increased by limited water supply (Table 2). PCA analyses supported the contribution of these micronutrients to variant differentiation (Figure 2). Similar findings were evident during analyses of fresh biomass-related data (Table 3, Figure 3). In contrast, moderate water reduction resulted in significantly decreased contents of Mn in radish root tubers grown within the same field experiment [19]. By analogy, water deficit-conditions have been shown to lead to lower Mn levels in leaves of cherry tomatoes [56].

3.13. Phenolic Compounds

In spinach dry biomass, total polyphenol content, including flavonoids, remained unchanged with RWS considering both the single datasets as well as the total dataset across all cultivation years. The levels of total flavonoids, as derived by summing up levels of individually quantitated flavonoids (Supplementary Tables S2 and S3), remained highly similar within a narrow range in all individual years (9.82–10.56 mg/g DM), while the amount of total polyphenol contents were marginally higher in 2016 (11.26–11.80 mg/g DM) and 2017 (11.38–11.42 mg/g DM) as compared to 2015 (10.09–10.31 mg/g DM).

In contrast to the quite stable dry biomass-related contents, fresh biomass-based contents displayed a much higher variability, particularly considering total flavonoids ranging from 54.45 to 86.93 mg/100 g FM. Fresh biomass-related levels of total flavonoids were significantly higher in reduced watered spinach than in well-watered spinach in 2017, which, however, was obviously based on significantly higher levels in dry biomass (Table 3). The levels found were in accordance with those of a previous study of Gil et al. [17], who found ca. 1000 mg/kg total flavonoids in fresh cut spinach. Because of generally higher contents of dry biomass in 2015, the fresh biomass-related levels of total flavonoids (86.52–86.93 mg/100 g FM) were substantially greater in 2015 as compared to those found in other years, reaching the lowest levels in 2016 (54.45–59.17 mg/100 g FM, Table 3). In accordance, strong seasonal variations in the levels of flavonoids such as quercetin and kaempferol were found in leafy vegetables such as lettuce (1.9–30 mg/kg quercetin) and endive (15–95 mg/kg kaempferol), based on fresh biomass-analyses [61]. Apart from fresh biomass-based variations, flavonoid levels in our study did not show any differences if compared on a dry biomass basis, even if there were strong differences in the global irradiation levels during the cultivation periods (Table 1, Supplementary Figures S1–S4). A rather low dependence or even independence of flavonoid contents from climatic factors such as temperature and global irradiation were reported previously in kale (*Brassica oleracea* var. *sabellica*) by Schmidt et al. [62].

Focusing on flavonoids, Bergquist et al. [22] observed the highest concentrations in baby spinach at early growth stages accompanied by high variations as associated with different sowing times. Furthermore, Hertog et al. [61] showed quite strong seasonal variations in the flavonoid levels of leafy vegetables such as lettuce and endive, which were 3 to 5 times higher in summer than in other seasons. In our study, the highest amounts of flavonoids in spinach fresh biomass were found in 2015 (86.52–86.93 mg/g FM), when cultivation took place in autumn with relatively low global irradiation in contrast to other years which had been sown at springtime (54.45–73.40 mg/ FM).

For individual flavonoids, RWS did not induce any clear effect in spinach. Univariate analyses indicated a dry biomass-related increase of some minor patuletin-derivatives in 2015 as well as an increase in the levels of a few more abundant flavonoids in 2016 upon moderate water limitation (Supplementary Table S3). For example, levels of specific patuletin- and spinatoside-derivatives (e.g., compounds **3**, **5** and **11**, Supplementary Table S2) were partly increased by RWS in 2015 and 2016. Among the other flavonoid-derivatives, the content of spinatoside, which represented one of the most abundant flavonoid-compounds in spinach according to a comparison of our mass spectral data to those reported previously by Aritomi et al. [63], tended to increase by RWS in 2016 ($p = 0.0677$). Apart from that, the study of Bergquist et al. [22] demonstrated that flavonoid profiles of baby spinach were highly similar during all growth stages, although the relative amounts of the individual flavonoids were prone to changes. Similar results were found in our study when comparing the different years, with hardly any difference in total flavonoids years, but individual flavonoids varied widely (Supplementary Table S3). The contents of individual flavonoids were almost the same when grown during early 2016 and 2017, while the proportions were markedly different when grown in late 2015. While an apparent methylenedioxyflavone-glucuronide (5,3′,4′-trihydroxy-3-methoxy-6:7-methylene-dioxyflavone-4′-β-D-glucuronide [21,22]) was the main flavonoid in 2016 and 2017 (30.0%–30.4%, Supplementary Table S4), the spinatoside accounted for the largest proportion in 2015 (20.2%–20.5%, Supplementary Table S4). A clear causal relationship between moderate limitations in water supply and changes in single or total flavonoids could not be established, which might indicate that the RWS effect on the levels of these compounds was rather low. Nevertheless, further studies in more controlled environments and, possibly, with more severe reductions in water supply, should be encouraged.

3.14. Carotenoids

In general, (all-*E*)-lutein was the major carotenoid found in spinach, followed by β-carotene and violaxanthin, which is in accordance with what had been expected in photosynthetically active plant tissues [33]. Moderately reduced water supply did not show clear effects on the contents of total carotenoids in spinach when evaluating the dry biomass-related dataset (Table 2). There was a slight increase of β-carotene, (9*Z*)-β-carotene and total carotenoids in 2015, not reaching statistical significance. In contrast, fresh biomass-related levels of all carotenoids significantly increased with RWS in all three cultivation years (Table 3). In contrast to the lack of effect of water supply, a seasonal variation was observed. For instance, total carotenoid levels in spinach dry biomass were ca. two-fold higher in 2017 (7.22–7.25 mg/g DM) than in 2015 and 2016 (3.03–4.11 mg/g DM). The high level of global irradiation in 2017 (756.9 MJ/m^2) had been expected to lead to comparably high levels of carotenoids, because increased radiation has been shown to enhance the accumulation of carotenoids in green vegetables [7]. This hypothesis would have explained the low carotenoid levels in 2015 with a global irradiation of 453.5 MJ/m^2, but not the low carotenoid levels in 2016 (3.03–3.05 mg/g DM) which were characterized by a high global irradiation (799.9 MJ/m^2). Carotenoid synthesis and accumulation must have been impacted by other, yet unknown, factors, such as the occurrence of multiple short-time drought events during the plant growth in 2016. By analogy to the levels of flavonoids, a clear effect of moderate water limitation and season on the levels of carotenoids was not observed in our open filed cultivation experiment.

4. Conclusions

In this study, the chemical composition of both dry and fresh biomass of spinach was shown to be strongly influenced by climatic conditions and/or water supply. The effects were highly dependent on the type of nutrient. Even moderately reduced water supply led to significant increases of dry biomass, which in turn often led to increased levels of numerous constituents and, thus, apparently enhanced the nutritional value of the vegetable product. For instance, fresh biomass-related levels of ascorbic acid, potassium, nitrogen, phosphorous as well as total flavonoids and carotenoids increased upon limiting water supply. Our results indicated that changes in levels of characteristic flavonoids might depend on seasonal variations, although further study is needed in this regard. Considering the composition of the dry biomass itself, we demonstrated that even mild water supply reductions led to significant increases of inositol, zinc and manganese levels, while malic acid, phosphate and chloride levels decreased. It is likely that such climate-related reductions in water supply will occur more frequently in the future, as is already occurring more and more due to the presumably accelerating climate change. In summary, our results indicate that the nutritional composition of spinach is sensitive to even moderately reduced water supply, but the overall quality of fresh spinach did not suffer regarding the levels of health-promoting constituents such as minerals, trace elements, flavonoids and carotenoids.

Supplementary Materials: The following are available online at http://www.mdpi.com/2311-7524/6/2/25/s1: Figure S1–S4: Daily mean weather conditions during the different spinach cultivation periods of 2015, 2016 and 2017: (1) Mean air temperature ($^\circ$C), (2) global irradiation sum (W/m^2), (3) daily mean relative humidity (%) and (4) daily mean evapotranspiration (mm). DAS: Days after sowing. Figure S5: HPLC-DAD chromatogram of spinach flavonoids at 350 nm. Flavonoid peaks assigned by comparing retention times, UV/Vis-spectra and mass spectral data to those reported by literature as shown in Table S2. Table S1: Proportion of single carotenoids on the total amount of carotenoids in spinach fresh biomass from three years (2015, 2016 and 2017). CTR: Control treatment with full water supply, RWS: Reduced water supply treatment. Table S2: Peak assignment to flavonoid-compounds as detected in spinach from years 2015, 2016 and 2017 by comparing retention time (RT), UV/Vis-spectra and negative ion m/z and important MS/MS-fragments to the literature. An exemplary chromatogram is shown in Figure S5. Table S3: Influence of reduced water supply on the dry biomass (DM)-related levels of single flavonoid compounds in spinach leaves from three years (2015, 2016 and 2017). Linear mixed model, t-test, *p*-values: < 0.1, < 0.05 (*), < 0.01 (**) and < 0.001 (***); CTR: Control treatment with full water supply, RWS: Reduced water supply treatment. Compound to peak number assignment is provided in Table S2. Table S4: Proportion of single flavonoids on the total amount of flavonoids in spinach dry biomass from three years (2015, 2016 and 2017). CTR: Control treatment with full water supply, RWS: Reduced water supply treatment. Compound to peak number assignment is provided in Table S2.

Author Contributions: Conceptualization, H.D. and J.Z.; methodology, C.S. and H.D.; software, M.F. and C.S.; formal analysis, C.S. and M.F.; investigation, C.S.; resources, H.D., R.S. and J.Z.; data curation, C.S.; writing—original draft preparation, C.S.; writing—review and editing, C.S., J.Z., H.D., M.F. and R.S.; visualization, C.S.; supervision, R.S., H.D., M.F. and J.Z.; project administration, H.D. and J.Z.; funding acquisition, H.D. and J.Z. All authors have read and agreed to the published version of the manuscript.

Funding: This research was funded by the LOEWE excellence cluster FACE2FACE from the Hessian State Ministry of Higher Education, Research and Arts (Wiesbaden, Germany).

Conflicts of Interest: The authors declare no conflict of interest.

References

1. Kundzewicz, Z.; Radziejewski, M.; Pińskwar, I. Precipitation Extremes in the Changing Climate of Europe. *Clim. Res.* **2006**, *31*, 51–58. [CrossRef]
2. Kang, Y.; Khan, S.; Ma, X. Climate change impacts on crop yield, crop water productivity and food security—A review. *Prog. Nat. Sci.* **2009**, *19*, 1665–1674. [CrossRef]
3. Chaves, M.M.; Maroco, J.P.; Pereira, J.S. Understanding plant responses to drought—From genes to the whole plant. *Funct. Plant Biol.* **2003**, *30*, 239. [CrossRef]

4. Cramer, G.R.; Urano, K.; Delrot, S.; Pezzotti, M.; Shinozaki, K. Effects of abiotic stress on plants: A systems biology perspective. *BMC Plant Biol.* **2011**, *11*, 163. [CrossRef]

5. Parida, A.K.; Dagaonkar, V.S.; Phalak, M.S.; Umalkar, G.V.; Aurangabadkar, L.P. Alterations in photosynthetic pigments, protein and osmotic components in cotton genotypes subjected to short-term drought stress followed by recovery. *Plant Biotech. Rep.* **2007**, *1*, 37–48. [CrossRef]

6. Koyama, R.; Itoh, H.; Kimura, S.; Morioka, A.; Uno, Y. Augmentation of Antioxidant Constituents by Drought Stress to Roots in Leafy Vegetables. *HortTechnology* **2012**, *22*, 121–125. [CrossRef]

7. Esteban, R.; Fleta-Soriano, E.; Buezo, J.; Míguez, F.; Becerril, J.M.; García-Plazaola, J.I. Enhancement of zeaxanthin in two-steps by environmental stress induction in rocket and spinach. *Food Res. Int.* **2014**, *65*, 207–214. [CrossRef]

8. Costa, J.M.; Ortuño, M.F.; Chaves, M.M. Deficit Irrigation as a Strategy to Save Water: Physiology and Potential Application to Horticulture. *J. Integr. Plant Biol.* **2007**, *49*, 1421–1434. [CrossRef]

9. Pérez-Bueno, M.L.; Pineda, M.; Barón, M. Phenotyping Plant Responses to Biotic Stress by Chlorophyll Fluorescence Imaging. *Front. Plant Sci.* **2019**, *10*, 1135. [CrossRef]

10. Taiz, L.; Zeiger, E. *Plant Physiology*, 5th ed.; Ainauer Associates Inc.: Sunderland, MA, USA, 2010; ISBN 978-0-87893-866-7.

11. Preece, J.E.; Read, P.E. The biology of horticulture. In *An Introductory Textbook*, 2nd ed.; John Wiley & Sons: Hoboken, NJ, USA, 2005; ISBN 0471465798.

12. Bergquist, S.A.M.; Gertsson, U.E.; Olsson, M.E. Influence of growth stage and postharvest storage on ascorbic acid and carotenoid content and visual quality of baby spinach (*Spinacia oleracea* L.). *J. Sci. Food Agric.* **2006**, *86*, 346–355. [CrossRef]

13. Toivonen, P.M.A.; Hodges, D.M. 10 Leafy Vegetables and Salads. In *Health-Promoting Properties of Fruit and Vegetables, [Enhanced Credo Edition]*; Terry, L.A., Ed.; CABI: Wallingford, UK; Cambridge, MA, USA; Boston, MA, USA, 2013; pp. 171–195. ISBN 978-1-78064-422-6.

14. Herrmann, K. *Inhaltsstoffe von Obst und Gemüse. 50 Tabellen und Übersichten, 97 Formeln*; Ulmer: Stuttgart, Germany, 2001; ISBN 3-8001-3139-0.

15. Gil, M.I.; Ferreres, F.; Tomás-Barberán, F.A. Effect of Modified Atmosphere Packaging on the Flavonoids and Vitamin C Content of Minimally Processed Swiss Chard (*Beta vulgaris* Subspecies *cycla*). *J. Agric. Food Chem.* **1998**, *46*, 2007–2012. [CrossRef]

16. Gil, M.I.; Castañer, M.; Ferreres, F.; Artés, F.; Tomás-Barberán, F.A. Modified-atmosphere packaging of minimally processed "Lollo Rosso" (*Lactuca sativa*). Phenolic metabolites and quality changes. *Z Lebensm Unters Forsch A* **1998**, *206*, 350–354. [CrossRef]

17. Gil, M.I.; Ferreres, F.; Tomás-Barberán, F.A. Effect of Postharvest Storage and Processing on the Antioxidant Constituents (Flavonoids and Vitamin C) of Fresh-Cut Spinach. *J. Agric. Food Chem.* **1999**, *47*, 2213–2217. [CrossRef]

18. Santamaria, P.; Elia, A.; Serio, F.; Todaro, E. A survey of nitrate and oxalate content in fresh vegetables. *J. Sci. Food Agric.* **1999**, *79*, 1882–1888. [CrossRef]

19. Schlering, C.B.; Dietrich, H.; Frisch, M.; Schreiner, M.; Schweiggert, R.; Will, F.; Zinkernagel, J. Chemical composition of field grown radish (*Raphanus sativus* L. var. *sativus*) as influenced by season and moderately reduced water supply. *J. Appl. Bot. Food Qual.* **2019**, 343–354. [CrossRef]

20. Singleton, V.L.; Rossi, J.A. Colorimetry of Total Phenolics with Phosphomolybdic-Phosphotungstic Acid Reagents. *Am. J. Enol. Vitic.* **1965**, *16*, 144.

21. Koh, E.; Charoenprasert, S.; Mitchell, A.E. Effect of Organic and Conventional Cropping Systems on Ascorbic Acid, Vitamin C, Flavonoids, Nitrate, and Oxalate in 27 Varieties of Spinach (*Spinacia oleracea* L.). *J. Agric. Food Chem.* **2012**, *60*, 3144–3150. [CrossRef]

22. Bergquist, S.A.M.; Gertsson, U.E.; Knuthsen, P.; Olsson, M.E. Flavonoids in baby spinach (*Spinacia oleracea* L.): Changes during plant growth and storage. *J. Agric. Food Chem.* **2005**, *53*, 9459–9464. [CrossRef]

23. Britton, G. UV/visible spectroscopy. In *Carotenoids: Volume 1B, Spectroscopy*; Britton, G., Liaaen-Jensen, S., Pfander, H., Eds.; Birkhäuser: Basel, Switzerland, 1995; pp. 13–62. ISBN 3-7643-2909-2.

24. Bates, D.; Maechler, M.; Bolker, B.; Walker, S. Fitting Linear Mixed-Effects Models Using {lme4}. *J. Stat. Softw.* **2015**, *67*, 1–48. [CrossRef]

25. R Core Team. *R: A Language and Environment for Statistical Computing*; R Foundation for Statistical Computing: Vienna, Austria, 2016.

26. RStudio Team. *RStudio: Integrated Development Environment for R.*; RStudio, Inc.: Boston, MA, USA, 2016.

27. Lenth, R.V. Least-Squares Means: The R Package lsmeans. *J. Stat. Softw.* **2016**, *69*, 1–33. [CrossRef]

28. Kuznetsova, A.; Brockhoff, P.B.; Christensen, R.H.B. lmerTest Package: Tests in Linear Mixed Effects Models. *J. Stat. Softw.* **2017**, *82*. [CrossRef]

29. Le, S.; Josse, J.; Husson, F. FactoMineR: An R Package for Multivariate Analysis. *J. Stat. Softw.* **2008**, *25*, 1–18. [CrossRef]

30. Kassambara, A.; Mundt, F. *Factoextra: Extract and Visualize the Results of Multivariate Data Analyses*, R package version 1.0.5; 2017. Available online: https://CRAN.R-project.org/package=factoextra (accessed on 27 March 2018).

31. Mercadante, A.Z.; Rodriguez-Amaya, D.B. Carotenoid composition of a leafy vegetable in relation to some agricultural variables. *J. Agric. Food Chem.* **1991**, *39*, 1094–1097. [CrossRef]

32. Schreiner, M. Vegetable crop management strategies to increase the quantity of phytochemicals. *Eur. J. Nutr.* **2005**, *44*, 85–94. [CrossRef]

33. Britton, G. Functions of Intact Carotenoids. In *Carotenoids: Volume 4: Natural Functions*; Britton, G., Liaaen-Jensen, S., Pfander, H., Eds.; Birkhäuser Basel: Basel, Switzerland, 2008; pp. 189–212. ISBN 978-3-7643-7499-0.

34. Schweiggert, R.; Ziegler, J.; Metwali, E.; Mohamed, F.; Almaghrabi, O.; Kadasa, N.; Carle, R. Carotenoids in mature green and ripe red fruits of tomato (*Solanum lycopersicum* L.) grown under different levels of irrigation. *Arch. Biol. Sci.* **2017**, *69*, 305–314. [CrossRef]

35. Radovich, T.J.K.; Kleinhenz, M.D.; Streeter, J.G. Irrigation Timing Relative to Head Development Influences Yield Components, Sugar Levels, and Glucosinolate Concentrations in Cabbage. *J. Am. Soc. Hortic. Sci.* **2005**, *130*, 943–949. [CrossRef]

36. Proietti, S.; Moscatello, S.; Giacomelli, G.A.; Battistelli, A. Influence of the interaction between light intensity and CO2 concentration on productivity and quality of spinach (*Spinacia oleracea* L.) grown in fully controlled environment. *Adv. Space Res.* **2013**, *52*, 1193–1200. [CrossRef]

37. Quick, W.P.; Chaves, M.M.; Wendler, R.; David, M.; Rodriguez, M.L.; Passaharinho, J.A.; Pereira, J.S.; Adcock, M.D.; Leegood, R.C.; Stitt, M. The effect of water stress on photosynthetic carbon metabolism in four species grown under field conditions. *Plant Cell Environ.* **1992**, *15*, 25–35. [CrossRef]

38. Cutler, J.M.; Rains, D.W. Effects of Water Stress and Hardening on the Internal Water Relations and Osmotic Constituents of Cotton Leaves. *Physiol. Plant.* **1978**, *42*, 261–268. [CrossRef]

39. Raschke, K. Stomatal Action. *Annu. Rev. Plant Physiol.* **1975**, *26*, 309–340. [CrossRef]

40. Lester, G.E.; Makus, D.J.; Hodges, D.M.; Jifon, J.L. Summer (subarctic) versus winter (subtropic) production affects spinach (*Spinacia oleracea* L.) leaf bionutrients: Vitamins (C, E, folate, K1, provitamin A), lutein, phenolics, and antioxidants. *J. Agric. Food Chem.* **2013**, *61*, 7019–7027. [CrossRef] [PubMed]

41. Da Silva, E.C.; Nogueira, R.J.M.C.; Silva, M.; Albuquerque, M. Drought Stress and Plant Nutrition. *Plant Stress* **2010**, *5*, 32–41.

42. Abdel Rahman, A.A.; Shalaby, A.F.; El Monayeri, M.O. Effect of moisture stress on metabolic products and ions accumulation. *Plant Soil* **1971**, *34*, 65–90. [CrossRef]

43. Sørensen, J.N.; Jørgensen, U.; Kühn, B.F. Drought Effects on the Marketable and Nutritional Quality of Carrots. *J. Sci. Food Agric.* **1997**, *74*, 379–391. [CrossRef]

44. Dubey, R.S.; Pessarakli, M. Physiological mechanisms of nitrogen absorption and assimilation in plants under stressful conditions. In *Handbook of Plant and Crop Physiology*, 2nd ed.; Pessarakli, M., Ed.; Dekker: New York, NY, USA, 2002; pp. 637–655. ISBN 0-8247-0546-7.

45. Morilla, C.A.; Boyer, J.S.; Hageman, R.H. Nitrate reductase activity and polyribosomal content of corn (*Zea mays* L.) having low leaf water potentials. *Plant Physiol.* **1973**, *51*, 817. [CrossRef]

46. Qiu, W.; Wang, Z.; Huang, C.; Chen, B.; Yang, R. Nitrate accumulation in leafy vegetables and its relationship with water. *J. Soil Sci. Plant Nutr.* **2014**, *14*, 761–768. [CrossRef]

47. Ćustić, M.; Poljak, M.; Čoga, L.; Ćosić, T.; Toth, N.; Pecina, M. The influence of organic and mineral fertilization on nutrient status, nitrate accumulation, and yield of head chicory. *Plant Soil Environ.* **2003**, *49*, 218–222. [CrossRef]

48. Anjana, S.U.; Iqbal, M. Nitrate accumulation in plants, factors affecting the process, and human health implications. A review. *Agron. Sustain. Dev.* **2007**, *27*, 45–57. [CrossRef]

49. Santamaria, P.; Gonnella, M.; Elia, A.; Parente, A.; Serio, F. Ways if reducing rocket salad nitrate content. *Acta Hortic.* **2001**, 529–536. [CrossRef]

50. Proietti, S.; Moscatello, S.; Colla, G.; Battistelli, Y. The effect of growing spinach (*Spinacia oleracea* L.) at two light intensities on the amounts of oxalate, ascorbate and nitrate in their leaves. *J. Hortic. Sci. Biotech.* **2004**, *79*, 606–609. [CrossRef]

51. He, M.; Dijkstra, F.A. Drought effect on plant nitrogen and phosphorus: A meta-analysis. *New Phytol.* **2014**, *204*, 924–931. [CrossRef] [PubMed]

52. Rouphael, Y.; Cardarelli, M.; Schwarz, D.; Franken, P.; Colla, G. Effects of Drought on Nutrient Uptake and Assimilation in Vegetable Crops. In *Plant Responses to Drought Stress: From Morphological to Molecular Features*; Aroca, R., Ed.; Springer: Berlin/Heidelberg, Germany, 2012; pp. 171–195. ISBN 978-3-642-32653-0.

53. Junjittakarn, J.; Pimratch, S.; Jogloy, S.; Htoon, W.; Singkham, N.; Vorasoot, N.; Toomsan, B.; Holbrook, C.C.; Patanothai, A. Nutrient uptake of peanut genotypes under different water regimes. *IJPP* **2013**, *7*, 677–692. [CrossRef]

54. Ashraf, M.; Shahbaz, M.; Qasim, A. Drought-induced modulation in growth and mineral nutrients of canola (*Brassica napus* L.). *Pak. J. Bot.* **2013**, *45*, 93–98.

55. Utrillas, M.J.; Alegre, L.; Simon, E. Seasonal changes in production and nutrient content of *Cynodon dactylon* (L.) Pers. subjected to water deficits. *Plant Soil* **1995**, *175*, 153–157. [CrossRef]

56. Sánchez-Rodríguez, E.; del Mar Rubio-Wilhelmi, M.; Cervilla, L.M.; Blasco, B.; Rios, J.J.; Leyva, R.; Romero, L.; Ruiz, J.M. Study of the ionome and uptake fluxes in cherry tomato plants under moderate water stress conditions. *Plant Soil* **2010**, *335*, 339–347. [CrossRef]

57. Hu, Y.; Schmidhalter, U. Drought and salinity: A comparison of their effects on mineral nutrition of plants. *J. Plant Nutr. Soil Sci.* **2005**, *168*, 541–549. [CrossRef]

58. Maathuis, F.J.M. Physiological functions of mineral macronutrients. *Curr. Opin. Plant Biol.* **2009**, *12*, 250–258. [CrossRef]

59. McLaughlin, S.B.; Wimmer, R. Tansley Review No. 104 Calcium physiology and terrestrial ecosystem processes. *New Phytol.* **1999**, *142*, 373–417. [CrossRef]

60. Pulupol, L.U.; Behboudian, M.H.; Fisher, K.J. Growth, Yield, and Postharvest Attributes of Glasshouse Tomatoes Produced under Deficit Irrigation. *HortScience* **1996**, *31*, 926–929. [CrossRef]

61. Hertog, M.G.L.; Hollman, P.C.H.; Katan, M.B. Content of potentially anticarcinogenic flavonoids of 28 vegetables and 9 fruits commonly consumed in the Netherlands. *J. Agric. Food Chem.* **1992**, *40*, 2379–2383. [CrossRef]

62. Schmidt, S.; Zietz, M.; Schreiner, M.; Rohn, S.; Kroh, L.W.; Krumbein, A. Genotypic and climatic influences on the concentration and composition of flavonoids in kale (*Brassica oleracea* var. *sabellica*). *Food Chem* **2010**, *119*, 1293–1299. [CrossRef]

63. Aritomi, M.; Komori, T.; Kawasaki, T. Flavonol glycosides in leaves of *Spinacia oleracea*. *Phytochemistry* **1985**, *25*, 231–234. [CrossRef]

© 2020 by the authors. Licensee MDPI, Basel, Switzerland. This article is an open access article distributed under the terms and conditions of the Creative Commons Attribution (CC BY) license (http://creativecommons.org/licenses/by/4.0/).

 *horticulturae*

Article

Container Type and Substrate Affect Root Zone Temperature and Growth of 'Green Giant' Arborvitae

Anthony L. Witcher [1,*], Jeremy M. Pickens [2] and Eugene K. Blythe [3]

[1] Department of Agricultural and Environmental Sciences, Otis L. Floyd Nursery Research Center, Tennessee State University, McMinnville, TN 37110, USA

[2] Department of Horticulture, Auburn University, Auburn, AL 36849, USA; jeremy.pickens@auburn.edu

[3] College of Agriculture, Auburn University, Auburn, AL 36849, USA; blythek@auburn.edu

* Correspondence: awitcher@tnstate.edu

Received: 20 December 2019; Accepted: 25 March 2020; Published: 7 April 2020

Abstract: Root zone temperature (RZT) in nursery containers commonly exceeds ambient temperature during the growing season, negatively impacting crop growth and quality. Black nursery containers absorb radiant heat resulting in excessive RZT, yet other types of containers and different substrates can moderate RZT. We conducted studies in Tennessee and Alabama to evaluate the effects of container type and substrate on RZT and growth of 'Green Giant' arborvitae (*Thuja standishii* × *plicata* 'Green Giant'). Trade gallon arborvitae were transplanted into black, white, or air pruning containers filled with pine bark (PB) or 4 PB: 1 peatmoss (v:v) (PB:PM). Plants grown in PB:PM were larger and had greater shoot and root biomass than plants grown in PB, likely due to increased volumetric water content. Plant growth response to container type varied by location, but white containers with PB:PM produced larger plants and greater biomass compared with the other container types. Root zone temperature was greatest in black containers and remained above 38 °C and 46 °C for 15% and 17% longer than white and air pruning containers, respectively. Utilizing light color containers in combination with substrates containing peatmoss can reduce RZT and increase substrate moisture content thus improving crop growth and quality.

Keywords: *Thuja standishii* × *plicata*; container production; nursery production; volumetric water content

1. Introduction

Container-grown nursery crops are subjected to extended periods of root zone heat stress throughout the growing season, negatively impacting crop growth and quality. The deleterious effects of high root zone temperature (RZT) have been observed historically in commercial container production. Self and Ward [1] observed less root growth of loquat (*Eriobotrya japonica*) on the south side of black metal cans where substrate temperature was 10 °C to 15 °C greater than ambient temperature (30 °C). In nursery containers, maximum RZT can reach 54 °C which can damage crops even when exposed for short periods [2,3]. At RZTs above 46 °C, direct injury to plants including immediate root cell damage can occur even with an exposure time under 30 min. Indirect injury to plants can occur at RZTs above 38 °C, including interruption of physiological mechanisms which may not present visible signs of damage but lead to reduced growth [4,5].

Many factors contribute to high RZT in container production. Nursery containers have a high surface-to-volume ratio, allowing excessive heat to be absorbed by the container and exchanged between the substrate through conduction [4]. Root zone temperature will vary within the container, with greater temperatures occurring in the region with direct exposure to the sun which will change throughout the growing season [3,4].

Container color, composition, or porosity can also affect RZT and plant growth. Dark-colored containers absorb more solar radiation than light-colored containers leading to excessive heat buildup in the substrate. Containers with a more porous exterior facilitate evaporation from the substrate and reduce heat exchange between the container wall and the substrate [6,7]. Markham et al. [3] evaluated the growth of red maple (*Acer rubrum*) and redbud (*Cercis canadensis*) along with RZT in conventional black containers and white painted containers. They reported RZT was 7.7 °C greater in black containers compared with white containers and that red maple had 2.5 times greater root density in white containers. In the same study, they noted white containers had little effect on redbud growth suggesting sensitivity to supraoptimal RZT varies by plant species.

Another alternative to conventional containers is air pruning containers which are available from several manufacturers and are designed with circular openings or long open slots to prevent root circling and improve root branching. Arnold and McDonald [8] evaluated the growth of five tree species in air pruning and conventional black containers. Although results were species specific, plant height, trunk diameter, and root and shoot biomass were similar or improved in the air pruning container compared to the conventional container. In the same study, it was also reported RZT near the exterior edges was 5 °C cooler in the air pruning container.

Container substrate porosity and moisture content are also factors in the rate of heat buildup or dissipation. Substrates with low air-filled porosity may improve heat energy diffusion through the substrate due to the physical connectivity of substrate particles. Water is an effective thermal conductor; thus, heat builds up more slowly in substrates with greater water content and minimizing temperature fluctuations in the substrate [4]. Pine bark (PB) is the most widely used substrate for nursery crops in the eastern United States and can be used alone or in combination with other components such as peatmoss (PM) or sand [9,10]. Pine bark typically has a high proportion of drainable air space, whereas PM has greater water holding capacity. Nevertheless, PB physical properties can vary by region and source due to processing methods and aging which could affect crop growth [11]. The combination of PM with PB would increase substrate water holding capacity and possibly reduce RZT.

Plant species vary in response to RZT and duration of exposure. Eastern arborvitae (*Thuja occidentalis*) is sensitive to high RZT and performs best in the United States Department of Agriculture (USDA) plant hardiness zones two to seven where summer temperatures are more moderate [5,12]. In a similar species such as 'Green Giant' arborvitae (*Thuja standishii* × *plicata* 'Green Giant'), root temperature sensitivity has not been reported but it is more heat tolerant and adapted to various soil types and climates including areas with long durations of high summer temperatures [13].

Although the benefits of light-colored containers and air pruning containers have been well documented, commercial adoption of these products remains low. Substrates with increased water retention properties can improve crop growth, yet most nursery producers continue to use PB as the sole substrate component (personal observation). Previous research has focused on differences among container type or among substrates, but the combined effects of container type and substrate have not been reported. Determining which factors (container or substrate) or combination of factors have the greatest impact on crop growth could increase adoption of these practices.

The objective of this research was to evaluate the combined effects of container type and substrate on RZT and growth of 'Green Giant' arborvitae.

2. Materials and Methods

Two separate studies were conducted concurrently at the Tennessee State University Otis L. Floyd Nursery Research Center, McMinnville, TN (USDA Plant Hardiness Zone 7a) and the Auburn University Ornamental Horticulture Research Center, Mobile, AL, USA (USDA Plant Hardiness Zone 8b). The studies were conducted at two locations due to potential differences in environmental conditions which may affect plant growth including temperature and rainfall. The average daily temperature in Tennessee (TN) ranged from 20.8 °C (September) to 26.7 °C (July) and ranged from

24 °C (May) to 27.3 °C (July) in Alabama (AL). Rainfall totaled 75.4 cm (TN) and 103.6 cm (AL) over the duration of each study.

Three different container types included black or white solid wall containers (11.3 L; PF1200; Nursery Supplies Inc., Kissimmee, FL, USA) and an air pruning container (10.5 L; #5 Rediroot; Nursery Source Inc., Boring, OR, USA). Two substrates were evaluated in combination with each container type (for a total of six treatments) and included PB and 4 PB: 1 peatmoss (v:v) (PB:PM). Pine bark was obtained from Morton's Horticultural Products (McMinnville, TN, USA) and from Longleaf Mulch (Semmes, AL, USA) for the TN and AL studies, respectively. Both substrates were amended (per 1 m^3) with 5.9 kg 18N-2.6P-6.6K controlled-release fertilizer (18-6-8 Nutricote® Total Type 180; Florikan USA, Sarasota, FL, USA), 3.6 kg dolomitic limestone, and 0.9 kg micronutrient granules (Micromax; ICL Specialty Fertilizers, Summerville, SC, USA). Trade gallon (2.4 L) 'Green Giant' arborvitae were transplanted on 19 April 2017 (AL) or 27 April 2017 (TN) into each treatment with 12 replicates (for a total of 72 individual experimental units) and plants were arranged on a gravel container pad in a randomized complete block design. To provide maximum container surface area to sunlight, plants were spaced 0.9 m apart.

Separate irrigation zones were used for each treatment to monitor and adjust irrigation application rates. Plants were irrigated daily using a modified dribble ring (15.2 cm diameter; Dramm Corp., Manitowoc, WI) fitted with a pressure-compensating emitter (8 L h^{-1}; Netafim USA, Fresno, CA). Irrigation application volume for each treatment was adjusted every two weeks to a target leaching fraction of 10% to 20%. Decagon 5TE (AL) and 5TM (TN) sensors and EM50 data loggers (Decagon Devices Inc., Pullman, WA) were used to measure and record RZT and volumetric water content (VWC; m^3·m^{-3}) every 15 min throughout both studies. Sensors (1 per container; n = 3) were positioned vertically approximately 4.3 cm from the south-facing container sidewall and placed midway between the substrate surface and bottom of the container. Plant height and diameter were measured at 0, 59, 138, and 166 days after planting (DAP) in AL and 0, 69, 95, 120, and 173 DAP in TN. Growth index was calculated [(height + width at widest point + perpendicular width) / 3] and increase in plant height and growth index was also reported (increase = final – initial). The studies were terminated at 166 (AL) and 173 (TN) DAP and plants were destructively harvested. Shoot dry weight (n = 12) and root dry weight (n = 4) were measured after samples were oven-dried at 70 °C for approximately 7 days. Substrate pH and electrical conductivity (EC) were recorded using the pour-through method [14] at 52, 97, 146, and 166 DAP (AL) and at 60, 95, and 120 DAP (TN). The percentage of time roots were exposed to temperatures above critical thresholds (38 °C and 46 °C) mentioned by Ingram et al. (2015) was calculated using the total number of data recordings during daylight hours. Substrate physical properties (n = 3) including air space, container capacity, total porosity, and bulk density were determined using porometer analysis [15].

Multi-factor data were analyzed with linear mixed models using the GLIMMIX procedure of SAS (Version 9.3; SAS Institute, Inc., Cary, NC, USA) by first testing for an interaction between treatment factors (container type and substrate). When there was an interaction between treatment factors, levels of container type were compared within each substrate. Porometer data were analyzed with linear models using the GLIMMIX procedure of SAS. *P*-values for all simultaneous comparisons were adjusted using the Tukey method to maintain an overall significance level of α = 0.05.

3. Results

3.1. Plant Growth

In the TN study, there was an interaction between container type and substrate for final plant height and height increase (Table 1). Both height and height increase were greatest for plants in the white container for PB:PM. White containers with PB:PM also produced plants with the greatest growth index (final and increase) (Table 2). Overall, plants grown in PB:PM produced more shoot and root dry weight (biomass) compared with plants grown in PB. White and black containers had similar shoot and root dry weight in PB but shoot dry weight was greatest in white containers for PB:PM. In the white container, shoot dry weight and root dry weight were 56% and 68% greater, respectively, for PB:PM compared with PB. In PB, final growth index and increase were similar for plants in black and white containers. Regardless of substrate, however, plants were shortest in the air pruning container.

Table 1. Plant height and height increase (n = 12) of 'Green Giant' arborvitae grown in different types of containers and substrates in Tennessee and Alabama.

		Plant Height (cm)			Plant Height (cm)		
		69 DAP [x]	173 DAP	Increase [w]	59 DAP	166 DAP	Increase
		Tennessee			Alabama		
		Significance of treatment factors					
Container (C) [z]		0.0055	<0.0001	<0.0001	0.1523	<0.0001	<0.0001
Substrate (S) [y]		0.8889	<0.0001	<0.0001	0.7328	0.0267	0.0053
C by S		0.1817	0.0006	0.0003	0.4744	0.9544	0.9757
		Least squares means for main effects					
Substrate	Container						
	Black	69.3 ab [v]	-	-	72.1 a	78.2 b	16.1 b
	White	71.8 a	-	-	72.9 a	92.7 a	31.3 a
	Air	66.9 b	-	-	69.3 a	75.1 b	14.0 b
PB		69.4 a	-	-	71.2 a	79.6 b	17.6 b
PB:PM		69.3 a	-	-	71.7 a	84.4 a	23.3 a
		Treatment least squares means grouped by substrate					
Substrate	Container						
	Black	67.8	85.8 a	27.3 a	70.9	75.8	13.4
PB	White	72.3	83.3 a	21.8 ab	73.9	90.7	28.6
	Air	68.2	75.0 b	14.9 b	68.7	72.3	10.8
	Black	70.7	99.2 b	41.0 b	73.3	80.7	18.8
PB:PM	White	71.4	114.8 a	56.0 a	71.8	94.7	34.1
	Air	65.7	88.2 c	32.5 c	70	77.9	17.2

[z] Container type: Black and White—standard solid wall (11.3 L; PF1200; Nursery Supplies Inc., Kissimmee, FL, USA); Air—air pruning (10.5 L; #5 Rediroot; NurserySource Inc., Boring, OR, USA). [y] Substrate: Pine bark alone (PB) or combined (v:v) with peatmoss (PB:PM; 4 pine bark: 1 peatmoss). [x] DAP = days after planting. [w] Increase = final plant height—initial plant height. [v] When the interaction term in the model is not significant ($P > 0.10$), main effects means for levels within each treatment factor followed by the same lower-case letter are not significantly different using the Tukey method for multiple comparisons ($\alpha = 0.05$). When the interaction term in the model is significant ($P \leq 0.10$), simple effects means (treatment means for container grouped within substrate) followed by the same lower-case letter are not significantly different using the Tukey method for multiple comparisons ($\alpha = 0.05$); otherwise, the treatment means are presented without letter groupings for informational purposes.

In the AL study after approximately two months of growth, plant height was similar between the two substrates and among the three types of containers (Table 1). At the end of the study, plant height was greater in PB:PM compared to PB and greatest in the white container (by over 18%). The plant height increase was also greatest in white containers and PB:PM. Plant growth index followed a similar trend with white containers and PB:PM producing larger plants throughout the study and the greatest growth index increase (Table 2). Although the white container with PB:PM tended to produce taller plants, shoot dry weight was similar among all container types in PB:PM and similar between black and white containers in PB. Container type did not have an effect on root dry weight, but root dry weight was 12% greater in PB:PM compared to PB.

Table 2. Plant growth index (n = 12), growth index increase and dry weight (shoot and root) of 'Green Giant' arborvitae grown in different types of containers and substrates in Tennessee and Alabama.

		Growth Index [x]			Shoot Dry wt (g)	Root Dry wt (g)	Growth Index			Shoot Dry wt (g)	Root Dry wt (g)
		69 DAP [w]	173 DAP	Increase [v]			59 DAP	166 DAP	Increase		
			Tennessee						Alabama		
					Significance of treatment factors						
Container (C) [z]		0.005	<0.0001	<0.0001	<0.0001	0.0084	0.55	<0.0001	<0.0001	0.0002	0.3203
Substrate (S) [y]		0.6342	<0.0001	<0.0001	<0.0001	<0.0001	0.0219	0.0002	0.0025	<0.0001	0.0452
C by S		0.4362	0.0002	<0.0001	<0.0001	0.0332	0.7646	0.6223	0.7424	0.0497	0.2689
					Least squares means for main effects						
Substrate	Container										
	Black	56.5 ab[u]	-	-	-	-	52.3 a	56.9 b	15.2 b	-	507 a
	White	58.1 a	-	-	-	-	53.0 a	63.3 a	22.4 a	-	560 a
	Air	54.4 b	-	-	-	-	51.7 a	55.2 b	13.8 b	-	534 a
PB		56.1 a	-	-	-	-	51.3 b	56.3 b	15.3 b	-	503 b
PB:PM		56.5 a	-	-	-	-	53.4 a	60.7 a	19.0 a	-	564 a
					Treatment least squares means grouped by substrate						
Substrate	Container										
	Black	56.5	68.9 a	26.6 a	330 a	629 a	51.3	55.1	12.8	162 a	508.5
PB	White	58.4	68.4 a	25.1 a	299 a	487 ab	52.3	61.5	21	170 a	506.5
	Air	53.4	61.9 b	21.4 b	191 b	366 b	50.2	52.2	12	120 b	494
	Black	56.4	74.9 b	33.5 b	383 b	698 a	53.4	58.8	17.7	195 a	504.5
PB:PM	White	57.8	83.8 a	42.2 a	466 a	820 a	53.7	65.1	23.7	201 a	614
	Air	55.4	72.6 b	31.8 b	361 b	653 a	53.3	58.1	15.5	186 a	574

[z] Container type: Black and White—standard solid wall (11.3 L; PF1200; Nursery Supplies Inc., Kissimmee, FL); Air—air pruning (10.5 L; #5 Rediroot; NurserySource Inc., Boring, OR). [y] Substrate: Pine bark alone (PB) or combined (v:v) with peatmoss (PB:PM; 4 pine bark: 1 peatmoss). [x] Growth index = (height + width at widest point + perpendicular width) / 3. [w] DAP = days after planting. [v] Increase = final growth index—initial growth index. [u] When the interaction term in the model is not significant ($P > 0.10$), main effects means for levels within each treatment factor followed by the same lower-case letter are not significantly different using the Tukey method for multiple comparisons ($\alpha = 0.05$). When the interaction term in the model is significant ($P \leq 0.10$), simple effects means (treatment means for container grouped within substrate) followed by the same lower-case letter are not significantly different using the Tukey method for multiple comparisons ($\alpha = 0.05$); otherwise, the treatment means are presented without letter groupings for informational purposes.

3.2. Substrate Chemical and Physical Properties

Substrate pH ranged from 5.3 to 6.4 (TN) and from 5.4 to 6.7 (AL) for all treatments in each study (data not shown), remaining within the recommended range of 4.5 to 6.5 [16] except at 52 DAP in AL. Substrate pH was only 0.3 units lower in PB:PM at 95 and 120 DAP in TN and the addition of PM had no effect on substrate pH in the AL study. Substrate EC varied among treatments in the TN study but no clear trend was observed (data not shown). Conversely, substrate and container type had no effect on substrate EC in the AL study except at 146 DAP with substrate EC being greatest in the black container.

Substrate container capacity increased 12% (TN) and 7% (AL) with the addition of PM compared to PB (Table 3). Substrate air space was greater for PB in the TN study, being reduced by 8% in PB:PM. Substrate air space was similar for substrates in the AL study, and air space was overall lower in PB from AL compared with TN.

Table 3. Physical properties (n = 3) of substrates for production of 'Green Giant' arborvitae in Tennessee and Alabama.

Substrate [z]	Air Space [y]	Container Capacity	Total Porosity	Bulk Density
		(% volume)		(g·cm^{-3})
		Tennessee		
PB	30.4 a [x]	42.8 b	73.2 a	0.233 a
PB:PM	21.8 b	54.7 a	76.5 a	0.226 a
		Alabama		
PB	25.4 a	41.1 b	66.6 a	0.267 a
PB:PM	21.7 a	48.9 a	70.6 a	0.259 a

[z] Substrate: Pine bark alone (PB) or combined (v:v) with peatmoss (PB:PM; 4 pine bark: 1 peatmoss). [y] Data obtained using the North Carolina State University porometer method (Fonteno and Harden, 2010). [x] Means followed by the same letter within a location are not significantly different (α = 0.05).

3.3. Root Zone Temperature and Volumetric Water Content

Substrate did not have an effect on the percentage of time at or above specific threshold temperatures (38 °C and 46 °C) in either study, but black containers produced the greatest RZT and remained above the critical thresholds far longer than the other container types (Table 4). In TN, RZT in the black container remained above 38° C for over 19% of the time which was 15% and 17% longer than white and air pruning containers, respectively. Although RZT in black containers only remained at or above 46 °C for 0.1% of the time, RZT in air pruning and white containers remained below 46° C throughout the study. In AL, there was an interaction between container type and substrate for a percentage of time at or above 38 °C. Root zone temperature in the black container remained above 38 °C for over 21% of the time (regardless of substrate), over 13% longer than the other container types. Root zone temperature in air pruning and white containers did not reach the 46 °C threshold but RZT in black containers remained at this threshold for nearly 2% of the time.

Volumetric water content was lowest in the air pruning container throughout most of the TN study (Table 5). Volumetric water content was similar for black and white containers in June and July, but black containers had the greatest VWC compared with the other container types in August and September. Substrate also affected VWC, with PB:PM maintaining greater VWC throughout the study. Similar trends were observed for VWC in the AL study, with VWC greater in black containers compared to the air pruning containers and VWC was also greater in PB:PM throughout the study (data not shown).

Table 4. Percent of time (n = 3) substrate temperature remained at or above critical thresholds (38 °C and 46 °C) during daylight hours for 'Green Giant' arborvitae grown in different types of containers and substrates in Tennessee and Alabama.

		38 °C (%)	46 °C (%)	38 °C (%)	46 °C (%)
		Tennessee		Alabama	
		Significance of treatment factors			
Container (C) [z]		<0.0001	0.0221	<0.0001	<0.0001
Substrate (S) [y]		0.5229	0.5144	0.7182	0.6128
C by S		0.397	0.5864	0.047	0.6464
		Least squares means for main effects			
Substrate	Container				
	Black	19.4 a [x]	0.1 a	-	1.93 a
	White	4.2 b	0.0 b	-	0.00 b
	Air	2.7 b	0.0 b	-	0.00 b
PB		7.9 a	0.0 a	-	0.85 a
PB:PM		8.3 a	0.0 a	-	0.43 a
		Treatment least squares means grouped by substrate			
Substrate	Container				
	Black	20.2	0.0	24.2 a	2.5
PB	White	5.1	0.0	8.4 b	0.0
	Air	2.5	0.0	5.5 c	0.0
	Black	18.8	0.1	21.6 a	1.3
PB:PM	White	3.2	0.0	7.2 b	0.0
	Air	2.9	0.0	7.8 b	0.0

[z] Container type: Black and White—standard solid wall (11.3 L; PF1200; Nursery Supplies Inc., Kissimmee, FL); Air—air pruning (10.5 L; #5 Rediroot; NurserySource Inc., Boring, OR). [y] Substrate: Pine bark alone (PB) or combined (v:v) with peatmoss (PB:PM; 4 pine bark: 1 peatmoss). [x] When the interaction term in the model is not significant (P > 0.10), main effects means for levels within each treatment factor followed by the same lower-case letter are not significantly different using the Tukey method for multiple comparisons (α = 0.05). When the interaction term in the model is significant ($P \leq 0.10$), simple effects means (treatment means for container grouped within substrate) followed by the same lower-case letter are not significantly different using the Tukey method for multiple comparisons (α = 0.05); otherwise, the treatment means are presented without letter groupings for informational purposes.

Table 5. Average daytime volumetric water content (n = 3) over a four-month period for 'Green Giant' arborvitae grown in different types of containers and substrates in Tennessee.

		Volumetric Water Content ($m^3 \cdot m^{-3}$)			
		June	July	August	September
		Significance of treatment factors			
Container (C)[z]		0.0274	0.0044	0.0006	0.0007
Substrate (S)[y]		0.0021	0.0226	0.0176	0.0183
C by S		0.0175	0.2421	0.7409	0.2288
		Least squares means for main effects			
Substrate	Container				
	Black	-	0.314 a [x]	0.320 a	0.312 a
	White	-	0.274 a	0.220 b	0.201 b
	Air	-	0.178 b	0.131 c	0.139 c
PB		-	0.214 b	0.180 b	0.176 b
PB:PM		-	0.283 a	0.247 a	0.240 a
		Treatment least squares means grouped by substrate			
Substrate	Container				
	Black	0.270 a	0.283	0.3	0.307
PB	White	0.319 a	0.266	0.187	0.159
	Air	0.152 b	0.117	0.093	0.105
	Black	0.367 a	0.346	0.341	0.316
PB:PM	White	0.323 a	0.286	0.269	0.265
	Air	0.342 a	0.238	0.169	0.173

[z] Container type: Black and White—standard solid wall (11.3 L; PF1200; Nursery Supplies Inc., Kissimmee, FL, USA); Air—air pruning (10.5 L; #5 Rediroot; NurserySource Inc., Boring, OR, USA). [y] Substrate: Pine bark alone (PB) or combined (v:v) with peatmoss (PB:PM; 4 pine bark: 1 peatmoss). [x] When the interaction term in the model is not significant (P > 0.10), main effects means for levels within each treatment factor followed by the same lower-case letter are not significantly different using the Tukey method for multiple comparisons (α = 0.05). When the interaction term in the model is significant ($P \leq 0.10$), simple effects means (treatment means for container grouped within substrate) followed by the same lower-case letter are not significantly different using the Tukey method for multiple comparisons (α = 0.05); otherwise, the treatment means are presented without letter groupings for informational purposes.

4. Discussion

Overall, arborvitae grew taller and produced more biomass in white containers with PB:PM. Plants in air pruning containers were smaller than those grown in black or white containers (regardless of substrate), but more growth occurred when PB:PM was used. The air pruning and white containers provided a lower RZT for plants throughout both studies, minimizing heat-related stress which likely led to reduced plant growth in the black container. Maximum RZT (data not shown) was reduced from 5 °C to 8 °C in the air pruning and white containers. Root zone temperature in the black container reached 46 °C (TN) and 51 °C (AL), while RZT in the air pruning and white containers reached a maximum of 41 °C (TN) and 43 °C (AL). These results support previous research documenting increased growth in white containers due to the deleterious effects of supraoptimal RZT. Root zone temperatures can commonly exceed 54 °C in container-grown crops, but RZT near 38 °C can cause indirect injury to plants leading to reduced shoot and root growth, increased water stress, interruption of physiological mechanisms (photosynthesis and respiration, and increased susceptibility to pathogens [3,4,7]. In our study, fewer roots were observed on the south-facing side of the solid wall containers, regardless of container color, suggesting supraoptimal RZT prevents root growth and development near the container sidewall even in light color containers.

Although air pruning containers had lower RZT, plant growth and biomass were consistently lower than plants in white containers. These results contradict previous work by Arnold and McDonald [8] where tree growth of several tree species was similar or superior in the air pruning container compared to the black container. In the present study, black and white containers had a 7% larger volume than the air pruning container which could have attributed to some of the reduced plant growth observed in the air pruning container. The physical design of the air pruning container likely also contributed to the reduced growth. The air pruning container was designed with open slots, aligned longitudinally around the container side wall, to prevent root circling and improve root branching and growth. The slots increase aeration which prevents buildup of heat, resulting in moderated RZT closer to ambient conditions. In air pruning containers, a larger portion of substrate surface is exposed to the outside environment resulting in more rapid drying out due to increased evaporation through solar radiation and air flow. Arnold and McDonald [8] also noted the air pruning container moderated RZT and that plants in the air pruning container dried out sooner. In the present study, VWC was lowest in the air pruning container which limited the amount of plant-available water and likely led to the reduced plant growth. Irrigation volume was calculated based on leaching fraction and applied once daily, thus the substrate in air pruning containers dried more rapidly after irrigation compared with the solid wall containers and likely never reached container capacity. Overall, black and white containers received more irrigation volume (13% and 38%, respectively; data not shown) compared to the air pruning containers corresponding to the observed differences in arborvitae growth. Therefore, growers using air pruning containers should utilize more frequent (cyclic) irrigation to maximize VWC throughout the day and reduce water-related stress especially in substrates with high air-filled porosity and lower water holding capacity.

Peatmoss increased substrate container capacity by up to 12% compared to PB alone resulting in PB:PM having at least 7% greater VWC throughout both studies. On average, plants grown in PB:PM received 90% more irrigation volume compared PB. The combined benefits of greater water retention and irrigation volume corresponded to superior plant growth in PB:PM. Peatmoss has greater container capacity and easily available water compared to PB, likely due to a higher proportion of macropores and a lower proportion of fine particles in PB [17]. The PB and PM for each study were obtained from different sources which led to slight differences in physical properties for each study. Pine bark physical properties can vary due to a number of factors including source, age, and processing method [11]. Nevertheless, the addition of PM provided PB:PM with more plant-available water resulting in overall improved plant growth at both locations.

Peatmoss typically has a lower inherent pH compared to PB. It has been shown that increasing the percentage of PM in PB does not increase cation exchange capacity (and thus nutrient retention) on a volumetric basis [9,18]. Johnson et al. [18] reported an increase in soluble salt level with an increasing proportion of PM in PB substrates despite no increase in CEC. In our studies, PB:PM was composed of 20% PM but substrate pH was not negatively affected and differences in EC between substrates were not observed. However, EC was generally lower for PB:PM in both studies and a lack of significance might be due to sample variation and small sample size (n = 4).

Root zone temperature was not affected by substrate in the present studies, yet substrate porosity and VWC may contribute to the rate of heat buildup and dissipation in a substrate. Amoroso et al. [19] reported RZT was greater in substrates irrigated to 100% container capacity compared to 30% container capacity. In their study, the substrate was composed of 80% PM which typically has smaller particle size and lower air space compared to PB. Martin and Ingram [20] suggested substrates with lower pore space combined with greater VWC (25% to 40%) could dissipate heat more effectively. As a result, substrate temperature would increase at a slower rate and maintain a lower RZT overall. Although PB:PM had greater VWC, there was very little difference in substrate total porosity compared to PB alone which may have minimized thermal dissipation in this study. Plants were irrigated once daily at 12 pm, thus VWC may have been too low during the hottest portion of the day for heat to effectively dissipate. Applying irrigation multiple times throughout the afternoon would increase VWC over a longer period and possibly improve heat dissipation from the substrate.

All the arborvitae plants grown in these studies were marketable, but plants grown in PB:PM grew significantly larger and were visually superior in quality (Figure 1). Plants in PB:PM benefited from overall greater VWC and which remained higher throughout the day compared with PB. Peatmoss is more expensive than PB and requires additional equipment for mixing into the substrate, but the added benefits (water retention and availability) could result in higher quality crops that reach a finished size more quickly. For example, plants in PB:PM were on average larger (height and growth index) than plants in PB that had been grown for 30 additional days (data not shown).

Figure 1. Representative 'Green Giant' arborvitae plants 173 days after planting (in Tennessee) in 100% pine bark (three plants on left) and 4 pine bark: 1 peatmoss (v:v; three plants on right) in standard black, standard white, and air pruning containers (left to right within each substrate).

White containers provided lower RZT that likely reduced plant stress resulting in slightly larger and potentially healthier plants. White plastic containers are commercially available from a number of manufacturers in a variety of sizes (up to 11 or 19 L). Modern white containers are high quality and typically manufactured with co-extruded black (interior) and white (exterior) plastic to prevent light diffusion through the container which was a problem with earlier products. Plants grown in air pruning containers tend to have better branched and more vigorous root systems that can improve transplant establishment and subsequent crop quality, but to prevent drying out they will require more frequent irrigation and possibly higher application rates compared with crops in traditional black containers.

The impact of abiotic factors on crop growth has been well documented and developing methods for reducing root zone stress would improve root development, crop quality, and transplant success [21]. The interaction of factors such as container type, substrate, and moisture content must be considered when evaluating alternative production practices. Growers typically use a single substrate/container type for all the different crop species in production. Light-colored containers can effectively reduce RZT which may be especially important in temperature-sensitive species, but we found that using substrates with higher VWC had a greater effect on overall crop growth especially when used in white containers. Although 'Green Giant' arborvitae is highly adaptable to soil type and is considered a heat-tolerant plant, we demonstrated increased growth by modifying the substrate and container type. Growers should consider conducting small-scale evaluations to determine if a particular substrate/container combination works in their production system. When conducting small trials, growers should place each substrate/container combination in separate irrigation zones and adjust irrigation volume based on plant needs to ensure moisture is not a limiting factor.

Author Contributions: Conceptualization, A.L.W. and J.M.P.; methodology, A.L.W. and J.M.P.; formal analysis, A.L.W. and E.K.B.; investigation, A.L.W. and J.M.P.; writing—original draft preparation, A.L.W.; writing—review and editing, A.L.W., J.M.P., and E.K.B.; funding acquisition, A.L.W. All authors have read and agreed to the published version of the manuscript.

Funding: This research was supported the National Institute of Food and Agriculture (NIFA), United States Department of Agriculture (USDA) Evans-Allen grant, under award number TENX-1519-CCOCP.

Acknowledgments: The authors wish to thank Terry Kirby for assistance with data collection and project maintenance.

Conflicts of Interest: The authors declare no conflicts of interest.

References

1. Self, R.L.; Ward, H.S. Effects of high soil temperature on root growth of loquat seedlings in nursery containers. *Plant Dis. Report.* **1956**, *40*, 957–959.
2. Martin, C.A.; Ingram, D.L.; Neil, T.A. Supraoptimal root-zone temperature alters growth and photosynthesis of holly and elm. *J. Arboric.* **1989**, *15*, 272–276.
3. Markham, J.W.; Bremer, D.J.; Boyer, C.R.; Schroeder, K.R. Effect of container color on substrate temperatures and growth of red maple and redbud. *HortScience* **2011**, *46*, 721–726. [CrossRef]
4. Ingram, D.L.; Ruter, J.M.; Martin, C.A. Review: Characterization and impact of supraoptimal root-zone temperatures in container-grown plants. *HortScience* **2015**, *50*, 530–539. [CrossRef]
5. Mathers, H.M. Summary of temperature stress issues in nursery containers and current methods of protection. *HortTechnology* **2003**, *13*, 617–624. [CrossRef]
6. Li, T.; Bi, G.; Harkness, R.L.; Denny, G.C.; Blythe, E.K.; Zhao, X. Nitrogen rate, irrigation frequency, and container type affect plant growth and nutrient uptake of Encore azalea 'Chiffon'. *HortScience* **2018**, *53*, 560–566. [CrossRef]
7. Nambuthiri, S.; Geneve, R.L.; Sun, Y.; Wang, X.; Fernandez, R.T.; Niu, G.; Bi, G.; Fulcher, A. Substrate temperature in plastic and alternative nursery containers. *HortTechnology* **2015**, *25*, 50–56. [CrossRef]
8. Arnold, M.A.; McDonald, G.V. Accelerator containers alter plant growth and the root-zone environment. *J. Environ. Hortic.* **1999**, *7*, 168–173.
9. Altland, J.E.; Locke, J.C.; Krause, C.R. Influence of pine bark particle size and pH on cation exchange capacity. *HortTechnology* **2014**, *24*, 554–559. [CrossRef]
10. Robbins, J.A.; Evans, M.R. Growing Media for Container Production in a Greenhouse or Nursery Part I. Available online: https://www.uaex.edu/publications/PDF/FSA-6097.pdf (accessed on 21 October 2019).
11. Altland, J.E.; Owen, J.S., Jr.; Jackson, B.E.; Fields, J.S. Physical and hydraulic properties of commercial pine-bark substrate products used in production of containerized crops. *HortScience* **2018**, *53*, 1883–1890. [CrossRef]
12. Gilman, E.F.; Watson, D.G. *Thuja occidentalis* White-Cedar. Available online: https://edis.ifas.ufl.edu/pdffiles/ST/ST62900.pdf (accessed on 21 October 2019).

13. Griffin, J.J.; Blazich, F.A.; Ranney, T.G. Propagation of Thuja x 'Green Giant' by stem cuttings: Effects of growth stage, type of cutting, and IBA treatment. *J. Environ. Hortic.* **1998**, *16*, 212–214. [CrossRef]

14. LeBude, A.V.; Bilderback, T.E. Pour-through Extraction Procedure: A Nutrient Management Tool for Nursery Crops. Available online: https://content.ces.ncsu.edu/the-pour-through-extraction-procedure-a-nutrient-management-tool-for-nursery-crops (accessed on 21 October 2019).

15. Fonteno, W.C.; Harden, C.T. North Carolina State University Horticultural Substrates Lab Manual. Available online: https://projects.ncsu.edu/project/hortsublab/pdf/porometer_manual.pdf (accessed on 15 December 2019).

16. Bilderback, T.; Boyer, C.; Chappell, M.; Fain, G.; Fare, D.; Gilliam, C.; Jackson, B.E.; Lea-Cox, J.; LeBude, A.V.; Niemiera, A.; et al. *Best Management Practices: Guide for Producing Nursery Crops*, 3rd ed.; Southern Nursery Association: Acworth, GA, USA, 2013.

17. Fields, J.S.; Fonteno, W.C.; Jackson, B.E.; Heltman, J.L.; Owen, J.S. Hydrophysical properties, moisture retention, and drainage profiles of wood and traditional components for greenhouse substrates. *HortScience* **2014**, *49*, 827–832. [CrossRef]

18. Johnson, C.R.; Midcap, J.T.; Hamilton, D.F. Evaluation of potting-media, fertilizer source, and rate of application on chemical composition and growth of *Ligustrum japonicum* Thumb. *Sci. Hortic.* **1981**, *14*, 157–163. [CrossRef]

19. Amoroso, G.; Frangi, P.; Piatti, R.; Fini, A.; Ferrini, F. Effect of mulching on plant and weed growth, substrate water content, and temperature in container-grown giant arborvitae. *HortTechnology* **2010**, *20*, 957–962. [CrossRef]

20. Martin, C.A.; Ingram, D.L. Evaluation of thermal properties and effect of irrigation on temperature dynamics in container media. *J. Environ. Hortic.* **1991**, *9*, 24–28.

21. Mathers, H.M.; Lowe, S.B.; Scagel, C.; Struve, D.K.; Case, L.T. Abiotic factors influencing root growth of woody nursery plants in containers. *HortTechnology* **2007**, *17*, 151–162. [CrossRef]

© 2020 by the authors. Licensee MDPI, Basel, Switzerland. This article is an open access article distributed under the terms and conditions of the Creative Commons Attribution (CC BY) license (http://creativecommons.org/licenses/by/4.0/).

 *horticulturae* 

Article

Effects of Non-Leguminous Cover Crops on Yield and Quality of Baby Corn (*Zea mays* L.) Grown under Subtropical Conditions

Atinderpal Singh [1], Sanjit K. Deb [1,*], Sukhbir Singh [1], Parmodh Sharma [2] and Jasjit S. Kang [3]

[1] Department of Plant and Soil Science, Texas Tech University, Box 42221, Lubbock, TX 79409, USA; atinderpal.singh@ttu.edu (A.S.); s.singh@ttu.edu (S.S.)

[2] PVR Technologies Inc., King of Prussia, PA 19406, USA; sharmap2@gmail.com

[3] Department of Agronomy, Punjab Agricultural University, Ludhiana, Punjab 141004, India; kangjs@pau.edu

* Correspondence: sanjit.deb@ttu.edu; Tel.: +1-806-834-1373; Fax: +1-806-742-0775

Received: 23 January 2020; Accepted: 13 March 2020; Published: 3 April 2020

Abstract: Effects of non-leguminous cover crops and their times of chopping on the yield and quality of no-till baby corn (*Zea mays* L.) were evaluated during two *kharif* seasons (May-August in 2014 and 2015) under subtropical climatic conditions of Punjab, India. The experiment was laid out in a split-plot design with four replications at Punjab Agricultural University's Research Farm. Three cover crops (pearl millet (*Pennisetum glaucum* L.), fodder maize (*Zea mays* L.), and sorghum (*Sorghum bicolor* L.)) and the control (no cover crop) were in the main plots and chopping time treatments (25, 35, 45 days after planting (DAP)) in the subplots. During both *kharif* seasons, the yield (cob and fodder yield) and dry matter accumulation of baby corn following cover crop treatments, especially pearl millet, were significantly ($p \leq 0.05$) higher than the control, and improved with increments in chopping time from 25 to 45 DAP. The effect of cover crops on baby corn quality (i.e., protein, starch, total soluble solids, crude fiber, total solid, and sugar content) did not differ among treatments, while increasing increments in chopping time had a significant effect on the protein and sugar content of baby corn. The use of cover crops and increment in chopping time helped in enhancing topsoil quality, especially available nitrogen; yet, the effect of cover crops and their times of chopping on topsoil organic carbon, phosphorus, and potassium did not differ among treatments. During both seasons, there was no significant interaction between cover crop and time of chopping among treatments with respect to baby corn yield and quality, as well as topsoil quality parameters.

Keywords: baby corn; non-leguminous cover crops; chopping; baby corn yield; baby corn quality; *kharif* season

---

1. Introduction

Maize (*Zea mays* L.), also called corn, is the third most important cereal crop in the world after rice and wheat as a source of calories and in terms of the value of production [1–3]. The importance of maize worldwide is driven by its multiple uses as human food, livestock feed, a variety of food and industrial products, and seed [4]. Maize is grown as a food crop (i.e., grain maize) in many tropical areas, particularly in Latin American, African, and Asian countries, including India. In 2014, India produced about 23.7 million tons of maize on 9.3 million ha [5]. The average maize yield in India was 2.56 t·ha^{-1} during 2015–2016, less than one-quarter of that obtained in the United States of America (USA), and less than half of that obtained in China and Brazil [6].

As a C4 plant, maize has a higher grain yield potential than other major cereal grains (i.e., wheat and rice) [7]. Maize can be directly consumed as food at different stages of crop development, from baby corn to mature grain. Baby corn is the young, fresh, and finger-like ears of fully-grown standard cultivars, which are harvested immediately after the silks emerge (i.e., within 2 or 3 days of silk emergence) and before pollination and fertilization [8,9]. In general, except for the length of time from the establishment to harvest, baby corn cultivation practices are similar to those of maize cultivation. Baby corn has increasingly gained popularity as a valued vegetable throughout the world. In India, baby corn has also emerged as a potential remunerative crop, especially among progressive farmers.

Maize, a heavy user of nutrients, requires more nitrogen (N) compared to other mineral nutrients. Shivay et al. [10] reported that increasing N application rate significantly increased leaf area index, dry matter accumulation, and net assimilation rate at different growth stages of maize. Cover crops (leguminous or non-leguminous cover crops) or green manures (particularly leguminous green manures) have the potential to fully or partially replace inorganic N fertilizer, particularly for high N-requiring cereal crops such as maize, and thereby promote the use of sustainable production practices. Leguminous cover crops are commonly used to provide N for use by subsequent crops [11,12]. Leguminous crops contribute N through symbiotic dinitrogen (N_2) fixation, reduce N fertilizer needs for subsequent crop, and increase soil N retention [13,14]. While leguminous cover crops are used as N sources to supplement or replace inorganic N fertilizer, non-leguminous cover crops have the potential to enhance soil organic matter by increasing biomass production and by scavenging nutrients, especially N leftover from previous crops [15]. Non-leguminous cover crops have been also reported to reduce nitrate leaching losses [16,17]. The use of legume-grass mixtures could combine the benefits of both, including N fixation, biomass production, and N scavenging [18,19].

The selection of cover crops for a given region requires, among others, knowledge of their growth potentials [20,21]. The major climatic variables affecting cover crop selection include temperature and rainfall [22]. Non-leguminous cover crops have become more important in tropical and subtropical areas, such as Punjab in India where crop residues in conventional systems are not enough to compensate for the loss of soil organic matter due to high rates of mineralization [23]. Additionally, the limited availability of farmyard manures could be overcome by using non-leguminous cover crops, especially in subtropical areas of Punjab by sowing them before *Kharif* baby corn (i.e., monsoon crop). In Punjab, these non-leguminous cover crops often include pearl millet (*Pennisetum glaucum* L.), fodder maize, and sorghum (*Sorghum bicolor* L.).

Cover crops could jump-start no-till, resulting in yield increases [3]. Hoorman et al. [3] reported that maize yields dropped slightly for the first five to seven years after switching to no-till because continuous conventional tillage oxidized the soil organic matter and soil productivity declined with time. However, long-term (i.e., seven to nine years), no-till practices improved soil health by getting microbes and soil fauna back into balance, restored nutrients lost by conventional tillage and increased organic matter levels, resulting in higher maize yields than conventionally tilled fields [3]. Hoorman et al. [3] suggested that cover crops could be an important part of a continuous no-till system for maintaining short-term as well as increasing long-term maize yields. The effects of leguminous cover crops on improved maize yield and enhanced soil N under no-till conditions have been repeatedly stressed in the literature [24–29]. Despite non-leguminous cover crops having beneficial effects on crop production, there remains a paucity of information about their effects on no-till maize or baby corn production, especially under subtropical conditions of Punjab, India.

In Punjab, planting *Kharif* non-leguminous cover crops has been often recommended in the second fortnight of April [30]. The benefits of non-leguminous cover crops as fodder crops could be obtained by cutting 50-day old crops before planting subsequent maize or baby corn crop [30]. Non-leguminous cover crops can be chopped before their flowering stages, and subsequent maize or baby corn crop can be planted under no-till conditions. The effects of leguminous cover crops such as sunn hemp (*Crotolaria juncea*), cowpea (*Vigna unguiculata*), and dhaincha (*Sesbania aculeata*) and their times of chopping on maize growth and yield have been evaluated in very few studies. Salaria [31] reported that the combination of leguminous cover crops (sunn hemp, cowpea, and dhaincha) and their times of chopping increased average maize grain yield by 15.3% over the control under subtropical climatic conditions. A significant interaction between leguminous cover crops and increment in chopping time indicated that chopping of cover crops at 45 days after planting (DAP) increased average maize grain yield by 12.9% and 24.6% over chopping at 35 DAP and 25 DAP, respectively [31]. Moreover, the manner in which leguminous cover crops have been used, various N levels as well as time of chopping, have improved available soil nitrogen (N), phosphorus (P), potassium (K) and organic carbon at harvest during a *kharif* no-till maize growing season [31]. In contrast, very little is known about the effects of non-leguminous cover crops and their times of chopping on the yield and quality of baby corn grown under tropical or subtropical climatic conditions. Therefore, the objective of this study was to evaluate the effects of non-leguminous cover crops (pearl millet, fodder maize, and sorghum) and increments in chopping time (25 DAP, 35 DAP, and 45 DAP) on the yield and quality of no-till baby corn during *kharif* seasons under the subtropical climatic conditions of Punjab, India.

2. Materials and Methods

2.1. Experimental Site

The field experiment was conducted at the Research Farm of the Department of Agronomy, Punjab Agricultural University, Ludhiana, India (latitude 30°53′58.27″ N, longitude 75°47′50.26″ E, at an altitude of 247 m above mean sea level) during two summer or *kharif* seasons (i.e., May–August in 2014 and 2015). The soil of the experimental field was sandy loam (coarse-loamy, calcareous, mixed, hyperthermic Typic Ustochrept) containing $3.3 \cdot g \cdot kg^{-1}$ organic carbon [32]. On average, within the 0–180 cm soil depths, soil pH, electrical conductivity (EC), soil water content at 30 kPa (i.e., soil water content at field capacity), and soil water content at 1500 kPa (i.e., soil water content at wilting point) were 8.0, $0.20 \cdot dS \cdot cm^{-1}$, 13.23%, and 7.61%, respectively [32]. As reported by Kukal and Sidhu [32], the soil in the experimental field was generally low in content of $KMnO_4$-extractable N ($152 \cdot kg \cdot ha^{-1}$), and medium in content of 0.5 N $NaHCO_3$-extractable P ($13.7 \cdot kg \cdot ha^{-1}$) and available K ($145 \cdot kg \cdot ha^{-1}$).

2.2. Weather Conditions

The experimental site was located in a region characterized by a subtropical, semi-arid climate with three distinct seasons: hot and dry summer (April–June) when *kharif* crops are grown, hot and humid monsoon (July–September), and cold winter (November–January) when *rabi* crops are grown. Weather data obtained from the meteorological observatory at Punjab Agricultural University, which was approximately 500 m from the experimental site, included maximum and minimum air temperatures, relative humidity, rainfall, and soil evaporation during two *kharif* seasons (2014–2015) (Figure 1). The weekly average maximum air temperature varied between 38.8 °C and 46.4 °C in 2014 (6 May–26 August) and between 38.9 °C and 48.2 °C in 2015 (6 May–26 August). The weekly average minimum air temperature ranged between 20.7 °C and 29.3 °C and 24.1 °C and 27.9 °C during 2014 and 2015, respectively. The weekly average relative humidity ranged from 43.9% to 84.4% in 2014 and from 29.6% to 81.1% in 2015. The experimental site received considerably less rainfall (161 mm) during the *kharif* season in 2014 compared to the *kharif* season in 2015 (197 mm). The weekly average soil evaporation rates varied from 24 mm to 80 mm and from 23.3 mm to 84 mm during 2014 and 2015, respectively.

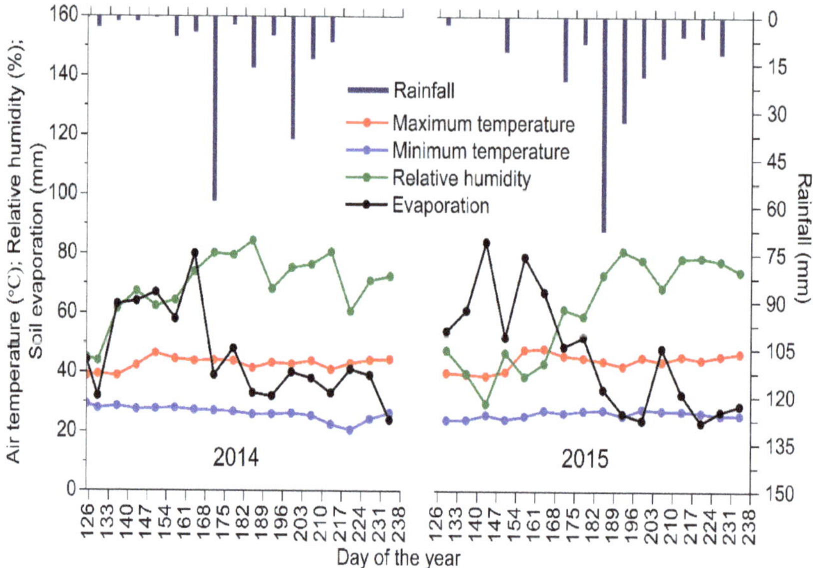

Figure 1. Weekly average weather data observed at the experimental site during both *kharif* seasons (6 May–26 August in 2014 and 2015).

2.3. Experimental Design and Agronomic Practices

The experimental field was divided into forty plots each 11 m × 3 m in size. All plots were subjected to no-till practices during both *kharif* seasons (2014–2015). The experiment was laid out in a split-plot randomized complete block design with four replications, which were comprised of three cover crops (i.e., non-leguminous pearl millet, fodder maize, and sorghum) and the control (no cover crop) in the main plots and cover crop chopping time treatments (i.e., 25 DAP, 35 DAP, and 45 DAP) in the sub-plots. All cover crops (pearl millet, fodder maize, and sorghum) were manually sown using the dibbling method on 5 May, 15 May, and 25 May during the *kharif* season in 2014, while all cover crops were sown on 9 May, 19 May, and 29 May during the *kharif* season in 2015. No fertilizers were applied to cover crops. The cover crops were chopped on 19 June in 2014 and on 22 June in 2015, corresponding to 25 DAP, 35 DAP, and 45 DAP at the time of chopping. The 25-, 35-, and 45-day old cover crops were chopped in situ using a chopper-cum-spreader and left uniformly on the soil surface. The chopper-cum-spreader used in this study was a tractor mounted flail type chopper, which was designed and developed by the Central Institute of Agricultural Engineering (CIAE), Bhopal, India [33]. The chopper cut cover crops above the ground level and chopped them into small pieces (i.e., 5–10 cm in length). The blades of the chopper were covered to prevent the spread of chopped cover crops from one plot to another. Baby corn was then planted on 19 June in 2014 and 22 June in 2015.

Baby corn (variety G-5414, Syngenta, India) was sown at a rate of 40·kg·ha^{-1} by the dibbling method at a row × plant spacing of 30 cm × 20 cm recommended by Dhaliwal and Kular for baby corn [30]. The recommended doses of N, P and K were used at rate of 120·kg·N·ha^{-1} as urea, 50·kg·P·ha^{-1} as diammonium phosphate and 30·kg·K$_2$O ha^{-1} as muriate of potash, respectively [30,34]. Half of the N and full doses of P and K were applied to baby corn as basal doses. The remaining dose of N was top-dressed at the knee-high stage of the baby corn. Prior to planting cover crops, pre-sowing flood irrigation was applied to all plots during both *kharif* seasons (2014–2015). Cover crops were sown when the soil water content was at field capacity in experimental plots. No post-sowing irrigation was applied to cover crops. As shown in Figure 1, a significant amount of rainfall during both seasons resulted in sufficient soil water content for cover crops. After chopping cover crops, all treatment plots had sufficient soil water content for the emergence of baby corn. A first irrigation was applied to baby corn at 8 DAP during both seasons. Additional irrigations were applied at two critical growth stages of water stress [35], i.e., at the knee-high stage and at the tassel emergence and silking stage, respectively.

2.4. Harvest of Baby Corn and Yield

To determine the dry matter accumulation, a plant sample of baby corn was collected at harvest from each treatment plot, sun-dried, and then oven-dried to a constant weight at 60 °C for 48–72 h. Generally, the dry matter accumulation provides an indicator of the growth and metabolic efficiency of the plant (i.e., an indicator of the crop yield) [31]. During both *kharif* seasons, baby corn cobs for each treatment plot was hand-harvested at two picking dates. The cobs were ready for first picking at 55 DAP when the silk length was about 2–4 cm. The second harvest was performed 5 days after the first harvest date. Baby corn can be marketed as green ears (with husk) and dehusked ears. All collected cobs, immediately after harvest, were dehusked by hand to remove their outer sheaths and weighed, and baby corn cob yield for each plot was recorded. After the completion of cob-picking, the crop was harvested, and green fodder (i.e., green stems and leaves of baby corn) yield for each plot was determined.

2.5. Quality Parameters of Baby Corn

The protein content (%) of baby corn cobs from individual treatments was estimated by multiplying its N content by the factor 6.25 [36]. The N content (%) was determined using the Kjeldahl distillation method [37]. The total soluble solids (TSS) (%) was determined with a hand-held digital refractometer (Erma, Tokyo, Japan) following the procedure described by Nelson and Sommers [38]. The total sugar content (%) of baby corn was estimated using the modified Nelson-Somogyi method [39,40]. The measurement of total solids content or dry matter content (% in relation to sample weight) of a baby corn cob was made by placing the sample (15 g) in a hot air oven at a temperature of 70 °C for 16–18 h until a constant mass was obtained. The dried sample was cooled down to room temperature in a desiccator, and the total solids content was determined as the remaining weight of the sample after drying.

The starch content (%) of a baby corn cob sample (i.e., about 5 g sample) was determined using the colorimetric method [41]. After the sugars present in the sample were extracted until a quantitative test with anthrone gave no green color, the sample was cooled and mixed with perchloric acid, solubilized, filtered, and diluted. The diluted solution was mixed with anthrone reagent and boiled until the reaction was completed. The solution was then allowed to cool, and its absorbance was measured at 630 nm in a spectrophotometer. The concentration of starch was calculated from a standard curve. A standard curve was prepared using 0.5, 1.0, and 1.5 mL of starch standard solution in a 100 mL volumetric flask. A 5 mL solution from each sample was used to develop the standard curve, i.e., a plot of absorbance versus concentration.

The crude fiber content (%) of baby corn cob was determined using the method described by Ranganna [41] and Horwitz [42]. About 2 g of baby corn cob sample was placed into the crucible, dissolved in 200 mL of sulfuric acid (H_2SO_4) solution, and boiled for 30 min in a digestion flask with the condenser. The hydrolyzed mixture was filtered, and the residue was rinsed with hot water to remove acid from the filtrate in the crucible. The residue from acid digestion was washed again in the flask with 200 mL of sodium hydroxide (NaOH) solution and boiled for 30 min. The hydrolyzed sample was again filtered, and the residue was rinsed with hot water to ensure that the crucible was free of alkalinity. The residue in the crucible was oven-dried at 105 °C until a constant mass was attained. The crucible containing dry residue was weighed and placed in a muffle furnace at 550 °C for 5 h. The crucible with ash was cooled to room temperature in a desiccator and weighed. The crude fiber content of baby corn cob sample was then calculated as:

$$\text{Crude fiber content } (\%) = \frac{\text{Weight of crucible with dry residue } - \text{ Weight of crucible with ash}}{\text{Weight of sample taken}} \times 100$$

2.6. Soil Analysis

Prior to planting cover crops in 2014, soil samples within the topsoil (0–20 cm soil depth) were collected from the experimental field to determine soil texture and soil chemical properties (Table 1). To determine soil chemical properties following cover crop treatments, soil samples (0–20 cm) were collected from all cover crop and control treatment plots immediately after harvesting baby corn in 2014 and 2015. The available soil nitrogen (N) was determined using the modified alkaline permanganate extraction method proposed by Subbaiah and Asija [43], the available soil phosphorus (P) using the sodium bicarbonate extraction method described by Olsen [44], and the available soil potassium (i.e., the ammonium acetate extractable K) using the method described by Merwin and Peech [45]. The chromic acid titration method [46] was used to determine the soil organic carbon.

Table 1. Selected soil chemical properties of the experimental field determined prior to planting non-leguminous cover crops during *kharif* season (May–August) in 2014.

Soil Depth (cm)	Particle Size Distribution (%)			Soil Texture [β]	Organic Carbon (%)	Nutrient Content (kg·ha^{-1})		
	Sand	Silt	Clay			Available N	Available P	Available K
0–20	74.3	19.8	5.9	Sandy Loam	0.30	179.6	20.1	156.7

[β] According to USDA (United States Department of Agriculture) classification.

2.7. Statistical Analysis

The analysis of variance (ANOVA) was performed using the General Linear Model (GLM) procedure of SAS software version 9.4 (SAS Institute, Cary, NC, USA). The appropriate error term was used to evaluate each factor and interaction. The main plot factors (cover crops) and sub-plot factors (times of chopping) were considered as fixed variables, and the data in 2014 and 2015 was considered as random variables. Differences among treatment means were compared using Fisher's protected least significance difference (LSD) test. Statistical significance was evaluated at $p \leq 0.05$.

3. Results and Discussion

3.1. Cover Crop Effects on Dry Matter Accumulation and Cob Yield of Baby Corn

The dry matter content of baby corn following non-leguminous cover crop treatments (pearl millet, fodder maize, and sorghum) was determined at harvest. The combination of cover crops and different times of chopping (25 DAP, 35 DAP, and 45 DAP) resulted in a higher amount of dry matter accumulation of baby corn compared to the control (no cover crop treatment) in 2014 and 2015 (Table 2). Similar enhanced dry matter content following cover crops and their times of chopping under no-till maize production was also reported by Salaria [31]. However, Salaria [31] evaluated leguminous cover crops (e.g., sunnhemp, cowpea, and dhaincha) and different N levels for their effects on dry matter yield of maize grown under subtropical climatic conditions.

Table 2. Effects of cover crops and their times of chopping (at different days after planting (DAP)) on the dry matter of baby corn, the cob yield of baby corn, and the green fodder yield of baby corn during both *kharif* seasons (May–August in 2014 and 2015).

Non-Leguminous Cover Crop	Dry Matter of Baby Corn (t·ha^{-1})				Baby Corn Cob Yield (t·ha^{-1})				Fodder Yield of Baby Corn (t·ha^{-1})			
	Chopping Time of Cover Crop				Chopping Time of Cover Crop				Chopping Time of Cover Crop			
	25 DAP	35 DAP	45 DAP	Mean	25 DAP	35 DAP	45 DAP	Mean	25 DAP	35 DAP	45 DAP	Mean
2014 Season:												
Pearl Millet	4.09	4.34	4.60	4.34 a [z]	1.39	1.41	1.57	1.45 a	20.01	20.24	21.29	20.52 a
Fodder Maize	4.19	4.37	4.45	4.33 a	1.25	1.36	1.52	1.37 a	17.95	19.05	21.09	19.37 ab
Sorghum	3.85	4.08	4.15	4.02 b	0.93	1.01	1.11	1.01 b	17.83	18.04	18.55	18.14 b
Mean	4.04 c	4.26 b	4.40 a	-	1.19 b	1.26 b	1.40 a	-	18.60 b	1.10 ab	20.31 a	-
Control (No Cover Crop)	-	-	-	3.56 c	-	-	-	0.98 b	-	-	-	18.93 ab
Cover Crop × Chopping Time	NS [y]	NS	NS		NS	NS	NS		NS	NS	NS	
2015 Season:												
Pearl Millet	4.18	4.38	4.65	4.40 a	1.47	1.46	1.59	1.50 a	20.07	20.31	21.31	20.56 a
Fodder Maize	4.22	4.39	4.48	4.36 a	1.30	1.40	1.52	1.40 a	18.05	19.06	21.13	19.41 ab
Sorghum	3.91	4.10	4.17	4.06 b	0.99	1.03	1.13	1.05 b	17.90	18.07	18.56	18.17 b
Mean	4.10 c	4.29 b	4.43 a	-	1.25 b	1.29 b	1.41 a	-	18.67 b	19.15 ab	20.33 a	-
Control (No Cover Crop)	-	-	-	3.58 c	-	-	-	1.01 b	-	-	-	19.20 ab
Cover Crop × Chopping Time	NS	NS	NS		NS	NS	NS		NS	NS	NS	

[z] Treatment means in columns for cover crops or rows for DAP across cover crops followed by the same letter are not significantly different. [y] NS: non-significant interaction at $p \leq 0.05$.

On average, the dry matter of baby corn in 2014 and 2015 (4.34 t·ha^{-1} and 4.40 t·ha^{-1}, respectively) was significantly ($p \leq 0.05$) higher following pearl millet cover crop treatment than following sorghum cover crop and control treatments. However, the dry matter of baby corn following fodder maize cover crop was similar to that observed following pearl millet cover crop in both years. In a review study, Dar et al. [47] reported that the dry matter yield of baby corn grown in India varied from 6.0 to 8.0 t·ha^{-1} during *rabi* and late-*rabi* seasons and from 8.0 to 9.0 t·ha^{-1} during *kharif* and summer seasons. The higher dry matter yield of baby corn reported by different studies was mainly influenced by various agronomic practices including closer crop geometries (i.e., closer row spacing) and higher rates of N application [47]. Eltelib et al. [48] observed that the dry matter yield of maize ranged from 3.7 to 12.2 t·ha^{-1} based on the amount of fertilizer application especially different levels of N application.

The maximum cob yield of baby corn was observed following pearl millet cover crop treatment in both years (Table 2). The cob yield of baby corn (1.45 t·ha^{-1} and 1.50 t·ha^{-1} in 2014 and 2015, respectively) was significantly ($p \leq 0.05$) higher following pearl millet than no cover crop (i.e., control) and sorghum cover crop. The cob yield of baby corn following fodder maize cover crop was statistically similar to the cob yield observed following pearl millet cover crop treatment in both years. On average, pearl millet cover crop treatment resulted in a relatively higher amount of green fodder yield of 20.52 t·ha^{-1} and 20.56 t·ha^{-1} in 2014 and 2015, respectively (Table 2). The green fodder yield of baby corn following sorghum was slightly lower than that of the control treatment in both years. However, the fodder yield of baby corn in both years was not significantly different between all cover crops and the control. Although information pertinent to the effect of non-leguminous and leguminous cover crops on cob and fodder yields of baby corn grown under subtropical climatic conditions is very limited, comparable results have been reported by several studies. For instance, as reported by Dar et al. [47], the yield of baby corn or sweet corn grown in India generally varied from 1.2 to 12.7 t·ha^{-1}, while the fodder yield of baby corn (sum of green fodder yield and dry fodder yield) ranged from 4.12 to 27.0 t·ha^{-1}. Baby corn intercropped with fodder legumes, such as maize, cowpea, clusterbean (*Cyamopsis tetragonoloba* L.), and pillipesara (*Phaseolus trilobus* L.), has been reported to produce 28.6 to 50.5 t·ha^{-1} green fodder and 5.1 to 8.8 t·ha^{-1} dry fodder yield of baby corn during *rabi* season [49]. The higher baby corn and fodder yield data reported in previous studies [47,49] were primarily attributed to high rates of N application, plant densities, and planting patterns.

In situ chopping of cover crops at 25 DAP, 35 DAP, and 45 DAP was likely to provide a large quantity of cover crop biomass, which could improve soil physical properties such as soil water retention and soil temperature. Cover crop residues could also provide additional nutrients for better growth responses of the subsequent crop [50]. Accordingly, compared to the control, the relatively higher dry matter accumulation of baby corn observed following all cover crop treatments was most likely due to the improved soil conditions. During both years, pearl millet cover crop was growing faster than sorghum and fodder maize. The relatively fast-growing deep root system of pearl millet [51,52] might scavenge more nutrients, resulting in more biomass production and dry matter accumulation of baby corn following this non-leguminous cover crop treatment (Table 2). The use of pearl millet as a cover crop to enhance biomass production and nutrients for subsequent crops has been reported in different studies (e.g., [53–56]). Schonbeck and Morse [53] reported that pearl millet cover crop could produce biomass from 7.0 to 12 t·ha^{-1}. Pearl millet cover crop has been also reported to improve N use efficiency by a succeeding maize crop [54], provide 60–80% of the potassium nutrient needed for the subsequent crop [55], and improve soil organic matter and inhibit soil-borne diseases [56].

Overall, chopping of cover crops at 45 DAP showed a significant ($p \leq 0.05$) effect on the dry matter accumulation of baby corn cob and green fodder yield of baby corn in both years (Table 2). The amount of dry matter accumulated in baby corn after chopping cover crops at 45 DAP (i.e., an average yield of 4.40 t·ha^{-1} and 4.43 t·ha^{-1} in 2014 and 2015, respectively) was significantly higher than the dry matter accumulation after chopping at 25 DAP and 35 DAP. Similarly, the average cob yield of baby corn observed after chopping cover crops at 45 DAP (i.e., 1.40 t·ha^{-1} and 1.41 t·ha^{-1} in 2014 and 2015, respectively) was significantly higher than those observed after chopping at 25 DAP and 35 DAP. The average green fodder yield of baby corn observed after chopping at 45 DAP (i.e., 20.31 t·ha^{-1} and t·ha^{-1} in 2014 and 2015, respectively) was slightly higher than those observed after chopping at 25 DAP and 35 DAP. However, the fodder yield of baby corn in both years was not significantly different among time of chopping treatments. Overall, the results suggested that chopping all non-leguminous cover crops, particularly pearl millet, at 45 DAP could enhance dry mass accumulation and cob and fodder yield of succeeding *kharif* baby corn under no-till practices. Another aspect to be noted in Table 2 is that during both *kharif* seasons, the interaction of cover crops and their times of chopping on the dry matter accumulation and cob and fodder yield were not significant among treatments.

3.2. Cover Crop Effects on Baby Corn Quality Parameters

3.2.1. Protein Content

The protein content of baby corn cob following non-leguminous cover crop treatments in 2014 and 2015 is presented in Table 3, suggesting that the average protein content of baby corn following cover crop treatments was slightly higher than that of no cover crop (i.e., control) treatment. On average, the higher amount of protein content of baby corn cob (12.15% and 12.26% in 2014 and 2015, respectively) was observed following pearl millet cover crop treatment. The reason might be attributed to the maximum amount of cover crop dry matter produced by pearl millet treatment that resulted in a higher amount of N available for use by subsequent baby corn, contributing to higher N uptake by baby corn. However, in both years, there was no statistically significant ($p \leq 0.05$) difference in the protein content of baby corn cob between all cover crops and the control. To the best of our knowledge, very little is known about the effect of non-leguminous cover crops on the protein content of baby corn; however, similar results of the protein content in baby corn intercropped with leguminous crops were reported by several studies. For instance, Kumar and Venkateswarlu [49] reported that the protein content of baby corn intercropped with fodder legumes (e.g., maize, cowpea, clusterbean, and pillipesara) varied from 7.01% to 8.73% during *rabi* season. The protein content of baby corn was significantly influenced by plant densities and fertilization practices, especially high rates of inorganic N levels in different studies (e.g., [57,58]). Hooda and Kawatra [59] reported that the protein content of baby corn (17.9%) was similar or slightly higher than vegetables like cabbage, bitter gourd, eggplant, French beans, and spinach. Comparable results (i.e., protein content varied between 10.3% and 12.96%) were also reported for sweet corn and maize [60,61]. The protein content of forage maize has been reported to vary from 3.67% to 9.06% under different levels of N application [48].

The average protein content of baby corn cob after chopping cover crops at 45 DAP was significantly ($p \leq 0.05$) higher than the protein content obtained after chopping at 25 DAP (Table 3). The protein content of baby corn cob was not significantly different after chopping cover crops at 35 DAP and 45 DAP. On average, the maximum protein content (12.60% and 12.68% in 2014 and 2015, respectively) was observed after chopping cover crops at 45 DAP. As mentioned earlier, the higher protein content of baby corn could be explained by the higher dry matter of cover crops and resulting higher amount of N associated with increment in chopping time from 25 DAP to 45 DAP. There was no interaction of cover crops and their times of chopping on the protein content of baby corn cob among treatments (Table 3).

Table 3. Effects of cover crops and their times of chopping (at different days after planting (DAP)) on the protein content, the starch content, and the crude fiber content of baby corn cob during both *kharif* seasons (May–August in 2014 and 2015).

Non-Leguminous Cover Crop	Protein Content (%)				Starch Content (%)				Crude Fiber (%)			
	Chopping Time of Cover Crop				Chopping Time of Cover Crop				Chopping Time of Cover Crop			
	25 DAP	35 DAP	45 DAP	Mean	25 DAP	35 DAP	45 DAP	Mean	25 DAP	35 DAP	45 DAP	Mean
2014 Season:												
Pearl Millet	11.26	12.59	12.61	12.15 a [z]	2.42	2.40	2.40	2.41 a	2.10	2.11	2.10	2.10 a
Fodder Maize	11.33	11.92	12.63	11.96 a	2.40	2.41	2.39	2.40 a	2.12	2.11	2.11	2.11 a
Sorghum	11.47	12.19	12.58	12.08 a	2.40	2.40	2.39	2.40 a	2.12	2.12	2.11	2.11 a
Mean	11.35 b	12.23 a	12.60 a	-	2.41 a	2.40 ab	2.39 b	-	2.11 a	2.11 a	2.10 a	-
Control (No Cover Crop)	-	-	-	11.79 a	-	-	-	2.41 a	-	-	-	2.10 a
Cover Crop × Chopping Time	NS [y]	NS	NS		NS	NS	NS		NS	NS	NS	
2015 Season:												
Pearl Millet	11.42	12.60	12.76	12.26 a	2.42	2.41	2.40	2.41 a	2.11	2.10	2.10	2.10 a
Fodder Maize	11.35	12.63	12.70	12.22 a	2.42	2.41	2.39	2.41 a	2.11	2.10	2.10	2.10 a
Sorghum	11.34	12.47	12.58	12.13 a	2.41	2.40	2.39	2.40 a	2.12	2.10	2.11	2.11 a
Mean	11.37 b	12.56 a	12.68 a	-	2.42 a	2.40 ab	2.39 b	-	2.11 a	2.10 a	2.10 a	-
Control (No Cover Crop)	-	-	-	12.02 a	-	-	-	2.41 a	-	-	-	2.10 a
Cover Crop × Chopping Time	NS	NS	NS		NS	NS	NS		NS	NS	NS	

[z] Treatment means in columns for cover crops or rows for DAP across cover crops followed by the same letter are not significantly different. [y] NS: non-significant interaction at $p \leq 0.05$.

3.2.2. Starch Content

The starch content of baby corn cob, as shown in Table 3, was not significantly ($p \leq 0.05$) different between all cover crops and the control in both years. On average, the maximum starch content (2.41% in both years) was observed following pearl millet cover crop treatment. There is a paucity of quantitative information on the starch content of baby corn following non-leguminous cover crop treatments. Nevertheless, evaluation of nutritional compositions of baby corn or sweet corn in several studies reported different starch content values for baby corn. For instance, Hooda and Kawatra [59] reported a starch content value of 15.6% for baby corn. Ugur and Maden [62] suggested that sweet corn would contain 10% to 11% starch. Generally, the average starch content of baby corn decreased with increment in chopping time from 25 DAP to 45 DAP. On average, the maximum starch content of baby corn (2.41% and 2.42% in 2014 and 2015, respectively) was observed after chopping cover crops at 25 DAP. As shown in Table 3, there was no interaction of cover crops and their times of chopping on the starch content of baby corn cob in both years.

3.2.3. Crude Fiber Content

The average crude fiber content of baby corn cob varied between 2.40% and 2.41% among cover crop and control treatments in both years (Table 3). The average crude fiber content was not significantly ($p \leq 0.05$) different between all cover crops and the control. A slightly lower amount of crude fiber was observed following pearl millet cover crop and no cover crop (control) treatments compared to fodder maize and sorghum cover crop treatments. The lower crude fiber content of baby corn in this study might be explained by the higher protein content of baby corn (Table 3), which generally decreased the deposition of lignin and cellulose [62]. Nutritional evaluation of baby corn for crude fiber content has yielded contrasting results in different studies. For instance, similar results of lower crude fiber contents of baby corn (4.53–5.89%) were reported by several studies (e.g., [57,59,63]). Shobha et al. [64] evaluated the quality of eleven maize genotypes at baby corn and grain maturity stages and observed lower crude fiber contents ranging from 1.96% to 2.40% among maize genotypes. In contrast, Kumar and Venkateswarlu [49] reported higher crude fiber content of baby corn (i.e., 23.76–25.71%) intercropped with fodder legumes (e.g., maize, cowpea, clusterbean, and pillipesara). Eltelib et al. [48] also reported that the crude fiber content of maize varied between 21.13% and 22.1%. The higher crude fiber contents of baby corn or maize in their studies were significantly influenced by high rates of N application.

The crude fiber content of baby corn cob was not significantly ($p \leq 0.05$) different after chopping cover crops at 25 DAP, 35 DAP, and 45 DAP in both years (Table 3). On average, the minimum amount of crude fiber content in baby corn cob (i.e., 2.10% in both years) was observed after chopping cover crops at 45 DAP. There was no statistically significant interaction between cover crops and their times of chopping among treatments with respect to the crude fiber content of baby corn cob.

3.2.4. Total Soluble Solids and Total Solid Content

There was no statistically significant ($p \leq 0.05$) difference in the total soluble solids (TSS) content of baby corn cob between all cover crops and the control in both years (Table 4). On average, the maximum amount of TSS (8.88 °Brix and 8.89 °Brix in 2014 and 2015, respectively) was observed following pearl millet cover crop treatment. Information about the effect of non-leguminous cover crops on the TSS content of baby corn is still limited; however, similar TSS values were reported by several studies that evaluated nutritional composition of baby corn. For instance, Joshi and Chilwal [58] reported that the TSS of baby corn varied from 8.1 to 9.5 °Brix. However, Khan et al. [65] found relatively higher TSS values in sweet corn cob ranging from 14.31 to 16.56 °Brix, which were attributed to agronomic practices associated with transplanting dates and higher N levels. Ugur and Maden [62] reported that average TSS values in sweet corn varied from 8.52 to 20.64 °Brix with the progression of the cultivation period among different sweet corn varieties. In contrast, Shobha et al. [64] observed relatively lower

TSS content values of maize at the baby corn stage, ranging from 5.06 to 5.86 °Brix among different maize genotypes.

Table 4. Effects of cover crops and their times of chopping (at different days after planting (DAP)) on the total soluble solids (TSS), the total solid content, and the sugar content of baby corn cob during both *kharif* seasons (May–August in 2014 and 2015).

Non-leguminous Cover Crop	TSS (°Brix)				Total Solid (%)				Sugar Content (%)			
	Chopping Time of Cover Crop				Chopping Time of Cover Crop				Chopping Time of Cover Crop			
	25 DAP	35 DAP	45 DAP	Mean	25 DAP	35 DAP	45 DAP	Mean	25 DAP	35 DAP	45 DAP	Mean
2014 Season:												
Pearl Millet	8.82	8.88	8.95	8.88 a [z]	14.93	14.95	15.60	15.16 a	6.8	7.1	7.6	7.1 a
Fodder Maize	8.71	8.67	8.80	8.72 a	14.91	14.80	15.05	14.92 a	6.7	6.9	7.2	6.9 a
Sorghum	8.67	8.77	8.97	8.80 a	14.93	14.94	14.94	14.94 a	6.8	7.0	7.2	7.0 a
Mean	8.73 a	8.77 a	8.90 a	-	14.92 a	14.89 a	15.20 a	-	6.8 b	7.0 b	7.3 a	-
Control (No Cover Crop)	-	-	-	8.84 a	-	-	-	15.04 a	-	-	-	6.9 a
Cover Crop × Chopping Time	NS [y]	NS	NS		NS	NS	NS		NS	NS	NS	
2015 Season:												
Pearl Millet	8.85	8.89	8.95	8.89 a	14.95	14.94	15.04	14.97 a	7.0	7.2	7.7	7.3 a
Fodder Maize	8.69	8.78	8.82	8.76 a	14.97	14.82	14.90	14.89 a	6.9	7.1	7.4	7.1 a
Sorghum	8.70	8.77	8.92	8.79 a	14.93	14.98	14.91	14.94 a	7.0	7.0	7.2	7.0 a
Mean	8.74 a	8.81 a	8.89 a	-	14.93 a	14.91 a	14.95 a	-	7.0 b	7.1 b	7.4 a	-
Control (No Cover Crop)	-	-	-	8.72 a	-	-	-	14.95 a	-	-	-	7.1 a
Cover Crop × Chopping Time	NS	NS	NS		NS	NS	NS		NS	NS	NS	

[z] Treatment means in columns for cover crops or rows for DAP across cover crops followed by the same letter are not significantly different. [y] NS: non-significant interaction at $p \leq 0.05$.

The TSS content of baby corn cob was not significantly ($p \leq 0.05$) different among the time of chopping treatments, i.e., after chopping cover crops at 25 DAP, 35 DAP, and 45 DAP in both years. The average amount of TSS was increased with increment in chopping time from 25 DAP to 45 DAP. For example, in 2014, the highest amount of TSS (8.90 Brix) was observed after chopping cover crops at 45 DAP, followed by chopping at 35 DAP (8.77 Brix), and the TSS was the lowest after chopping cover crops at 25 DAP (Table 4). There was no statistically significant interaction between cover crops and their times of chopping among treatments with respect to the TSS of baby corn cob.

As shown in Table 4, there was no statistically significant ($p \leq 0.05$) difference in the total solid content (i.e., dry matter content) of baby corn cob between all cover crops and the control in both years. To the best of our knowledge, there is almost no information on total solids content or dry matter content of baby corn cob following leguminous or non-leguminous cover crops. On average, the maximum total solids content of baby corn cob (15.16% and 14.97% in 2014 and 2015, respectively) was observed following pearl millet treatment, followed by no cover crop (control) treatment (15.04% and 14.95% in 2014 and 2015, respectively). The total solids content of baby corn cob was not significantly different among the time of chopping treatments, i.e., after chopping cover crops at 25 DAP, 35 DAP, and 45 DAP. The total solids content of baby corn observed after chopping cover crops at each time was statistically similar, with the highest amount after chopping cover crops at 45 DAP in both years. There was no interaction of cover crops and their times of chopping on the total solids content of baby corn among treatments (Table 4).

3.2.5. Sugar Content

Similar to the starch content of baby corn, there was no statistically significant ($p \leq 0.05$) difference in the sugar content of baby corn cob between all cover crops and the control in both years (Table 4). On average, the maximum sugar content of baby corn cob (7.1% and 7.3% in 2014 and 2015, respectively) was observed following pearl millet treatment. Like other baby corn quality parameters discussed earlier, evaluation of nutritional composition of baby corn or sweet corn in several studies reported different total sugar content values for baby corn. Shobha et al. [64] evaluated the total sugar content among different maize genotypes at the baby corn stage and reported that the total sugar content ranged from 0.40% to 0.89%. The total sugar content of baby corn has been also reported to vary from 0.002% to 2.3% (e.g., [57,61]). Prajwal Kumar et al. [63] reported relatively lower total sugar content of baby corn, ranging from 0.021% to 0.025%. In contrast, Rosli and Anis [66] reported that baby corn contained a significantly higher total sugar content of 10.7–21.48%.

In both years, the average sugar content of baby corn cob increased with increment in chopping time from 25 DAP to 45 DAP (Table 4). Notably, on average, the sugar content of baby corn after chopping cover crops at 45 DAP was significantly ($p \leq 0.05$) higher than the sugar content observed after chopping cover crops at 25 DAP and 35 DAP. There was no interaction of cover crop treatments and their times of chopping on the sugar content of baby corn among treatments (Table 4).

3.3. Cover Crop Effects on Soil Quality Parameters

3.3.1. Soil Organic Carbon

The amount of soil organic carbon content in different treatment plots in the topsoil (0–20 cm), presented in Table 5, was determined after the harvest of baby corn in both years. All cover crop treatments resulted in slightly higher average soil organic carbon content (i.e., ranging from 0.32% to 0.35% during 2014–2015) compared to the control (i.e., ranging from 0.31% to 0.32% during 2014–2015) (Table 5). Sharma et al. [67] suggested that although cover crops were highly decomposable, increased soil organic matter following cover crops was only confined to the topsoil (0–20 cm). However, in this study, there was no statistically significant ($p \leq 0.05$) difference in soil organic carbon content between all cover crops and the control in both years. Among only cover crop treatments, the soil organic carbon content was slightly higher following pearl millet and sorghum treatments than that observed following fodder maize treatment. The enhanced organic carbon content in the topsoil (0–20 cm) following non-leguminous cover crops (i.e., pearl millet, fodder maize, and sorghum) was most likely due to increased biomass or dry matter accumulation produced by these cover crop treatments under no-till practices [15,68,69]. It is worth noting that the soil organic carbon content is a good indicator of soil quality [38,70].

Table 5. Effects of cover crops and their times of chopping (at different days after planting (DAP)) on the organic carbon content in soil at harvest during both *kharif* seasons (May–August in 2014 and 2015).

Non-Leguminous Cover Crop	Organic Carbon (%)			
	Chopping Time of Cover Crop			
	25 DAP	35 DAP	45 DAP	Mean
2014 Season:				
Pearl Millet	0.32	0.32	0.37	0.34 a [z]
Fodder Maize	0.31	0.31	0.36	0.32 a
Sorghum	0.34	0.35	0.36	0.35 a
Mean	0.32 b	0.32 b	0.36 a	-
Control (No Cover Crop)	-	-	-	0.32 a
Cover Crop × Chopping Time	NS [y]	NS	NS	
2015 Season:				
Pearl Millet	0.35	0.34	0.37	0.35 a
Fodder Maize	0.35	0.33	0.34	0.34 a
Sorghum	0.35	0.35	0.35	0.35 a
Mean	0.35 ab	0.34 b	0.36 a	-
Control (No Cover Crop)	-	-	-	0.31 a
Cover Crop × Chopping Time	NS	NS	NS	

[z] Treatment means in columns for cover crops or rows for DAP across cover crops followed by the same letter are not significantly different. [y] NS: non-significant interaction at $p \leq 0.05$.

Generally, improved soil organic carbon following various cover crop treatments have been reported in numerous studies (e.g., [13,50,67,71,72]). Sainju et al. [68] reported that a non-leguminous cover crop (rye) was better than legumes (hairy vetch and crimson clover) in increasing soil organic carbon. Several studies observed that the use of both leguminous and non-leguminous cover crops and conservation tillage practices increased soil organic carbon content under maize production systems (e.g., [69,73]). On average, the soil organic carbon after chopping cover crops at 45 DAP was significantly ($p \leq 0.05$) higher than the soil organic carbon observed after chopping cover crops at 25 DAP and 35 DAP in both years (Table 5). In both years, there was no statistically significant interaction between cover crops and their times of chopping among treatments with respect to the soil organic carbon.

3.3.2. Available Soil Nitrogen

The amount of available soil nitrogen (N) in different treatment plots in the topsoil (0–20 cm), which was determined after the harvest of baby corn in both years, is presented in Table 5. The use of pearl millet, fodder maize, and sorghum cover crops improved available topsoil N content at harvest in 2014 and 2015 (Table 6) compared to N values observed at the beginning of this study (Table 1).

On average, all cover crop treatments generally resulted in significantly ($p \leq 0.05$) higher available soil N as compared to the control (Table 6). Sharma et al. [67] also reported that most of the changes in soil chemical properties (e.g., soil N) following cover crops appeared to be confined in the topsoil (0–20 cm). Among cover crop treatments, the amount of available soil N following pearl millet (196.7 kg·ha^{-1} and 196.4 t·ha^{-1} in 2014 and 2015, respectively) was significantly higher than the available soil N observed following fodder maize and sorghum cover crops.

The enhanced soil N content following various cover crops have been observed in numerous studies (e.g., [67,68,72]). Substantial changes in soil total N content were primarily due to long-term use of cover crops, which increased total soil N through additions of fixed N or prevention of N losses (e.g., [67,74]). The evaluation of the effects of leguminous and non-leguminous cover crops on soil N has yielded contrasting results in different studies for different crops. For instance, Kuo et al. [20] observed that leguminous cover crops, particularly hairy vetch, were more effective than non-leguminous cover crops (i.e., rye and annual ryegrass) in increasing soil inorganic N levels. In contrast, Sainju et al. [68] suggested that a non-leguminous cover crop (rye) was much more effective than legumes (hairy vetch and crimson clover) in increasing N availability in the soil.

Compared to legumes, the use of non-leguminous cover crops has been reported to reduce the loss of nitrate through leaching [16,17,75–77]. McCracken et al. [16] observed that a non-leguminous cover crop (rye) was much more effective than a leguminous cover crop (hairy vetch) in reducing nitrate leaching. A direct evaluation of the efficacy of non-leguminous cover crops in scavenging N was not examined in this study; however, the non-leguminous cover crop has been recognized for its potential as a scavenger of soil N [15,78]. Moreover, all non-leguminous cover crops, particularly pearl millet, were more likely to reduce nitrate leaching. Overall, as shown in Table 6, the contribution from non-leguminous cover crops to N availability in the topsoil (0–20 cm) suggested that inorganic fertilizer nutrients could be reduced in *kharif* no-till baby corn production.

In both years, the average amount of available soil N was increased with an increase in the time of chopping of cover crops from 25 DAP to 45 DAP (Table 6). On average, the available soil N after chopping cover crops at 45 DAP was significantly ($p \leq 0.05$) higher than the soil N observed after chopping cover crops at 25 DAP. However, the available soil N was not significantly different among increment in chopping time treatments 35 DAP and 45 DAP in both the years. There was no interaction of cover crops and their times of chopping on the available soil N among treatments.

Table 6. Effects of cover crops and their times of chopping (at different days after planting (DAP)) on the available nitrogen (N) content, the phosphorus (P) content, and the potassium (K) content in soil at harvest during both *kharif* seasons (May–August in 2014 and 2015).

Non-Leguminous Cover Crop	Available N (kg·ha^{-1})				Available P (kg·ha^{-1})				Available K (kg·ha^{-1})			
	Chopping Time of Cover Crop				Chopping Time of Cover Crop				Chopping Time of Cover Crop			
	25 DAP	35 DAP	45 DAP	Mean	25 DAP	35 DAP	45 DAP	Mean	25 DAP	35 DAP	45 DAP	Mean
2014 Season:												
Pearl Millet	193.4	192.1	204.8	196.7 a [z]	20.4	20.1	23.0	21.2 a	155.5	162.7	170.2	162.8 a
Fodder Maize	183.3	191.7	194.3	189.7 b	21.2	22.5	26.0	23.2 a	158.8	151.9	161.9	157.5 a
Sorghum	179.6	186.7	186.1	184.1 c	22.1	22.6	22.7	22.4 a	156.2	161.0	159.9	159.0 a
Mean	185.4 b	190.1 ab	195.0 a	-	21.2 b	21.7 b	23.9 a	-	156.8 b	158.5 ab	164.0 a	-
Control (No Cover Crop)	-	-	-	181.5 c	-	-	-	22.6 a	-	-	-	158.1 a
Cover Crop × Chopping Time	NS [y]	NS	NS		NS	NS	NS		NS	NS	NS	
2015 Season:												
Pearl Millet	194.0	193.2	202.2	196.4 a	21.0	22.1	23.1	22.0 a	157.1	163.3	171.3	163.9 a
Fodder Maize	183.0	187.6	195.7	188.7 b	22.0	22.5	24.8	23.1 a	159.0	155.4	162.6	159.0 b
Sorghum	185.3	187.6	187.3	186.7 b	23.1	22.9	23.8	23.2 a	157.7	163.0	162.5	161.0 ab
Mean	187.3 b	189.4 ab	195.0 a	-	22.0 b	22.5 ab	23.9 a	-	157.9 b	160.5 b	165.4 a	-
Control (No Cover Crop)	-	-	-	182.8 c	-	-	-	22.7 a	-	-	-	158.6 b
Cover Crop × Chopping Time	NS	NS	NS		NS	NS	NS		NS	NS	NS	

[z] Treatment means in columns for cover crops or rows for DAP across cover crops followed by the same letter are not significantly different. [y] NS: non-significant interaction at $p \leq 0.05$.

3.3.3. Available Soil Phosphorus

The amount of available soil phosphorus (P) at different treatment plots in the topsoil (0–20 cm), which was determined after the harvest of baby corn in both years, is presented in Table 6. There was no statistically significant ($p \leq 0.05$) difference in the average available soil P between all cover crops and the control in 2014 and 2015. The average available soil P among all cover crop and control treatments ranged from 21.2 to 23.2 kg·ha^{-1} during 2014–2015. The use of cover crop treatments did not markedly increase available topsoil P content at harvest (Table 6) compared to p values observed at the beginning of this study (Table 1). Several studies showed that soil P content could be conserved or maintained and/or enhanced (primarily in the topsoil) by various cover crop species [67,79]. Cover crops could also accumulate P near the soil surface due to the deposition of crop residue [67]. However, cover crops have been shown to have relatively little effect on soil P availability, even though cover crops increased crop dry matter accumulation and recycled a large amount of P to the soil surface [80]. Cover crops could improve P uptake of succeeding crops by converting unavailable native P and residual fertilizer P to chemical forms that are more available to succeeding crops, resulting in lower soil P concentrations [67].

In both years, the average amount of available soil P was increased with an increase in the time of chopping of cover crops from 25 DAP to 45 DAP (Table 6). On average, the available soil P after chopping cover crops at 45 DAP (23.9 kg·ha^{-1} in both years) was significantly ($p \leq 0.05$) higher than the soil P observed after chopping cover crops at 25 and 35 DAP. As shown in Table 6, there was no statistically significant interaction between cover crops and their times of chopping among treatments with respect to the available soil P.

3.3.4. Available Soil Potassium

As shown in Table 6, the amount of available soil potassium (K) in the topsoil (0–20 cm), which was determined after the harvest of baby corn in both years, was not significantly ($p \leq 0.05$) different between all cover crops and the control. The average available soil K among all cover crop and control treatments ranged from 158.1 to 163.9 kg·ha^{-1} during 2014–2015. The amount of available soil K following pearl millet treatment was higher than the available soil K observed following sorghum and fodder maize cover crops, and no cover crop (i.e., control), particularly in 2015 *kharif* season when the amount of available soil K following pearl millet cover crop was significantly higher than following sorghum, fodder maize, and control treatments. In both years, the average amount of available soil K was increased with increment in chopping time from 25 DAP to 45 DAP. On average, the available soil K after chopping cover crops at 45 DAP (164.0 kg·ha^{-1} and 165.4 kg·ha^{-1} in 2014 and 2015, respectively) was significantly higher than after chopping cover crops at 25 DAP and 35 DAP. Similar to the available soil N and P content, there was no interaction of cover crops and their times of chopping on the available soil K among treatments in both years. The use of cover crops treatments did not markedly increase topsoil K content at harvest (Table 6) compared to K values observed at the beginning of this study (Table 1). The enhanced soil K content following various cover crops have been observed in several studies (e.g., [81,82]). Studies also suggested that cover crops could accumulate K at the soil surface due to deposition of crop residue and lack of surface-applied fertilizers [81,82]. However, the succeeding crop in its growing season could take up soil K at a much higher rate than the addition of K by cover crops, resulting in lower soil K concentrations [82].

4. Conclusions

Effects of three non-leguminous cover crops (pearl millet, fodder maize, and sorghum) and their times of chopping on the yield and quality of no-till baby corn were evaluated during two *kharif* seasons (during 2014–2015) under subtropical climatic conditions of Punjab, India. During both *kharif* seasons, the yield (cob and green fodder yield) and dry matter accumulation of baby corn following cover crop treatments were significantly higher than the control (no cover crop) and improved with increment in chopping time from 25 DAP to 45 DAP. Among cover crop treatments, the yield (cob and green fodder yield) and dry matter accumulation of baby corn following pearl millet cover crop were significantly higher compared to fodder maize and sorghum cover crop and control treatments. Chopping of cover crops at 45 DAP showed significantly higher yield and dry matter accumulation of baby corn over chopping at 25 DAP and 35 DAP. The effect of cover crops on baby corn quality parameters (i.e., protein, starch, crude fiber, total soluble solids (TSS), total solid, and sugar content) was not significant among treatments during both *kharif* seasons, while increment in chopping time (from 25 DAP to 45 DAP) had a significant effect on the protein and sugar content of baby corn cob. The use of cover crops and increment in chopping time generally helped in enhancing topsoil quality at harvest, especially available soil N. However, the effect of cover crops and their times of chopping on other topsoil quality parameters (i.e., organic carbon content, and available soil P and K) did not differ among treatments. During both *kharif* seasons, there was no significant interaction between cover crops and their times of chopping among treatments with respect to baby corn yield and quality as well as topsoil quality parameters. Based on the results during two *kharif* seasons, it is suggested that non-leguminous cover crops and their times of chopping evaluated in this study could be used for sustainable maize crop production system to improve baby corn growth and yield, baby corn quality, and topsoil quality. However, long-term evaluation of these non-leguminous cover crops and increment in chopping time on *kharif* baby corn yield and quality, as well as soil quality under subtropical climatic conditions, is needed.

Author Contributions: A.S. and J.S.K. conceived and designed the experiments. Funding acquisition and project administration were performed by J.S.K., A.S., and J.S.K. performed the experiments and collected the data. A.S., J.S.K., and S.K.D. analyzed the data. S.S. and P.S. provided technical assistance when the data was analyzed. S.K.D. and A.S. wrote the manuscript. All authors have read and agreed to the published version of the manuscript.

Funding: This research received no external funding.

Acknowledgments: We are grateful to the Department of Agronomy, Punjab Agricultural University (PAU), Ludhiana for providing field and laboratory facilities for the experiment. We gratefully acknowledge the funding support from the "Farm Fresh Foods" (Ladhowal, Punjab) to carry out this research during 2014–2015. We are also thankful to graduate students and technical staff of the Department of Agronomy, PAU, for their assistance during the field experiment. The constructive comments from three anonymous reviewers have improved this manuscript and are greatly appreciated.

Conflicts of Interest: The authors declare no conflict of interest.

References

1. Dowswell, C.R.; Paliwal, R.L.; Cantrell, R.P. *Maize in the Third World*; Routledge (Taylor & Francis): New York, NY, USA, 1996.
2. Verheye, W. Growth, and production of maize: Traditional low-input cultivation. In *Land Use, Land Cover and Soil Sciences*; Verheye, W.H., Ed.; UNESCO-EOLSS Publishers: Oxford, UK, 2010; pp. 1–23.
3. Hoorman, J.J.; Islam, R.; Sundermeier, A.; Reeder, R. Using cover crops to convert to no-till. *Crop. Soils* **2009**, *42*, 9–13.
4. Naik, G. Closing the yield gap: Maize in India. In *Technology, Adaptation, and Exports: How Some Developing Countries Got It Right*; Chandra, V., Ed.; The World Bank: Washington, DC, USA, 2006; pp. 275–300.
5. FAO. FAOSTAT. Available online: http://www.fao.org/faostat/en/#data/QC (accessed on 2 August 2017).
6. Federation of Indian Chambers of Commerce and Industry (FICCI). *Maize Vision 2022: A Knowledge Report*; PricewaterhouseCoopers (PwC) Private Limited: Mumbai, India, 2018. Available online: http://ficci.in/spdocument/22966/India-Maize-Summit.pdf (accessed on 12 February 2020).
7. Zhong, H.; Sticklen, M.B. Genetic engineering of corn: Sustainability of shoot tip meristem in genetic transformation. In *Transgenic Crops I*; Bajaj, Y.P.S., Ed.; Springer: Berlin, Germany, 2000; pp. 37–59.
8. Hallauer, A.R. *Specialty Corns*, 2nd ed.; CRC Press: Boca Raton, FL, USA, 2001.
9. Almeida, I.P.d.C.; Negreiros, M.Z.d.; Barbosa, Z. Baby corn, green ear, and grain yield of corn cultivars. *Hortic. Bras.* **2005**, *23*, 960–964. [CrossRef]

10. Shivay, Y.; Singh, R.; Pal, M.; Pal, M. Productivity and economics of maize as influenced by intercropping with legumes and nitrogen levels. *Ann. Agric. Res.* **2001**, *22*, 576–582.

11. Smith, M.S.; Frye, W.W.; Varco, J.J. Legume winter cover crops. In *Advances in Soil Science*; Stewart, B.A., Ed.; Springer: New York, NY, USA, 1987; pp. 95–139.

12. Fageria, N.K.; Baligar, V.C.; Bailey, B.A. Role of cover crops in improving soil and row crop productivity. *Commun. Soil Sci. Plant Anal.* **2005**, *36*, 2733–2757. [CrossRef]

13. Drinkwater, L.E.; Wagoner, P.; Sarrantonio, M. Legume-based cropping systems have reduced carbon and nitrogen losses. *Nature* **1998**, *396*, 262–265. [CrossRef]

14. Yadvinder, S.; Bijay, S.; Ladha, J.K.; Khind, C.S.; Gupta, R.K.; Meelu, O.P.; Pasuquin, E. Long-term effects of organic inputs on yield and soil fertility in the rice-wheat rotation. *Soil Sci. Soc. Am. J.* **2004**, *68*, 845–853.

15. Clark, A. *Managing Cover Crops Profitably*, 3rd ed.; Handbook Series Book 9; Sustainable Agriculture Research and Education (SARE): College Park, MD, USA, 2007.

16. McCracken, D.V.; Smith, M.S.; Grove, J.H.; Blevins, R.L.; MacKown, C.T. Nitrate leaching as influenced by cover cropping and nitrogen source. *Soil Sci. Soc. Am. J.* **1994**, *58*, 1476–1483. [CrossRef]

17. Meisinger, J.; Hargrove, W.; Mikkelsen, R.; Williams, J.; Benson, V. Effects of cover crops on groundwater quality. In *Cover Crops for Clean Water*; Hargrove, W., Ed.; Soil and Water Conservation Society: Ankeny, IA, USA, 1991; pp. 793–799.

18. Ranells, N.N.; Wagger, M.G. Nitrogen release from grass and legume cover crop monocultures and bicultures. *Agron. J.* **1996**, *88*, 777–882. [CrossRef]

19. Dabney, S.M.; Delgado, J.A.; Meisinger, J.J.; Schomberg, H.H.; Liebig, M.A.; Kaspar, T.; Mitchell, J.; Reeves, W. Using cover crops and cropping systems for nitrogen management. In *Advances in Nitrogen Management for Water Quality*; Delgado, J.A., Follett, R.F., Eds.; Soil and Water Conservation Society: Ankeny, IA, USA, 2010; pp. 230–281.

20. Kuo, S.; Sainju, U.; Jellum, E. Winter cover cropping influence on nitrogen in the soil. *Soil Sci. Soc. Am. J.* **1997**, *61*, 1392–1399. [CrossRef]

21. Sharma, P.; Singh, A.; Kahlon, C.S.; Brar, A.S.; Grover, K.K.; Dia, M.; Steiner, R.L. The role of cover crops towards sustainable soil health and agriculture—A review paper. *Am. J. Plant Sci.* **2018**, *9*, 1935–1951. [CrossRef]

22. Dabney, S.; Delgado, J.; Reeves, D. Using winter cover crops to improve soil and water quality. *Commun. Soil Sci. Plant Anal.* **2001**, *32*, 1221–1250. [CrossRef]

23. Florentín, M.A.; Peñalva, M.; Calegari, A.; Derpsch, R.; McDonald, M. *Green Manure/Cover Crops and Crop Rotation in Conservation Agriculture on Small Farms*; Integrated Crop Management Vol.12-2010; Food and Agriculture Organization (FAO) of the United Nations: Rome, Italy, 2011.

24. Mitchell, W.; Tell, M. Winter-annual cover crops for no-tillage corn production. *Agron. J.* **1977**, *69*, 569–573. [CrossRef]

25. Ebelhar, S.; Frye, W.; Blevins, R. Nitrogen from legume cover crops for no-tillage corn. *Agron. J.* **1984**, *76*, 51–55. [CrossRef]

26. Decker, A.M.; Clark, A.J.; Meisinger, J.J.; Mulford, F.R.; McIntosh, M.S. Legume cover crop contributions to no-tillage corn production. *Agron. J.* **1994**, *86*, 126–135. [CrossRef]

27. Dou, Z.; Fox, R.; Toth, J. Tillage effect on seasonal nitrogen availability in corn supplied with legume green manures. *Plant Soil* **1994**, *162*, 203–210. [CrossRef]

28. Cline, G.R.; Silvernail, A.F. Effects of cover crops, nitrogen, and tillage on sweet corn. *HortTechnology* **2002**, *12*, 118–125. [CrossRef]

29. Muthukumar, V.; Velayudham, K.; Thavaprakaash, N. Growth and yield of baby corn (Zea mays l.) as influenced by plant growth regulators and different time of nitrogen application. *Res. J. Agric. Biol. Sci.* **2005**, *1*, 303–307.

30. Dhaliwal, H.S.; Kular, J.S. *Package of Practices for the Crops of Punjab: Kharif 2014*; Punjab Agricultural University: Ludhiana, India, 2014; Volume 31.

31. Salaria, A. Productivity of No-till Maize (*Zea mays* L.) as Influenced by Leguminous Cover Crops and Nitrogen Levels. Ph.D. Thesis, Punjab Agricultural University, Ludhiana, India, 2016.

32. Kukal, S.; Sidhu, A. Percolation losses of water in relation to pre-puddling tillage and puddling intensity in a puddled sandy loam rice (*Oryza sativa* L.) field. *Soil Till. Res.* **2004**, *78*, 1–8. [CrossRef]

33. Central Institute of Agricultural Engineering (CIAE). *Tractor Mounted Fodder Harvester: A Suceess Story*; All India Coorninated Research Project on Farm Implements and Machinery; Extension Bulletin No. CIAE/FIM/2004/45; Central Institute of Agricultural Engineering: Bhopal, India, 2004.

34. Singh, M.; Singh, R.; Singh, S.; Yadav, M.; Singh, V. Integrated nutrient management for higher yield, quality, and profitability of baby corn (Zea mays). *Indian J. Agron.* **2010**, *55*, 100.

35. Mahajan, G.; Sharda, R.; Kumar, A.; Singh, K. Effect of plastic mulch on economizing irrigation water and weed control in baby corn sown by different methods. *Afr. J. Agric. Res.* **2007**, *2*, 19–26.

36. Jones, D.B. *Factors for Converting Percentages of Nitrogen in Foods and Feeds into Percentages of Proteins*; Circular No. 183; United States Department Of Agriculture: Washington, DC, USA, 1941.

37. Howitz, E. *Official Methods of Analysis of AOAC International*; AOAC (Association of Official Agricultural Chemists) International: Rockville, MD, USA, 2000.

38. Nelson, D.W.; Sommers, L.E. Total carbon, organic carbon, and organic matter. In *Methods of Soil Analysis, Part 3 Chemical Methods*; Sparks, D.L., Page, A.L., Helmke, P.A., Loeppert, R.H., Soltanpour, P.N., Tabatabai, M.A., Johnston, C.T., Sumner, M.E., Eds.; Soil Science Society of America and American Society of Agronomy: Madison, WI, USA, 1996; pp. 961–1010.

39. Asana, R.; Saini, G. Rapid method for determination of reducing and total sugar. *J. Biochem.* **1962**, *22*, 23–25.

40. Somogyi, M. A new reagent for the determination of sugars. *J. Biolog. Chem.* **1945**, *160*, 61–68.

41. Ranganna, S. *Handbook of Analysis and Quality Control for Fruit and Vegetable Products*, 2nd ed.; Tata McGraw-Hill: New Delhi, India, 1986.

42. Horwitz, W.; Chichilo, P.; Reynolds, H. *Official Methods of Analysis of the AOAC*; AOAC (Association of Official Agricultural Chemits) International: Rockville, MD, USA, 1970.

43. Subbaiah, B.V.; Asija, G.L. A rapid procedure for estimation of available nitrogen in soil. *Curr. Sci.* **1956**, *25*, 259–260.

44. Olsen, S.R. *Estimation of Available Phosphorus in Soils by Extraction with Sodium Bicarbonate*; United States Department of Agriculture: Washington, DC, USA, 1954.

45. Merwin, H.D.; Peech, M. Exchangeability of soil potassium in the sand, silt, and clay fractions as influenced by the nature of the complementary exchangeable cation. *Soil Sci. Soc. Am. J.* **1951**, *15*, 125–128. [CrossRef]

46. Walkley, A.; Black, I.A. An examination of the degtjareff method for determining soil organic matter, and a proposed modification of the chromic acid titration method. *Soil Sci.* **1934**, *37*, 29–38. [CrossRef]

47. Dar, E.A.; Yousuf, A.; Bhat, M.A.; Poonia, T. Growth, yield and quality of baby corn (*Zea mays* L.) and its fodder as influenced by crop geometry and nitrogen application—A review. *Bioscan. Int. Quart. J. Life Sci.* **2017**, *12*, 463–469.

48. Eltelib, H.A.; Hamad, M.A.; Ali, E.E. The effect of nitrogen and phosphorus fertilization on growth, yield, and quality of forage maize (*Zea mays* L.). *J. Agron.* **2006**, *5*, 515–518.

49. Kumar, T.K.; Venkateswarlu, B. Baby corn (*Zea mays* L.) performance as vegetable-cum-fodder in intercropping with legume fodders under different planting patterns. *Range Manag. Agroforest.* **2013**, *34*, 137–141.

50. Reddy, P.P. Cover/green manure crops. In *Sustainable Intensification of Crop Production*; Springer Nature: Singapore, 2016; pp. 55–67.

51. Hannaway, D.B.; Larson, C. *Forage Fact Sheet: Pearl Millet (Pennisetum americanum)*; Oregon State University: Corvallis, OR, USA, 2004.

52. Payne, W.A. Optimizing crop use in sparse stands of pearl millet. *Agron. J.* **2000**, *92*, 808–814. [CrossRef]

53. Schonbeck, M.; Morse, R. Cover Crops for All Seasons: Expanding the Cover Crop Toolbox for Organic Vegetable Producers. Virginia Association for Biological Farming Information Sheet Number 3-06. May 2006. Available online: https://www.sare.org/content/download/69536/985534/Cover_crops_for_all_seasons.pdf?inlinedownload=1 (accessed on 10 January 2020).

54. Rosolem, C.A.; Pace, L.; Crusciol, C.A. Nitrogen management in maize cover crop rotations. *Plant Soil* **2004**, *264*, 261–271. [CrossRef]

55. Rosolem, C.A.; Calonego, J.C.; Foloni, J.S. Potassium leaching from millet straw as affected by rainfall and potassium rates. *Commun. Soil Sci. Plant Anal.* **2005**, *36*, 1063–1074. [CrossRef]

56. Wang, G.S.; Noite, K. Summer Cover Crop Use in Arizona Vegetable Production Systems. Arizona Cooperative Extension AZ1519. July 2010. Available online: https://extension.arizona.edu/sites/extension.arizona.edu/files/pubs/az1519.pdf (accessed on 10 January 2020).

57. Ramachandrappa, B.K.; Nanjappa, H.V.; Shivakumar, H.K. Yield and quality of baby corn (*Zea mays* L.) as influenced by spacing and fertilization levels. *Acta Agron. Hung.* **2004**, *52*, 237–243. [CrossRef]

58. Joshi, G.; Chilwal, A. Determination of quality of baby corn (*Zea mays* L.) under the effect of integrated nutrient management. *Int. J. Chem. Stud.* **2018**, *6*, 3244–3247.

59. Hooda, S.; Kawatra, A. Nutritional evaluation of baby corn (Zea mays). *Nutr. Food Sci.* **2013**, *43*, 68–73. [CrossRef]

60. Rahman, N.A.; Rosli, W.I.W. Nutritional compositions and antioxidative capacity of the silk obtained from immature and mature corn. *J. King Saud Univ. Sci.* **2014**, *26*, 119–127. [CrossRef]

61. Budak, F.; Aydemir, S.K. Grain yield and nutritional values of sweet corn (*Zea mays var. saccharata*) in produced with good agricultural implementation. *Nutr. Food Sci. Int. J.* **2018**, *7*, 555710.

62. Ugur, A.; Maden, H.A. Sowing and planting period on yield and ear quality of sweet corn. *Cienc. Agrotechnol.* **2015**, *39*, 112–116.

63. Prajwal Kumar, G.K.; Lalitha, B.S.; Shamshad Begum, S.; Somashekar, K.S. Effect of food nutrition on nutritional quality of baby corn (*Zea mays* L.) and dehydrated baby corn enriched products. *Int. J. Pure App. Biosci.* **2018**, *6*, 107–113. [CrossRef]

64. Shobha, D.; Sreeramasetty, T.A.; Puttarama, N.; Gowda, K.P. Evaluation of maize genotypes for physical and chemical composition at silky and hard stage. *Karnataka J. Agric. Sci.* **2010**, *23*, 311–314.

65. Khan, A.A.; Hussain, A.; Ganai, M.A.; Sofi, N.R.; Hussain, S.T. Yield, nutrient uptake and quality of sweet corn as influenced by transplanting dates and nitrogen levels. *J. Pharmacog. Phytochem.* **2018**, *7*, 3567–3571.

66. Rosli, W.I.W.; Anis, J.C. The potential of *Zea mays* ears and it extracts as an alternative food nutritive ingredients. *APCBEE Procedia* **2012**, *2*, 141–147. [CrossRef]

67. Sharma, V.; Irmak, S.; Padhi, J. Effects of cover crops on soil quality: Part I. soil chemical properties—Organic carbon, total nitrogen, pH, electrical conductivity, organic matter content, nitrate-nitrogen, and phosphorous. *J. Soil Water Conserv. Soc.* **2018**, *73*, 637–651. [CrossRef]

68. Sainju, U.M.; Singh, B.P.; Whitehead, W.F. Long-term effects of tillage, cover crops, and nitrogen fertilization on organic carbon and nitrogen concentrations in sandy loam soils in Georgia, USA. *Soil Till. Res.* **2002**, *63*, 167–179. [CrossRef]

69. Mazzoncini, M.; Sapkota, T.B.; Bàrberi, P.; Antichi, D.; Risaliti, R. Long-term effect of tillage, nitrogen fertilization and cover crops on soil organic carbon and total nitrogen content. *Soil Till. Res.* **2011**, *114*, 165–174. [CrossRef]

70. Bauer, A.; Black, A. Quantification of the effect of soil organic matter content on soil productivity. *Soil Sci. Soc. Am. J.* **1994**, *58*, 185–193. [CrossRef]

71. Wyland, L.J.; Jackson, L.E.; Schulbach, K.F. Soil plant nitrogen dynamics following incorporation of a mature rye cover crop in a lettuce production system. *J. Agric. Sci.* **1995**, *124*, 17–25. [CrossRef]

72. Kuo, S.; Sainju, U.; Jellum, E. Winter cover crop effects on soil organic carbon and carbohydrate in soil. *Soil Sci. Soc. Am. J.* **1997**, *61*, 145–152. [CrossRef]

73. Olson, K.R.; Ebelhar, S.A.; Lang, J.M. Cover crop effects on crop yields and soil organic carbon content. *Soil Sci.* **2010**, *175*, 89–98. [CrossRef]

74. Kaspar, T.C.; Singer, J.W. The use of cover crops to manage soil. In *Soil Management: Building a Stable Base for Agriculture*; Hartfield, J.L., Sauer, T.J., Eds.; American Society of Agronomy and Soil Science Society of America: Madison, WI, USA, 2011; pp. 321–337.

75. Kaspar, T.C.; Jaynes, D.B.; Parkin, T.B.; Moorman, T.B. Rye cover crop and gamagrass strip effects on NO_3 concentration and load in tile drainage. *J. Environ. Qual.* **2007**, *36*, 1503–1511. [CrossRef]

76. Tonitto, C.; David, M.; Drinkwater, L. Replacing bare fallows with cover crops in fertilizer-intensive cropping systems: A meta-analysis of crop yield and n dynamics. *Agric. Ecosyst. Environ.* **2006**, *112*, 58–72. [CrossRef]

77. Strock, J.S.; Porter, P.M.; Russelle, M.P. Cover cropping to reduce nitrate loss through subsurface drainage in the Northern U.S. corn belt. *J. Environ. Qual.* **2004**, *33*, 1010–1016. [CrossRef] [PubMed]

78. Wagger, M.G.; Mengel, D.B. The role of nonleguminous cover crops in the efficient use of water nitrogen. In *Cropping Strategies for Efficient Use of Water and Nitrogen*; Hargrove, W.L., Ed.; ASA-CSSA-SSSA (American Society of Agronomy-Crop Science Society of America-Soil Science Society of America): Madison, WI, USA, 1988; pp. 115–127.

79. Janegitz, M.C.; Martins, A.R.H.; Rosolem, C.A. Cover crops and soil phosphorus availability. *Commun. Soil Sci. Plant Anal.* **2017**, *48*, 1240–1246. [CrossRef]

80. Teles, A.P.B.; Rodrigues, M.; Herrera, W.F.B.; Soltangheisi, A.; Sartor, L.R.; Withers, P.J.A.; Pavinato, P.S. Do cover crops change the lability of phosphorus in a clayey subtropical soil under phosphate fertilizers? *Soil Use Manag.* **2017**, *33*, 34–44. [CrossRef]

81. Eckert, D.J. Chemical attributes of soils subjected to no-till cropping with rye cover crops. *Soil Sci. Soc. Am. J.* **1991**, *55*, 405–409. [CrossRef]

82. Sharma, V.; Irmak, S.; Padhi, J. Effects of cover crops on soil quality: Part II. soil exchangeable bases (potassium, magnesium, sodium, and calcium), cation exchange capacity, and soil micronutrients (zinc, manganese, iron, copper, and boron). *J. Soil Water Conserv. Soc.* **2018**, *73*, 652–668. [CrossRef]

© 2020 by the authors. Licensee MDPI, Basel, Switzerland. This article is an open access article distributed under the terms and conditions of the Creative Commons Attribution (CC BY) license (http://creativecommons.org/licenses/by/4.0/).

 *horticulturae*

Article

Effect of Stand Reduction at Different Growth Stages on Yield of Paprika-Type Chile Pepper

Israel Joukhadar * and Stephanie Walker

Extension Plant Sciences, New Mexico State University, P.O. Box 30003, MSC 3AE, Las Cruces, NM 88003, USA;
swalker@nmsu.edu
* Correspondence: icalsoya@nmsu.edu

Received: 16 December 2019; Accepted: 24 February 2020; Published: 5 March 2020

Abstract: Paprika-type chile (*Capsicum annuum* L.) crops are susceptible to plant population losses through pest activity, disease, and extreme weather events such as hail storms. This study was conducted to determine the influence of intensity and timing of plant population reductions on the final harvested yield of paprika-type chile so that informed decisions can be made regarding continuing or ending a damaged field. Two trials, one per year, were conducted in southern New Mexico. 'LB-25', a standard commercial cultivar, was direct seeded on 29 March 2016 and 4 April 2017. Plants were thinned at three different growth stages; early seedling, first bloom, and peak bloom. Plants were thinned to four levels at each phenological stage; 0% stand reduction (control; ~200,000 plants/ha), 60% stand reduction (~82,000 plants/ha), 70% stand reduction (~60,000 plants/ha), and 80% stand reduction (~41,000 plant/ha). In both years, the main effects of stand reduction had a significant impact on harvested yield, emphasizing the percentage of stand reduction has more of an impact on yield than timing in paprika-type red chile. Consistently, an 80% stand reduction in paprika-type chile significantly reduced fresh red chile yield by 26% to 38%.

Keywords: *Capsicum annuum*; heat units; plant population density; hail damage

---

1. Introduction

Crop hail damage can cause considerable economic loss during the growing season [1]. Physical crop injury can be divided into two main categories; defoliation and stand reduction [2]. Many researchers have simulated stand reduction in crops such as cotton (*Gossypium hirsutum* L.), corn (*Zea mays* L.), soybeans (*Glycine max* L.), and wheat (*Triticum aestivum* L.) by cutting and removing a specific number of plants from the field [2–4]. Conversely, for many vegetable crops including paprika-type chile, little to no research of a similar type has been conducted [1].

Paprika-type red chile (*Capsicum annuum* L.) is a specialty crop important in the southwest region of the United States with a total of 5382 ha of red chile harvested in New Mexico, Arizona, and Texas in 2016 [5,6]. Stand reductions due to pests and extreme weather events have been identified as threats to chile farmers in both Arizona and New Mexico. In response to these threats, the United States Department of Agriculture Risk Management Agency has started pilot programs to insure chile crops [7]. For example, in 2016, there were 136 hail events in New Mexico, 29 hail events in Arizona, and over 500 hail events in Texas [8], causing both defoliation and stand reduction damage to crops. To adequately insure and provide coverage for losses in chile, both farmers and insurance companies must have information on how chile yield changes due to stand reduction caused by pests or extreme weather events at different growth stages.

In the southwest US, New Mexico-type green and red chile are the two most prominent chile products. New Mexico-type red chile is harvested when fruit are at a mature red stage and are partially dried on the plant [9]. Paprika-type chile is a subset of red chile distinctive for fruit exhibiting very low heat level and high carotenoid content [10]. Carotenoids are extracted and used as a natural dye in a variety of food and cosmetic products [11]. Paprika-type chile is also ground into powder and used as a spice [12]. New Mexico is the only state in the southwest to categorize chile production, and in 2016, the highest harvested category of chile was paprika-type chile at 1416 ha [13]. Paprika-type chile was selected for this study due to its importance, not only in the southwest, but to the food industry all over the world. Throughout the world, red paprika-type chile is used as a culinary spice and the extracted pigments of paprika-type red chile are used a natural food colorant in many food products.

All of the previously reported research on the effect of plant population losses in paprika-type chile has been done at one growth stage, leaving a gap in the knowledge about responses during different growth stages. As Cavero et al. [14] found, paprika yield increased as plant density increased from 13,333 to 200,000 plants per hectare when the plants were thinned during the ten to twelve leaves growth stage. Although this illustrated that there is an impact on paprika-type red chile when plant populations change, how they respond to such changes over the season has not been explored. On the other hand, a paprika-type chile field with a high plant population density of 322,335 plants per hectare experienced a 60% yield reduction [10]. It has been reported that removing plants from fields with high plant populations at specific growth stages can be beneficial due to reduction in competition for light [15]. Pariossien and Flynn [10] reported the best planting density for paprika-type red chile to be 98,800 plants per hectare.

In other crops such as soybeans, when plant populations were reduced at the early growth stage, there were no significant changes in seed yield, but when stand reduction occurred in the later growth stage seed yield was decreased [2]. Similar results were found when sunflowers (*Helianthus annuus* L.) underwent stand reduction at early and late growth stages. Sunflower stand losses of 25% during later growth stages significantly reduced yield, while no reductions in yield occurred when stand losses of 25% occurred at an early growth stage [16]. Many crops can compensate for stand reduction losses early in the season.

The goal of this study was to understand how a simulation of population losses by stand reduction at different growth stages affected the yield of paprika-type red chile. Obtaining this knowledge will give farmers more insight into their yield expectations after a stand reduction event caused by pests and/or extreme weather events at any growth stage. The specific objectives were to determine how four levels of stand reduction simulating hail damage at three growth stages affect the yield components. Our hypothesis was that paprika-type red chile, an indeterminate crop, would recover from stand reduction early in the growing season.

2. Materials and Methods

Field experiments were conducted during 2016 and 2017 at the New Mexico State University Leyendecker Plant Science Research Center in La Mesa, NM, USA [LPSRC (lat. 32.16° N; long. 106.46° W; elevation 1186 m)]. The soil at LPSRC was a Glendale clay loam [17]. Fertilization during both years consisted of total nitrogen (Helena Chemicals, Collierville, TN, USA) at 168.1 kg·ha^{-1} and total phosphorus at 112.1 kg·ha^{-1}. All phosphorus and a quarter of the nitrogen were broadcast preplant as ammonium phosphate and the remaining nitrogen was delivered throughout the season in the irrigation water as urea and ammonium nitrate.

2.1. Field Cultivation

The field was plowed, disced, laser-leveled, and listed before planting. 'LB-25' (Biad Chili Co., Leasburg, NM, USA), a common commercial paprika-type red chile cultivar, was planted at a rate of 5.6 kg·ha^{-1} on 29 March 2016 and 4 April 2017 with metalaxyl fungicide (Ridomil Gold; Syngenta, Greensboro, NC, USA) at 146 mL·ha^{-1} banded into the planting bed during the direct seeding of the 'LB-25'. A two-way factorial treatment structure in a randomized complete block design with four replications for a total of 48 plots was used. The first factor, stand reduction, had four levels, and the second factor, growth stage, had three levels, and each were combined and randomized in the field plot. Each plot consisted of three rows, with a total area of 13.8 m^2 (3.0 m between row spacing × 4.6 m length). The field was 662.24 m^2 (13.8 m^2 × 48 plots) surrounded by one row (north and south) or plot (east and west) borders of paprika-type red chile plants. All plots were hand-weeded weekly each season. The field was furrow irrigated once every 10–14 days and irrigation ended on 16 September 2016 and 1 September 2017 when the crop was at a mature red growth stage.

2.2. Stand Reduction

At three different growth stages, plants were thinned to four levels of stand reduction. When plants were thinned, two plants were left in a clump [18] at different spacing intervals to achieve desired plant counts per plot. When describing stand reduction treatments, a row is one of the three rows within a plot with an area of 4.6 m^2 (1.0 m × 4.6 m). Each of the three rows in a plot were thinned to one of the specified treatments. The four stand reductions treatments were: control with no thinning and ~64 plants per row, 60% stand reduction with 35.7-cm spacing and ~25 plants per row, 70% stand reduction with 45.7-cm spacing and ~19 plants per row, and 80% stand reduction with 66.0-cm spacing and ~13 plants per row. The densities achieved in 2016 for each stand reduction level were 209,974 plants·ha^{-1} (control, no thinning), 82,021 plants·ha^{-1} (60% stand reduction), 62,336 plants·ha^{-1} (70% stand reduction), and 42,651 plants·ha^{-1} (80% stand reduction). The densities achieved in 2017 for each stand reduction level were 200,131 plants·ha^{-1} (control, no thinning), 82,021 plants·ha^{-1} (60% stand reduction), 59,055 plants·ha^{-1} (70% stand reduction), and 39,370 plants·ha^{-1} (80% stand reduction).

2.3. Growth Stages

Stand reduction treatments occurred at pre-determined growth stages based on heat units accumulated after planting (HUAP). HUAP values were calculated using the method described by Brown [19] and Silvertooth et al. [20] (Tables 1 and 2) using 12 °C as the base temperature. Using heat unit systems in a phenology model for crops relates plant growth to local weather and climate conditions [19] and take into account day to day changes in temperature [20]. Daily weather data such as maximum temperatures, minimum temperatures, mean temperatures, and precipitation were collected from the LPSRC weather station, La Mesa, NM, USA [21,22] (Tables 1 and 2).

Table 1. Growing season (29 March–25 October 2016) weather data ^{z}: weekly total precipitation, daily maximum, minimum, and mean temperatures, and calculated heat units accumulated after planting.

Week	Maximum Temperature (°C)	Minimum Temperature (°C)	Mean Daily Temperature (°C)	Total Weekly Precipitation (cm)	Heat Units Accumulated after Planting ^{y} (HUAP)
1	22.4	3.8	13.0	0.0	16.8
2	25.9	7.2	17.0	0.3	69.8
3	23.7	5.4	15.3	0.6	102.3
4	29.1	9.1	19.8	0.0	190.6
5	23.6	5.4	15.7	0.0	227.7
6	27.8	9.6	19.1	0.0	309.7
7	31.2	10.8	21.8	0.0	423.8
8	29.3	8.9	19.7	0.0	510.9
9	30.3	8.2	20.1	0.3	603.2
10	32.0	16.2	23.9	0.3	743.7
11	35.8	18.9	26.8	0.0	920.3
12	36.4	14.2	25.9	0.0	1085.5
13	35.9	19.8	27.3	0.5	1267.6
14	34.8	18.3	26.3	0.1	1437.6
15	34.7	22.2	27.5	0.0	1622.1
16	37.9	19.1	28.5	0.1	1820.4
17	38.3	20.1	29.2	0.1	2026.2
18	36.1	20.7	27.6	0.3	2213.5
19	35.7	19.4	27.1	0.3	2394.1
20	33.2	18.7	25.2	0.6	2550.5
21	32.8	16.4	24.0	0.4	2692.0

Table 1. *Cont.*

Week	Maximum Temperature (°C)	Minimum Temperature (°C)	Mean Daily Temperature (°C)	Total Weekly Precipitation (cm)	Heat Units Accumulated after Planting [y] (HUAP)
22	32.1	16.2	23.5	4.1	2826.6
23	31.1	18.9	24.2	1.6	2970.9
24	29.3	17.4	22.4	0.7	3092.2
25	33.2	11.8	21.5	0.0	3201.7
26	29.8	14.7	21.6	0.8	3313.2
27	28.2	13.1	19.8	2.3	3401.3
28	26.5	11.2	18.3	0.0	3471.3
29	31.1	8.3	18.3	0.0	3540.7
30	29.0	7.2	16.7	0.0	3597.8
Season [w]	31.2	13.7	22.2	13.1	3597.8

[z] Precipitation and temperature collected from LPSRC Weather Station, La Mesa, NM (2016). [y] Total weekly calculated heat units accumulated after planting; GDD (Growing Degree Days based on Fahrenheit scale) = mean daily temperature °F-32 °F; for paprika-type red chile with a base temperature of 55 °F, if maximum temperature exceeds 86 °F then maximum temperature is set at 86 °F in DDF equation; if minimum temperature is below 55 °F then minimum temperature set at 55 °F in DDF equation; HUAP = cumulative DDF. 32 °F = 0 °C. [w] Mean temperatures, total precipitation during growing season, total heat units accumulated during growing season.

Table 2. Growing season (4 April–17 October 2017) weather data [z]: weekly total precipitation, daily maximum, minimum, mean temperatures, and calculated heat units accumulated after planting.

Week	Maximum Temperature (°C)	Minimum Temperature (°C)	Mean Daily Temperature (°C)	Total Weekly Precipitation (cm)	Heat units Accumulated after Planting [y] (HUAP)
1	26.2	4.9	16.4	0.0	46.1
2	30.3	8.8	19.8	0.0	134.8
3	29.8	7.5	19.9	0.0	224.7
4	24.0	6.5	16.3	0.2	274.1
5	30.8	9.3	21.2	0.0	380.4
6	27.9	10.3	19.6	0.0	466.9
7	27.9	9.3	18.9	0.0	543.8
8	32.8	12.6	23.7	0.0	681.6
9	30.9	14.5	22.4	0.3	802.8
10	35.5	17.3	26.7	0.1	977.8
11	38.3	12.6	25.9	0.0	1143.1
12	37.8	20.4	29.1	0.1	1342.6
13	36.6	18.3	27.4	0.0	1526.3
14	36.8	19.8	28.4	0.0	1722.9
15	33.4	19.3	25.9	2.3	1894.2
16	32.3	19.0	24.1	8.4	2037.3
17	34.2	19.5	25.7	0.3	2214.2
18	33.3	18.9	25.7	0.0	2377.4
19	34.4	20.4	26.6	1.1	2551.3
20	32.6	17.1	24.1	2.6	2694.0
21	32.4	17.8	24.5	1.3	2842.1
22	32.6	14.6	23.3	0.0	2974.6
23	32.4	15.7	24.0	0.1	3116.1
24	34.2	14.0	23.5	0.0	3251.2
25	31.9	13.4	21.8	0.1	3364.9
26	27.4	14.6	20.5	1.0	3462.3
27	30.1	12.3	20.6	0.4	3560.9
28	27.3	9.0	17.5	0.1	3629.1
Season [w]	31.9	14.2	23.0	18.3	3629.1

[z] Precipitation and temperature collected from LPSRC Weather Station, La Mesa, NM (2017). [y] Total weekly calculated heat units accumulated after planting; GDD (Growing Degree Days based on Fahrenheit scale) = mean daily temperature °F-32 °F; for paprika-type red chile with a base temperature of 55 °F, if maximum temperature exceeds 86 °F then maximum temperature is set at 86 °F in DDF equation; if minimum temperature is below 55 °F then minimum temperature set at 55 °F in DDF equation; HUAP = cumulative DDF. 32 °F = 0 °C. [w] Mean temperatures, total precipitation during growing season, total heat units accumulated during growing season.

The targeted growth stages for the stand reduction treatments were early seedling stage at 700 HUAP, first bloom at 1400 HUAP, and peak bloom at 2000 HUAP [20]. Although HUAP values were used to determine phenological growth stages, we observed that early seedling stage was characterized by the plants having about 30 true leaves, 60–70 days after planting. First bloom was when anthesis began on each plant and peak bloom when more than 60% anthesis was observed. Due to inclement weather and scheduling constraints, stand reduction events did not occur at the exact targeted number of HUAPs for each growth stage. The actual HUAPs and dates for each stand reduction event in 2016 were: early seedling stage on 1 June 2016 at 623 HUAP, first bloom on 27 June 2016 at 1268 HUAP, and peak bloom on 19 July 2016 at 1849 HUAP. The actual HUAPs and dates for each stand reduction event in 2017 were: early seedling stage on 31 May 2017 at 717 HUAP, first bloom on 26 June 2017 at 1398 HUAP, and peak bloom on 17 July 2017 at 1894 HUAP.

2.4. Harvest

The plots were harvested on 17 October 2016 at 3598 HUAP and on 25 October 2017 at 3629 HUAP. The harvested sample area was 3.1 m^2 (3.04 m × 1.01 m) taken from the middle section of the middle row of each plot. In 2017, due to labor constraints, the sample size was reduced to 1.5 m^2 (1.52 m × 1.01 m). All fruit within a sample area was hand-harvested into plastic bags and then removed from the field for sorting.

2.5. Yield Data Collection

Harvested material was sorted into the following categories: (1) fresh red yield, (2) fresh green yield, (3) unmarketable yield, (4) immature yield. Fruit classified as red were fruits with more than 50% red color. Fruit classified as green were fruits with more than 50% green color. Fruit classified as unmarketable yield were fruits with blemishes and/or discoloration from disease covering over 40% of the fruit. Immature fruit were fruits under 7.6-cm and had a malleable pericarp. Immature yield was nominal, so data were not included in this report. All of the sorted material was weighed (SVI-100E; Sartorius Stedim North America, Bohemia, NY, USA). Fresh red yield was put in a drier at 54.4 °C until fruit were completely dehydrated and then weighed for a dry red yield. In 2016, red yield subsamples in the drier were overcome with mold and had to be discarded.

2.6. Data Analysis

Additionally, this study was designed to measure and compare the interaction of stand reduction and growth stage on various yield components. Analysis was conducted on each year separately due to environmental variation between the years. Response variables analyzed in 2016 and 2017 were: fresh red yield, green yield, unmarketable yield, and plant counts. Additionally, dry red yield was analyzed in 2017, but not in 2016 due to the mold growth noted above. Response variable data were analyzed by analysis of variance (ANOVA) using SAS (version 9.4; SAS Institute, Cary, NC, USA). Tukey's significant difference test ($p \leq 0.05$) was used to separate means when interactions between stand reduction level and growth stage were significant. When interactions were not significant, ANOVA was conducted on the main effects of stand reduction levels. If statistically significant differences were detected in the main effects, then Tukey's significant difference test ($p \leq 0.05$) was used to separate means.

3. Results

3.1. Weather Differences

There were two major differences in weather patterns between the 2016 and 2017 growing seasons. First, 2017 had an overall higher average minimum temperature for the entire season. In 2016, the season average minimum temperature was 13.7 °C, 0.5 °C cooler than in 2017. The higher minimum temperatures 2017 increased the growth rate of the plants, so they matured at a faster rate. Due to this, the 2017 season was 28 weeks long and the 2016 season was 30 weeks long. Second, 2017 had 5.2 cm more total precipitation during the growing season. Much of the precipitation recorded in 2017 occurred in the month of July 2017; it fell at a fast rate, leaving the field with standing water for over a week from 17 July through 24 July 2017.

3.2. Yield Components

Growth stage by stand reduction interactions were not statistically significant for all of the yield components measured in 2016 and 2017 (Tables 3 and 4). So significant stand reduction main effects were evaluated. In 2016, stand reduction had a significant impact on fresh red fruit yield and plant counts (Table 3). The 0%, 60%, and 70% stand reduction plots had on average 36% more fresh red fruit yield than the 80% stand reduction plots (Figure 1A). As expected, the 0% stand reduction plots had over two and a half times more plants than the 80% and 70% stand reduction plots (Figure 1B). In 2017, stand reduction had an effect on fresh red fruit yield, dry red fruit yield, and plant counts (Table 4). The 60% stand reduction plots in 2017 had 45% more fresh red yield than the 0%, 70%, and 80% stand reduction plots (Figure 2A). The 60% stand reduction plots also had 83% more dry red yield than the 80% stand reduction plots (Figure 2B). When evaluating the plant counts, the 0% stand reduction plots had over four times the number of plants as the 80% stand reduction plots (Figure 2C).

Table 3. Yield and plant counts of paprika-type red chile with four stand reduction levels at three growth stages harvested on 25 October 2016.

Stand Reduction Level [z]	Growth Stage at Reduction [y]	Fresh Red Fruit [x]	Green Fruit [w]	Unmarketable Fruit [v]	Plant Count [u]
0%	Early seedling	21.5	5.6	1.8	60.0
0%	First bloom	22.4	7.8	2.6	68.0
0%	Peak bloom	20.7	5.9	1.5	64.3
60%	Early seedling	21.5	8.1	1.8	26.3
60%	First bloom	20.8	6.5	2.8	24.0
60%	Peak bloom	19.6	5.6	1.2	30.5
70%	Early seedling	18.9	7.5	1.0	18.0
70%	First bloom	21.9	9.3	2.0	22.3
70%	Peak bloom	20.1	6.2	1.3	22.8
80%	Early seedling	15.0	8.4	2.7	17.8
80%	First bloom	15.1	8.5	2.5	15.8
80%	Peak bloom	15.8	7.6	0.7	15.8
Significance					
Growth Stage (GS)		NS [t]	NS	NS	NS
Stand Reduction Level (SL)		***	NS	NS	***
GS X SL		NS	NS	NS	NS

[z] Percent of plant population reduced from standard population of ~200,000 plants per hectare. [y] Growth stage characterized by heat units accumulated after planting (HUAP) during stand reduction events for 2016; early seedling = 623 HUAP, first bloom = 1268 HUAP, peak bloom = 1849 HUAP. [x] All chile yields were harvested in kg per 3.1 m^2; reported in tons per hectare. Fresh red yield were fruits at the mature red stage; means of n = 4. [w] Green yield were fruits with more than 50% green color; means of n = 4. [v] Unmarketable yield were fruits with more than 40% disease caused discoloration and/or blemishes; means of n = 4. [u] Number of counted plants in each row per plot; means off n = 4. [t] NS, *** Nonsignificant or significant at $p \leq 0.001$, respectively.

Table 4. Yield and plant counts of paprika-type red chile with four stand reduction levels three growth stages harvested on 17 October 2017.

Stand Reduction Level [z]	Growth Stage [y]	Fresh Red Fruit [x]	Dry Red Fruit [w]	Green Fruit [v]	Unmarketable Fruit [u]	Plant Count [t]
0%	Early seedling	10.5	2.3	1.6	1.8	56.8
0%	First bloom	8.1	2.4	1.5	1.2	64.8
0%	Peak bloom	11.5	3.1	3.4	1.4	62.8
60%	Early seedling	18.1	4.0	2.6	2.1	24.5
60%	First bloom	12.3	2.7	2.0	1.6	23.5
60%	Peak bloom	14.2	3.6	1.5	1.9	25.5
70%	Early seedling	14.1	3.1	2.8	1.5	19.0
70%	First bloom	12.1	2.5	2.5	1.3	14.0
70%	Peak bloom	8.8	1.7	1.4	1.3	14.0
80%	Early seedling	9.1	2.1	1.3	2.0	9.5
80%	First bloom	9.0	1.8	2.5	1.3	11.5
80%	Peak bloom	9.3	1.8	1.8	1.4	12.5
Significance						
Growth Stage (GS)		NS [s]	NS	NS	NS	NS
Stand Reduction Level (SL)		*	*	NS	NS	***
GS X SL		NS	NS	NS	NS	NS

[z] Percent of plant population reduced from standard population of ~200,000 plants per hectare. [y] Growth stage characterized by heat units accumulated after planting (HUAP) during stand reduction events for 2017; early seedling= 717 HUAP, first bloom = 1398 HUAP, peak bloom = 1894 HUAP. [x] All chile yields were harvested in kg per 3.1 m^2; reported in tons per hectare. [w] Weight of dehydrated fresh red fruit yield, means of n = 4. [v] Green yield were fruits with more than 50% green color; means of n = 4. [u] Unmarketable yield were fruits with more than 40% disease caused discoloration and/or blemishes; means of n= 4. [t] Number of counted plants in each row per plot; means off n = 4. [s] NS, *, *** Nonsignificant or significant at $p \leq 0.05$, or 0.001, respectively.

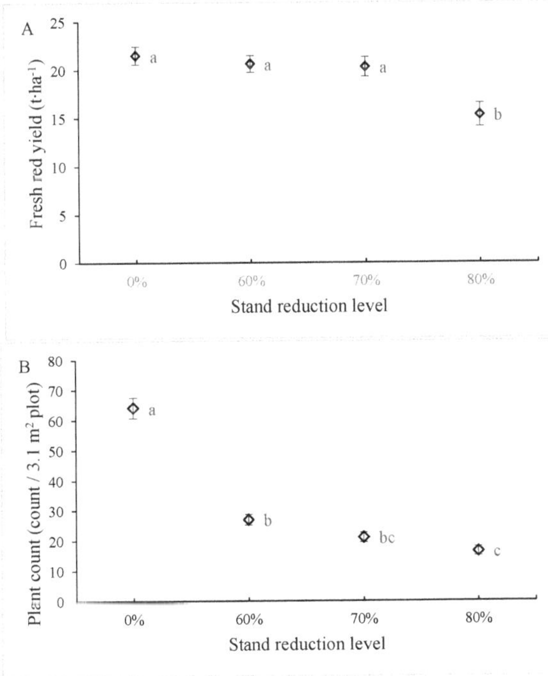

Figure 1. Stand reduction effects on fresh red yield (**A**) and plant counts (**B**) of paprika-type chile in 2016. Mean values of yield component measurements ± SE; all values are means of n = 12. Means separated by Tukey's test, $p \leq 0.05$. Means with common letter do not differ significantly. Yield reported in tons per hectare.

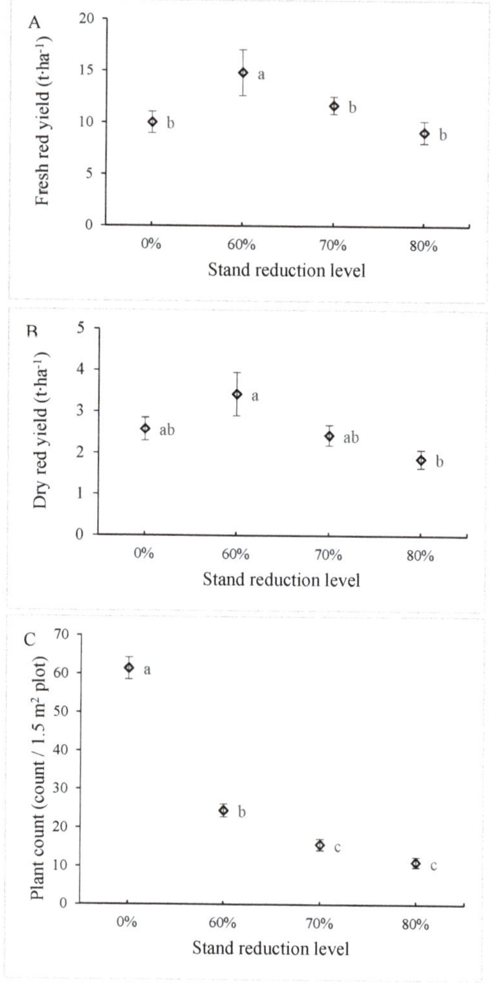

Figure 2. Stand reduction effects on the fresh red yield (**A**), dry red yield (**B**), and plant counts (**C**) of paprika-type chile in 2017. Mean values of yield component measurements ± SE; all values are means of n = 12. Means separated by Tukey's test, $p \leq 0.05$. Means with common letter do not differ significantly. Yield reported in tons per hectare.

4. Discussion

We found that the timing of stand reductions for paprika-type chile did not impact marketable red yields at the end of the season. Studies conducted in soybeans and sunflowers showed that yield was significantly impacted by the growth stage during which a stand reduction occurs. When sunflower plant populations were reduced in early growth stages they were able to recover yield, but stand losses in later growth stages resulted in yield reductions [16]. Similar results were found in soybeans when stand losses occurred in the early growth stages and yield was not affected due to plant compensation [2]. Yet, we found paprika-type chile yield was not affected by the growth stage during which stand reduction occurred. This could be due to our methodology of thinning the plots to clumps of 2 to 3 plants [18]. This standard practice, long employed by red chile growers in New Mexico, may provide protection from yield losses by increasing interplant competition. Interplant competition driven by clumped plants may increase vigorous plant growth earlier in the season [9] producing robust plants by midseason that are able to compensate for plants lost. Additionally, we may not have decreased plant populations at optimal growth stages to have an impact on yield components. Our 70% and 80% stand reduction plots did not have statistically different plant counts in either 2016 or 2017; perhaps a 90% stand reduction plot was necessary.

In 2017, our control plots with 0% stand reduction had less fresh and dry red yields. Reports have shown that chile grown in dense populations will yield less due to a decrease in plant light reception [10,23]. Our lower yields in 2017 may suggest that some thinning might be necessary to ensure each plant has access to light and enough space to adequately grow.

Traditionally, when evaluating how crops respond to stand reductions due to pest and/or extreme weather damage, two variables are taken into consideration: growth stage and extent of crop loss [16,24,25]. Our yield components were not significantly affected by the growth stage during the stand reduction event. The percentage of stand losses had a greater impact on the fresh red yield and dry red yield of paprika-type chile. According to our results, percentage of crop loss is a better predictor of end of season crop loss than the growth stage during which the stand reduction occurs. Therefore, insurance adjusters and farmers can estimate paprika-type chile crop losses based on percentage of stand losses instead of growth stage. Fresh red yield will be significantly reduced by 26% to 38% when plant populations are reduced by 80%. Cavero et al. [14] had comparable yield loss results, indicating paprika-type chile has some capacity to recover and compensate for stand reduction losses. This has been observed in other indeterminate crops such as lentils (*Lens culinaris* L.) that can compensate for stand reduction caused by hail damage anytime during the season [26]. Our data shows that a farmer could lose up to 70% of their paprika-type chile stand (a remaining plant population of at least 60,000 plants per hectare) due to hail damage and experience minimal to no impact on their yields.

Author Contributions: I.J., conception or design of the work, and the acquisition, analysis, or interpretation of data for the work, wrote initial draft of manuscript, and completed edits. S.W., conception or design of the work, co-authored initial draft of manuscript, and completed edits. All authors have read and agreed to the published version of the manuscript.

Funding: This research received no additional funding.

Conflicts of Interest: The authors declare no conflict of interest.

References

1. Sij, J.W.; Ott, J.P.; Olson, B.L.; Baughman, T.A. Growth and yield response to simulated hail damage in guar. *Agron. J.* **2005**, *97*, 1636–1639. [CrossRef]
2. Teigen, J.B.; Vorst, J.J. Soybean response to sand reduction and defoliation. *Agron. J.* **1975**, *67*, 813–816. [CrossRef]
3. McGinty, J.; Morgan, G.; Mott, D. Cotton response to simulated hail damage and stand loss in central Texas. *J. Cotton Sci.* **2019**, *23*, 1–6.
4. Vorst, J.V. Assessing hail damage to corn. In *National Corn Handbook*; Iowa State University: Ames, IA, USA, 1990; Volume 1, p. 4.
5. Soto-Ortiz, R.; Silvertooth, J.C.; Galadima, A. *Crop Phenology for Irrigated Chiles (Capsicum Annuum L.) in Arizona and New Mexico*; University of Arizona Vegetable Report; University of Arizona: Tucson, AZ, USA, 2006.
6. United States Department of Agriculture. Vegetables: 2016 Summary. 2017. Available online: http://usda.mannlib.cornell.edu/usda/current/VegeSumm/VegeSumm-02-22-2017_revision.pdf (accessed on 2 January 2018).
7. United States Department of Agriculture. Risk Management Agency State Profiles. 2017. Available online: https://www.rma.usda.gov/pubs/state-profiles.html (accessed on 3 January 2018).
8. National Oceanic and Atmospheric Administration. Storm Event Database. 2016. Available online: https://www.ncdc.noaa.gov/stormevents/ (accessed on 3 January 2018).
9. Walker, S.J.; Funk, P.A. Mechanizing chile peppers: Challenges and advances in transitioning harvest of New Mexico's signature crop. *HortTechnology* **2014**, *3*, 281–284. [CrossRef]

10. Paroissien, M.; Flynn, R. Plant spacing/plant population for machine harvest. In *New Mexico Chile Task Force. Rpt 13*; New Mexico State Univeristy: Las Cruces, NM, USA, 2004.

11. Wolf, I.; Aper, Y. Mechanization of paprika harvest. In Proceedings of the First International Conference on Fruit, Nut and Vegetable Harvesting Mechanization, Bet Dagan, Israel, 5–12 October 1983; American Society of Agricultural and Biological Engineers: St. Joesph, MI, USA, 1984; Volume 5, pp. 265–275.

12. Bosland, P.W.; Votava, E.J. *Pepper: Vegetable and Spice Capsicums*; CABI Publishers: Oxon, UK, 2000.

13. New Mexico Department of Agriculture. 2016 New Mexico Chile Production. 2016. Available online: https://www.nass.usda.gov/Statistics_by_State/New_Mexico/Publications/Special_Interest_Reports/NM_Chile_Production_03012017.pdf (accessed on 2 January 2018).

14. Cavero, J.; Ortega, R.G.; Guitierrez, M. Plant density affects yield, yield components, and color of direct-seeded paprika pepper. *HortScience* **2001**, *36*, 76–79. [CrossRef]

15. Donald, C.M. Competition among crop and pasture plants. *Adv. Agron.* **1963**, *15*, 1–118.

16. Miller, J.F.; Roath, W.W. Compensatory response of sunflower to stand reduction applied at different plant growth stages. *Agron. J.* **1982**, *74*, 119–121. [CrossRef]

17. Natural Resources Conservation Service. Web Soil Survey. 2017. Available online: https://websoilsurvey.sc.egov.usda.gov/App/WebSoilSurvey.aspx (accessed on 20 October 2017).

18. Bosland, P.W.; Walker, S.J. Growing chiles in New Mexico. In *New Mexico State Cooperative Extension Services Guide H-230*; Mew Mexico State University: Las Cruces, NM, USA, 2004.

19. Brown, P.W. *Heat Units. University of Arizona Cooperative Extension Bulletin AZ1602*; University of Arizona: Tucson, AZ, USA, 2012.

20. Silvertooth, J.C.; Brown, P.W.; Walker, S.J. Crop growth and development for irrigated chile (Capsicum annuum). In *New Mexico Chile Task Force Report 32*; New Mexico State University: Las Cruces, NM, USA, 2010.

21. New Mexico Climate Center. Leyendecker II PSRC Weather Data. 2016. Available online: https://weather.nmsu.edu/ziamet/request/station/nmcc-da-5/data/ (accessed on 5 January 2018).

22. New Mexico Climate Center. Leyendecker II PSRC Weather Data. 2017. Available online: https://weather.nmsu.edu/ziamet/request/station/nmcc-da-5/data/ (accessed on 5 January 2018).

23. Decoteau, D.R.; Graham, H.A.H. Plant spatial arrangement affects growth, yield, and pod distribution of cayenne peppers. *HortScience* **1994**, *29*, 149–151. [CrossRef]

24. McGregor, D.I. Effect of plant density on development and yield of rapeseed and its significance to recovery from hail injury. *Can. J. Plant Sci.* **1987**, *67*, 43–51. [CrossRef]

25. Rangarajan, A.; Ingall, B.A.; Orzolek, M.D.; Otjen, L. Moderate defoliation and plant population losses did not reduce yield or quality of butternut squash. *HortTechnology* **2003**, *13*, 463–468. [CrossRef]

26. Bueckert, R.A. Simulated hail damage and yield reduction in lentil. *Can. J. Plant Sci.* **2011**, *91*, 117–124. [CrossRef]

© 2020 by the authors. Licensee MDPI, Basel, Switzerland. This article is an open access article distributed under the terms and conditions of the Creative Commons Attribution (CC BY) license (http://creativecommons.org/licenses/by/4.0/).

MDPI

St. Alban-Anlage 66

4052 Basel

Switzerland

Tel. +41 61 683 77 34

Fax +41 61 302 89 18

www.mdpi.com

Horticulturae Editorial Office

E-mail: horticulturae@mdpi.com

www.mdpi.com/journal/horticulturae

www.ingramcontent.com/pod-product-compliance
Lightning Source LLC
LaVergne TN
LVHW071014170726
843515LV00006B/1135